AF361949

Thermal Fluid Dynamics and Control in Aerospace

Thermal Fluid Dynamics and Control in Aerospace

Editors

Dan Zhao

Chenzhen Ji

Hexia Huang

Basel • Beijing • Wuhan • Barcelona • Belgrade • Novi Sad • Cluj • Manchester

Editors

Dan Zhao
University of Canterbury
Christchurch, New Zealand

Chenzhen Ji
Tongji University
Shanghai, China

Hexia Huang
Nanjing University of
Aeronautics and Astronautics
Nanjing, China

Editorial Office
MDPI
St. Alban-Anlage 66
4052 Basel, Switzerland

This is a reprint of articles from the Special Issue published online in the open access journal *Aerospace* (ISSN 2226-4310) (available at: https://www.mdpi.com/journal/aerospace/special_issues/L37IU3181D).

For citation purposes, cite each article independently as indicated on the article page online and as indicated below:

Lastname, A.A.; Lastname, B.B. Article Title. *Journal Name* **Year**, *Volume Number*, Page Range.

ISBN 978-3-0365-8558-1 (Hbk)
ISBN 978-3-0365-8559-8 (PDF)
doi.org/10.3390/books978-3-0365-8559-8

Contents

About the Editors

Dan Zhao

Prof. Dan Zhao is the Director of Master Engineering Studies at the University of Canterbury, New Zealand. He has been awarded with fellowships from Engineering New Zealand, the European Academy of Sciences and Arts, the European Academy of Sciences, as well as the ASEAN Academy of Engineering and Technology. Prof. Zhao obtained his PhD and Master's degrees from the University of Cambridge and the University of Manchester (UMIST), respectively. As of now, he has successfully secured USD 17.2M funding from industries and governments, including the Singapore Prime Minister's Office and the New Zealand Ministry of Business, Innovation and Employment. His research expertise and interests include applying theoretical, numerical, and experimental approaches in order to study CO_2-free combustion science and technology; fabric drying; aeroacoustics; thermoacoustics; UAV aerodynamics; propulsion; energy harvesting; and renewable energy and fuel (ammonia and hydrogen).

Chenzhen Ji

Dr. Chenzhen Ji is a junior professor in the School of Mechanical Engineering at Tongji University. His research interests focus on clean combustion, combustion instability and control, aeroacoustics and computational fluid dynamics. He obtained his Ph.D. degree from Nanyang Technological University in Singapore and later worked as a postdoctoral researcher at Imperial College London, UK, from 2018 to 2020. He has led and participated in over 10 national-level scientific research projects in China, Singapore and the United Kingdom. So far, he has published over 40 journal papers, including over 30 SCI papers, with 2 of them selected as "ESI highly cited paper". He has participated in two chapters of professional books, and presented reports at more than 20 academic international conferences. In 2017, he was awarded the Best Paper Award at the 13th International Conference on Heat Transfer Fluids and Thermodynamics (HEFAT2017, Slovenia).

Hexia Huang

Dr. Hexia Huang is an associate professor who received their bachelor's and doctorate degrees from Nanjing University of Aeronautics and Astronautics in 2011 and 2017, respectively. Huang is a postdoctoral at Tsinghua University and a member of the American Institute of Aeronautics and Astronautics (AIAA), the Chinese Society of Aeronautics and Astronautics, and the Chinese Society of Astronautics. Huang was selected for the Young Talented Lift Scientists Sponsorship Program and "Changkong Star" Scholar Program of Nanjing University of Aeronautics and Astronautics. Huang has been engaged in the research of inlets of aerospace propulsion systems for a long time and has conducted a large amount of research in the fields of the internal flow aerodynamics of inlets, shock wave/boundary layer interaction, the aerodynamic design and flow mechanism of high-speed inlets, flow control technologies, inlet/engine/nozzle integration, and advanced internal flow measurement technologies. Huang has undertaken over 20 scientific research projects, including at the National Natural Science Foundation of China, the key program of the National Major Project, the Natural Science Foundation of Jiangsu, and the Aeronautics Power Foundation, among others. Huang has provided technical services for developing many types of inlets, and his inlet design software has been promoted and applied in many research institutes. Huang has published over 30 academic papers in the Physics of Fluids and the AIAA Journal, among others, and has authorized more than 30 invention patents.

Preface

Thermal fluid dynamics and control are critical scientific issues for aircraft design and aerospace propulsion systems, involving aerodynamics, fluid dynamics, thermal science and engineering, and posing significant challenges to academic researchers and engineers. In recent years, various novel aerospace propulsion systems have been proposed and developed rapidly. Thermal fluid dynamics and their efficient control are the key to improving the propulsion efficiency, controlling the combustion stability, reducing emissions, and expanding the engine operating boundaries. There is an urgent need to develop new theories of thermal fluid dynamics, reveal the coupling mechanism of multiple physical fields, such as fluid/combustion/acoustics/thermal/structure, and propose some new methods for thermal fluid control.

This Special Issue published recent advances in the inlet, compressor, combustor, turbine and nozzle of aero-engines. The inlet is an aerodynamic interface between the aircraft and the engine. Sun et al. [1] numerically assessed the effect of a horizontal periodic gust on the internal flow of the inlet at a low flight Mach number of 0.235. They found that the gust changes the time-averaged flow structure, and the internal flow oscillates significantly. The results demonstrated that the gust has unfavorable impacts on the aerodynamic performance of the inlet, with total pressure distortion and increases in the total pressure lost. Luo et al. [2] analyzed the unsteady shock motion in a hypersonic inlet of a rotating detonation engine. The transient three-dimensional shock structure, surface pressure distribution and the unsteady shock/boundary layer interaction are discussed in detail. The flow structure shows a spatial–temporal self-similarity property, and a theoretical model of the inclination angles of the MSW with an accuracy of 3% is developed and validated. To enhance the performance of the centrifugal compressor, Zhang et al. [3] proposed a self-circulating casing treatment and explored its underlying control mechanism. The results showed that the area of high relative total pressure loss in the blade tip passage and the flow loss reduced, resulting in an improvement of 20.22% for the stall margin. As the circumferential coverage ratios increased, the self-circulating control effect improved significantly. The efficient atomization of liquid fuel in scramjet combustors was one of the key issues. Zhao et al. [4] numerically investigated the effect of an incident shock on the atomization characteristics of kerosene with supersonic crossflow. Due to the interaction of the incident shock, the Sauter mean diameter of the jet reduced, and its distribution became more uniform. As the incident shock moved upstream, it had a reduced impact on the penetration depth of the kerosene jet, but the trajectory of the jet became closer to the wall. Tai et al. [5] used dielectric barrier discharge plasma actuation to enhance the turbulent mixing of fuel droplets and oxidant air in a ramjet combustor. The primary control mechanics of such an actuator involve the acceleration of the fluid, and the recirculation zone behind the evaporative V-groove flame holder was rearranged. Adjusting the actuation dimension, actuation intensity, and actuation position can achieve a relatively optimal turbulent mixing. As the performance of an axial turbine becomes poor at lower-than-nominal regimes, Nicoara et al. [6] proposed a specific fluid injection to overcome such shortcomings. Compared to the uncontrolled state, the power generated by the turbine could be improved by as much as 30%, which validated its effectiveness. An exhausted nozzle is one of the major sources of infrared radiation in aero-engines. The serpentine nozzle, characterized by its curved shape, was a practical configuration with which to minimize infrared radiation. Ahmed et al. [7] numerically addressed the impact of the cone mixer angle and aft-deck. With the increase in the mixer cone angle, the jet velocity and the temperature dropped, reducing the infrared radiation. The increase in the length of the aft-deck reduced the jet temperature also. A trapezoid and triangle aft-deck had a superior infrared radiation performance. Heat dissipation is critical for aerospace equipment, its battery, and electronic equipment. Zhao et al. [8] presented a comprehensive review regarding the micro-channel oscillating heat pipe (MCOHP) used in aerospace spacecraft for heat dissipation. The effects of heat flow on the thermal performance of MCOHPs were summarized in detail. The choice of heat flow work fluids, such as nano-fluids, gases, single liquids, mixed liquids, surfactants and self-humidifying fluids, was revealed to have a significant effect on the thermal performance of MCOHPs, as well as influences the gravity and flow characteristics. The heat flow pattern also affected the flow mode and mass transfer efficiency of the oscillating heat pipe.

Moreover, this Special Issue collected and published recent research addressing aircraft design and engineering. The formation of ice on aircraft wings has become a research hotspot in recent years, involving aerodynamics, two-phase flow and fluid–structure interactions, among others. Lu et al. [9] conducted numerical simulations in order to study the effect of blockages on the icing of various airfoils. Their simulation results showed that, with an increase in blockages, the flow velocity above the stagnation point of the airfoil rose, which led to the distribution of pressure coefficients and stronger heat transfer capacities. Meanwhile, it was found that the position of icing moved forward and the angle of the upper ice horn became smaller. Han et al. [10], from the Shanghai Aircraft Design and Research Institute, performed normal droplet and supercooled large droplet icing wind tunnel tests. They found that compared with normal droplets, the ice horn height of supercooled large droplets decreased with the increase in the droplet particle size. The range and height of the rough element ice shape after the main ice horn of the supercooled large droplets were significantly larger and higher than those of the normal droplets. Flight mechanics is a crucial research direction in aircraft design. Liu et al. [11] performed experimental investigations on the flight control of a V-tail configuration for a wind tunnel aircraft model, and they proposed a flight control law in order to control the aircraft model's static stability, height, and pitch angle in the wind tunnel. Their results indicated that the proposed experimental method, via a full model of the aircraft with twin V-tails and a novel longitudinal flight control law, was effective.

The editors of this Special Issue would like to thank each one of these authors for their contributions and for making this Special Issue a success. Additionally, the guest editors would like to thank the reviewers and the Aerospace editorial office, particularly Ms. Linghua Ding.

References

[1] Sun, S.; Wu, Z.; Huang, H.; Bangga, G.; Tan, H. Aerodynamic Response of a Serpentine Inlet to Horizontal Periodic Gusts. *Aerospace* **2022**, *9*, 824. https://doi.org/10.3390/aerospace9120824

[2] Luo, Z.; Huang, H.; Tan, H.; Liang, G.; Lv, J.; Wu, Y.; Li, L. On the Self-Similarity in an Annular Isolator under Rotating Feedback Pressure Perturbations. *Aerospace* **2023**, *10*, 188. https://doi.org/10.3390/aerospace10020188

[3] Zhang, H.; Wang, H.; Li, Q.; Jing, F.; Chu, W. Mechanism Underlying the Effect of Self-Circulating Casings with Different Circumferential Coverage Ratios on the Aerodynamic Performance of a Transonic Centrifugal Compressor. *Aerospace* **2023**, *10*, 312. https://doi.org/10.3390/aerospace10030312

[4] Zhao, Y.; Wu, J.; Mu, X. Atomization Characteristics of Kerosene in Crossflow with an Incident Shock Wave. *Aerospace* **2023**, *10*, 30. https://doi.org/10.3390/aerospace10010030

[5] Tai, Z.; Chen, Q.; Niu, X.; Lin, Z.; Yang, H. Plasma Actuation for the Turbulent Mixing of Fuel Droplets and Oxidant Air in an Aerospace Combustor. *Aerospace* **2023**, *10*, 77. https://doi.org/10.3390/aerospace10010077

[6] Nicoara, R.E.; Crunteanu, D.E.; Vilag, V.A. Axial Turbine Performance Enhancement by Specific Fluid Injection. *Aerospace* **2023**, *10*, 47. https://doi.org/10.3390/aerospace10010047

[7] Ahmed, H.M.A.; Ahn, B.-G.; Lee, J. Numerical Simulations on the Performance of

Two-Dimensional Serpentine Nozzle: Effect of Cone Mixer Angle and Aft-Deck. *Aerospace* **2023**, *10*, 76. https://doi.org/10.3390/aerospace10010076

[8] Zhao, X., Su, L., Jiang, J., Deng, W., Zhao, D. A review of working fluids and flow state effects on thermal performance of micro-channel oscillating heat pipe for aerospace heat dissipation. *Aerospace* **2023**, *10*, 179. https://doi.org/10.3390/aerospace10020179

[9] Lu, D., Lu, Z., Han, Z., Xu, X., Huang, Y. Numerical Investigation on the Effect of Blockage on the Icing of Airfoils. *Aerospace* **2022**, *9*, 587. https://doi.org/10.3390/aerospace9100587

[10] Han, Z., Si, J., Wu, D. Contrast Icing Wind Tunnel Tests between Normal Droplets and Supercooled Large Droplets. *Aerospace* **2022**, *9*, 844. https://doi.org/10.3390/aerospace9120844

[11] Liu, J., Qian, W., Bai, Y., Xu, X. Numerical and Experimental Research on Flight Control of a V-Tail Configuration for the Wind Tunnel Model of Aircraft. *Aerospace* **2022**, *9*, 792. https://doi.org/10.3390/aerospace9120792

Dan Zhao, Chenzhen Ji, and Hexia Huang
Editors

Article

Aerodynamic Response of a Serpentine Inlet to Horizontal Periodic Gusts

Shu Sun [1], Zhenlong Wu [2,3,*], Hexia Huang [2], Galih Bangga [4,5] and Huijun Tan [2]

[1] College of Civil Aviation, Nanjing University of Aeronautics and Astronautics, Nanjing 210016, China
[2] College of Energy and Power Engineering, Nanjing University of Aeronautics and Astronautics, Nanjing 210016, China
[3] Integrated Energy Institute, Nanjing University of Aeronautics and Astronautics, Nanjing 210016, China
[4] Institute of Aerodynamics and Gas Dynamics, University of Stuttgart, 70569 Stuttgart, Germany
[5] DNV Services UK Limited, Bristol BS2 0PS, UK
* Correspondence: zhenlongwu@nuaa.edu.cn

Abstract: Gust is a common atmospheric turbulence phenomenon encountered by aircraft and is one major cause of several undesired instability problems. Although the response of aircraft to the incoming gust has been widely investigated within the subject of external-flow aerodynamics in the past decades, little attention is paid to its effects on the internal flow within aircraft engines. In this paper, a newly implemented Field Velocity Method (FVM) in OpenFOAM is used to simulate the flow field and aerodynamic responses of a serpentine inlet exposed to non-stationary horizontal sinusoidal gusts. Validations are performed on the results obtained based on the baseline Computational Fluid Dynamics (CFD) solver and the gust modeling method. Finally, the flow field and aerodynamic characteristics of the serpentine inlet under horizontal sinusoidal gust conditions are comprehensively investigated. It is found that the gusts not only significantly change the flow structure but also play an unfavorable role in the total pressure distortion of the serpentine inlet. This finding shows the necessity to consider gust effects when designing and evaluating the performance of aircraft engines.

Keywords: gust; aerodynamics; serpentine inlet; computational fluid dynamics

Citation: Sun, S.; Wu, Z.; Huang, H.; Bangga, G.; Tan, H. Aerodynamic Response of a Serpentine Inlet to Horizontal Periodic Gusts. *Aerospace* **2022**, *9*, 824. https://doi.org/10.3390/aerospace9120824

Academic Editor: Sergey Leonov and Kung-Ming Chung

Received: 20 October 2022
Accepted: 12 December 2022
Published: 14 December 2022

Publisher's Note: MDPI stays neutral with regard to jurisdictional claims in published maps and institutional affiliations.

1. Introduction

Gust is one of the most common and important atmospheric phenomena for flight and has played an important role in many catastrophic flight accidents in the past decades [1–3]. Investigations of aircraft aerodynamic responses to atmospheric disturbances like gusts have demonstrated the importance of gusts in aerospace engineering because they are generally unexpected, giving the aircraft control less time to respond. Taking aircraft lifting surfaces such as wing and horizontal tail as examples, the unsteady loads caused by gusts not only decrease the fatigue lifetime of the structure but also affect the flight performance and passengers' comfort level [4]. Severe atmospheric gusts can lead to serious effects on the flight path and propulsion system [5,6]. Therefore, gust loads have to be considered in the load cases during aircraft development and all the representative airworthiness regulations such as the FAR-25 and the CS-25 have made detailed specifications for the determination of critical gust loads [4]. Several artificial gust generators have been developed to experimentally investigate these phenomena, which include rotating slotted cylinders [7], oscillating plates [8], wind vanes [9], and active grids [10].

Using CFD (Computational Fluid Dynamics) methods, it is possible to prescribe the gust velocity using the Field Velocity Method (FVM) [11,12] and the Split Velocity Method (SVM) [13–16] for simulating the airfoil lift response to a gust of arbitrary shapes. The main differences between FVM and SVM are twofold. First, SVM requires special grid refinement for gust grid transport while FVM does not. Second, SVM can account for the mutual interaction of the gust and the research object [6] since the gust is resolved in the flow field, whereas the gust shape is unaffected from the start to the end in FVM [17]. A wide range of reduced-order models [18–21] was also developed for the efficient estimation of airfoil aerodynamic force responses to gusts [22–24]. These simple models may be combined with a CFD method to enhance the accuracy of the result and to provide a direct coupling with the aeroelastic equations of motions for predicting displacement responses [25].

It is well known that the engine inlet is usually located at the very front of the propulsion system and plays an important role in deciding the performance of the engine itself. Therefore, any disturbances in the incoming air may cause significant flow field deterioration, such as distortion and separation in the inlet. Although first noticed as early as in the 1960s [26], hardly any emphasis has been placed on investigating the effects of gusts on engine inlets, and only a handful of numerical studies on gust-inlet interactions were carried out so far [27–29]. Kozakiewicz and Frant [30,31] numerically studied the impact of changes in speed, gust angle, and sideslip angle on the development of the intake vortex for both a single and a double fuselage-shielded inlet. A common conclusion can be drawn that low-speed gust is the most dangerous because low-speed gusts contribute to the formation of the inlet vortex regardless of the gust angles for a wide range of sideslip angles [31]. However, the gusts in the aforementioned study were indeed steady crosswinds, rather than unsteady oscillating gusts as prescribed in the existing airworthiness regulations. Halwas and Aggarwal [28] numerically investigated the effect of side gust on the performance of a supersonic inlet. They found that the shock structure, flow characteristics, and inlet performance all are strongly affected by the presence of gust. The shock structures and the external and internal flows become asymmetric. The shock/boundary-layer interaction becomes stronger, resulting in strong boundary-layer separation, secondary flow, and a separated flow region in the diffuser section. In addition, the inlet performance generally deteriorates, resulting in significant reductions in the total pressure recovery and the mass flow ratio and increases in the flow distortion. However, the adverse effects of the shock waves and their interaction with the boundary layer can be effectively removed by using a bleed system [29]. Übelacker, Hain and Kähler [32] experimentally investigated the interaction between the gust generated by a pitching airfoil and the leading-edge separation bubble of a through-flow nacelle. It was found that the gust significantly influences the inlet separation region by changing the angle of attack.

From the above discussions, it is easily found that only limited gust studies have been performed concerning the inlet internal flow consideration compared to the extensive studies on the wing external flow aspect. Moreover, it is seen from the small amount of open literature that the more common unsteady gusts in nature, such as oscillating and transporting gusts, were rarely involved. The present work is motivated by these considerations. In this paper, we present new insights into the flow characteristics and mechanisms of an ultra-compact serpentine inlet subjected to periodic horizontal gusts via computational fluid dynamics (CFD) on the platform of the open-source package OpenFOAM®. The aims of this study are primarily threefold. First, this study is thus motivated to provide a primary insight into the effects of gusts on internal flows and attract more attention to this relevant field. Second, all existing studies about gust effects on engine inlets [28–31] investigated the effect of side gusts with changing sideslips but with a constant magnitude. The present study investigates unsteady sinusoidal gusts, which may reveal different mechanisms regarding gust-inlet interactions in a real environment. Finally, most of the existing gust modeling methods are based on in-house codes. This study implements the FVM within the open-source OpenFOAM toolbox, providing better access for researchers in the field.

Considering the streamwise direction of the main flow inside the diffuser, horizontal gusts are deemed to more significantly affect the inlet flow and thus are adopted by this study. The unsteady solver for the inlet aerodynamics is validated by comparing the calculation results with the previous wind-tunnel measurement data [33]. Both the flow field characteristics and aerodynamic performance responses of the inlet are inspected to reveal the underlying mechanism for the gust-inlet interaction. The following content of this paper is structured as follows. Section 2 presents the relevant theories and details of the numerical setup. Section 3 presents and analyzes the results of the effects of periodic gusts on the serpentine inlet. Finally, a conclusion is made in Section 4 to summarize the results and findings obtained from the present study.

2. Theory and Methodology

2.1. Model Description

The inlet model in the present work is an ultra-compact serpentine inlet integrated with the forebody of a flying-wing stealth aircraft, which has been studied previously by the Inlet Research Group of Nanjing University of Aeronautics and Astronautics (NUAA) [33,34], as shown in Figure 1a. The main geometric parameters of the model are tabulated in Table 1. The entrance of the top-mounted serpentine inlet is smoothly merged with the forebody of the aircraft. The inlet is ultra-compact for the sake of the rigorous restrictions of the aircraft size and weight, with a length of 2.3 D for the compact diffuser, where D = 65 mm is the diameter of the aerodynamic interface plane (AIP). Section 1-1 shows half of the trapezoidal shape of the entrance section and Section 2-2 the unique concave shape of the middle section. The typical cross-sectional shape of the middle section of a traditional design is also shown in Figure 1a for contrast. Furthermore, the vertical offset between the entrance center and the exit center of the diffuser is as high as 0.66 D, which is highly prone to boundary layer separations at the aft part of the top surface and the fore part of the bottom surface. Therefore, a large internal bump is introduced into the top surface to suppress massive flow separations and decrease the distortion index at the AIP (Figure 1b). More descriptions of the configuration of the inlet model are accessible in our previous investigation [33]. Experimental measurements were previously done in the NH-1 high-speed wind tunnel of NUAA and will be reused here for the current numerical model validation.

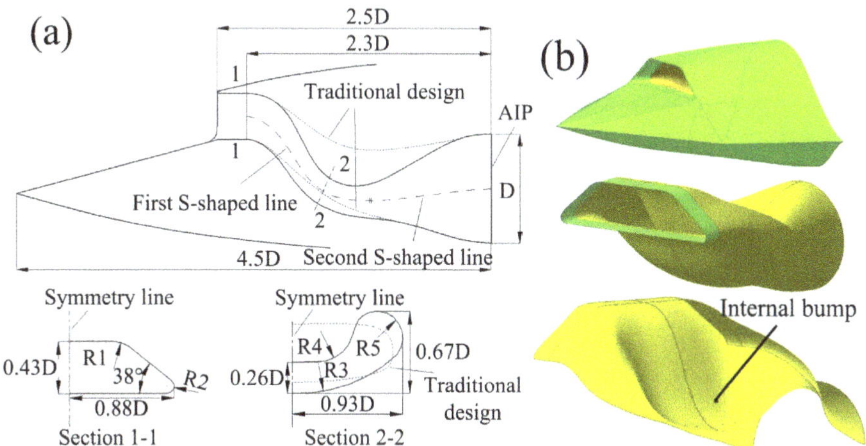

Figure 1. (**a**) Geometry and (**b**) experimental setup of the serpentine inlet of NUAA Inlet Research Group [33].

Table 1. Main geometric parameters of the ultra-compact serpentine inlet.

Parameter	Value
Diameter of the AIP	D (65 mm)
Distance between the forebody tip and the inlet	2.0 D
Total length of the inlet	2.5 D
Length of the diffuser	2.3 D
Vertical offset of the diffuser	0.66 D

2.2. Performance Parameter Definition

Two common parameters, i.e., total pressure recovery σ and circumferential total pressure distortion index, DC_{60} are employed in this paper to characterize the inlet performance at the aerodynamic interface plane (AIP) between the inlet and the engine. σ is defined as

$$\sigma = p^*_{avg360}/p^*_{\infty} \tag{1}$$

where p^*_{avg360} is the mass flow weighted average total pressure at the AIP, and p^*_{∞} is the total pressure of the freestream air. And DC_{60} is calculated as

$$DC_{60} = \left(p^*_{min60}/p^*_{avg360} \right)/q_{avg360} \tag{2}$$

where p^*_{min60} is the minimum value of the mass flow weighted average total pressure over any $60°$ sector around the center of the AIP, q_{avg360} is the mass flow weighted average dynamic pressure at the AIP.

In addition, the static pressure coefficient Cp is used to discuss the static pressure distribution on the wall of the inlet, which is defined as

$$Cp = (p - p_{\infty})/q_{\infty} \tag{3}$$

where p and p_{∞} are the static pressures on the inlet wall surface and of the freestream air respectively. q_{∞} is the dynamic pressure of the freestream air.

2.3. Governing Equations

The unsteady three-dimensional (3D) compressible Reynolds averaged Navier-Stokes (URANS) equations are solved for both investigations: no-gust and gust cases. Meanwhile, the turbulence is modeled by the Menter $k - \omega$ Shear Stress Transport (SST) model [35] with a fully turbulent boundary layer assumed. The governing equations including the mass, momentum, energy, and turbulence equations can be written in a common vector form as:

$$\frac{\partial}{\partial t} \int_{\Omega} \vec{W}\, d\Omega + \oint_{\partial\Omega} \left(\vec{F}_c - \vec{F}_v \right) dS = \int_{\Omega} \vec{Q}\, d\Omega \tag{4}$$

where $\vec{W}$ is the vector of conservative variables, $\vec{F}_c$ is the vector of convective fluxes, $\vec{F}_v$ is the vector of viscous fluxes, and $\vec{Q}$ is the source term comprising all volume sources due to body forces, volumetric heating, and turbulence. Ω is the control volume bounded by the closed surface $\partial\Omega$, and dS denotes the surface element. The concrete expressions of all the variables can be found in Blazek's book [36].

The velocity correction is applied throughout the flow field. A horizontal sinusoidal gust can be perceived as a superposition of a sinusoidal time-variant velocity field parallel to the freestream. As opposed to SVM, which is also named as Resolved Gust Approach (RGA) in [25,37,38], the superposed velocity field is modeled by modifying the grid time metrics without actually moving the grid. Therefore, there is no need for a high grid resolution in the whole domain to transport a gust from the inflow boundary to the aircraft to minimize the numerical losses [17]. Mathematically, FVM can be explained by considering the general flow velocity $\vec{V}$ in the computational domain, which can be rewritten as

$$\vec{V} = (u - x_{\tau})i + (v - y_{\tau})j + (w - z_{\tau})k \tag{5}$$

where u, v, and w are the components of the velocity along the coordinate directions, and x_{τ}, y_{τ}, and z_{τ} are the grid time metrics components. For the flow over a stationary body, these components are zero.

The velocity field in the presence of a horizontal gust can be expressed as

$$\vec{V} = (u - x_\tau + u_g)i + (v - y_\tau)j + (w - z_\tau)k \tag{6}$$

Thus, the modified time metrics are

$$\tilde{x}_\tau i + \tilde{y}_\tau j + \tilde{z}_\tau k = (x_\tau - u_g)i + y_\tau j + z_\tau k \tag{7}$$

The type of gust of interest in the present study is horizontal and sinusoidal, i.e., only the velocity component in the horizontal (streamwise) direction changes with the time variable in a sinusoidal fashion. The gusty inflow velocity profile simulated in this study is illustrated in Figure 2. It consists of only one full period of sinusoidal oscillation followed by the mean velocity for $t > T$. The oscillation amplitude, frequency, and mean velocity of the horizontal sinusoidal gusts of interest in this study are $\tilde{u}_g$, f, and u_0, respectively.

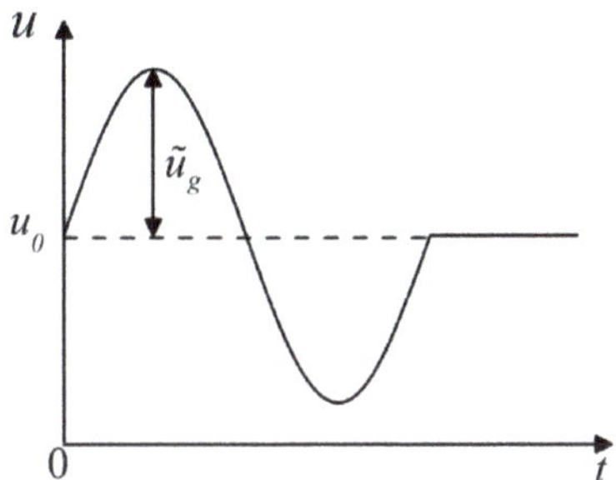

Figure 2. Definition of the gust velocity profile of interest in this study.

The horizontal component of the gusty inflow velocity can be expressed as

$$u = u_0 + \tilde{u}_g \cdot \sin(2\pi f \cdot t) \tag{8}$$

where $t = \left[0 : \frac{T}{N} : \frac{(N-1)T}{N}\right]$ is the vector of time, discretizing one period of gust T into N equal intervals. N decides the resolution of the discretized gust shape. A larger N implies that the discretized gust shape is more approximate to the physically continuous one.

Discretization of the gust velocity was done by using Matlab. As for a complete running of a gust case, an unsteady calculation of the no-gust case was first carried out by the compressible solver 'rhoPimpleFoam', of which the result was used to initialize the flow field for the subsequent gust computation. Then, the gust velocity, which is read from an external velocity document, is superposed with the mean velocity of the flow field for calculation of the inlet response, as illustrated in Figure 3.

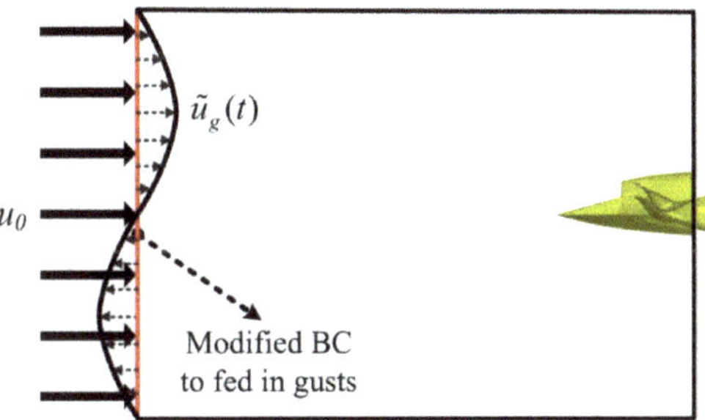

Figure 3. Illustration of the implementation of the current FVM gust modeling method.

2.4. Solution Algorithm

All the velocity, turbulence, and energy terms were discretized by the second-order linear-upwind scheme, as well as the pressure terms by the second-order linear differencing scheme. The dual time stepping approach was used for the temporal discretization together with a second-order backward Euler differencing scheme. Within each time step, the linearized algebraic equations were solved using a generalized Geometric-algebraic Multi-grid (GAMG) convergence acceleration method with the Gauss-Seidel smoother. In addition, the PIMPLE algorithm [39], which is a combination of the pressure-implicit split operator (PISO) and the semi-implicit method for pressure-linked equations (SIMPLE) algorithms, was used to solve the pressure-velocity equation [40]. The time step size is adjustable so that the Courant number remains below one, to satisfy the CFL condition. Solutions were computed on a 2-CPU workstation with a total of 128 "AMD-EPYC-7742" processors. All the calculations were performed using a Linux-compiled version of the OpenFOAM code (Version 1912) for multiple parallel computations. The convergence criterion was based upon a decrease in solution residual error of at least four orders of magnitude from the starting conditions.

2.5. Mesh and Boundary Condition

Due to the symmetry of the geometric model and the flow field characteristics of all simulations in this study, half of the model is adopted for numerical construction. The computational domain is set as 14 D, 16 D, and 10 D in the x, y, and z directions, which can sufficiently eliminate the boundary effect on the computed results, as shown in Figure 4. The origin of the coordinate system concerning the geometry is at the leading of the forebody. The internal duct was also extended linearly by 1.0 D from the AIP to the actual exit of the diffuser. For mesh generation, a fully structured multi-block mesh strategy is adopted using the ANSYS ICEM CFD software [41]. The fine mesh, which will be adopted for the whole study, in total has 51 blocks and 4.20 million cells. The internal diffuser mesh employs an O-grid topology and has 201, 120, and 31 nodes in the streamwise, circumferential, and radial directions, respectively. The height of the first layer adjacent to all the walls is set to be 1×10^{-5} m, which satisfies the requirement of the normalized wall distance $y^+ \sim 1$.

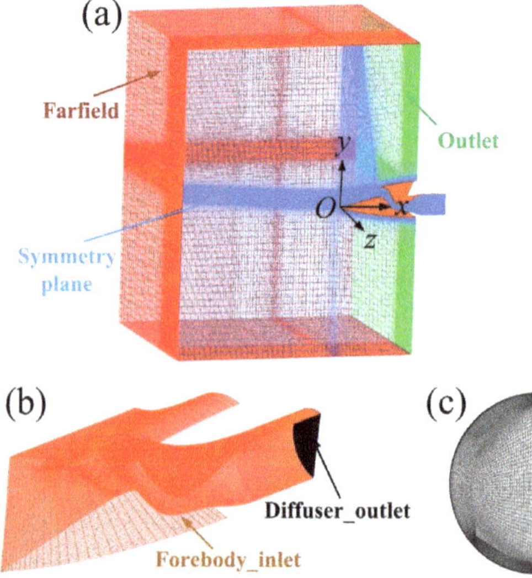

Figure 4. Computational grid used in the current study: (**a**) Global view of the whole computational domain; (**b**) Close view of the surface mesh on the forebody-inlet integration; (**c**) Close view of the O-type mesh at the AIP.

The boundary conditions are also shown in Figure 4. At the symmetry plane, colored in blue, no fluid can enter or exit the domain, and the normal gradients of the scalar variables and velocity components tangential to the symmetry plane are zero, which can be expressed as

$$\vec{v} \cdot \vec{n} = 0, \, \vec{n} \cdot \nabla \phi = 0, \quad \vec{n} \cdot \nabla \left(\vec{v} \cdot \vec{t} \right) = 0 \tag{9}$$

where ϕ stands for the scalar variables including p, T, K, and ω. $\vec{t}$ denotes the unit tangential vector. The outlet of the domain (in green) and the outlet of the diffuser are applied with the Neumann boundary condition given by

$$\nabla \vec{v} = 0 \tag{10}$$

and all the walls are treated as no-slip and impermeable walls, i.e.,

$$\vec{v} \cdot \vec{t} = 0, \, \vec{v} \cdot \vec{n} = 0 \tag{11}$$

A list of the boundary condition settings for the computational domain concerning the main variables in OpenFOAM is shown in Table 2.

Table 2. Boundary condition settings for the computational domain.

Boundary	k	ω	p	T	$\vec{V}$
Farfield	inletOutlet	inletOutlet	freestreamPressure	inletOutlet	freestreamVelocity
Outlet	inletOutlet	inletOutlet	fixedValue	inletOutlet	zeroGradient
Symmetry plane	symmetry	symmetry	symmetry	symmetry	symmetry
Forebody_inlet	kqRWallFunction	omegaWallFunction	zeroGradient	zeroGradient	noSlip
Diffuser_outlet	inletOutlet	inletOutlet	fixedValue	inletOutlet	zeroGradient

3. Results and Discussions

3.1. Mesh Independence Examination

For the CFD simulation, the first important consideration is to guarantee the results are independent of the mesh used, i.e., to conduct a mesh independence examination. To this end, a set of grids sequentially named as coarse-, fine-, and dense-mesh is checked at the freestream Mach number of M = 0.7 and the non-dimensional static pressure at the AIP $p_{\text{AIP}}/p_0 = 1.18$ where p_0 is the static pressure of the freestream air. Table 3 tabulates the number of cells, the calculated total pressure recovery σ_{AIP}, the absolute error relative to the result by the dense mesh, as well as the previous wind-tunnel experimental measurement [33]. All three grids obtain satisfactory results with good accuracy relative to the experimental measurement. Enhancing the mesh resolution from the coarse mesh to the fine one has reduced the prediction error by an order of magnitude.

Table 3. Inlet performance calculated with different resolved meshes at M = 0.7 and $p_{\text{AIP}}/p_0 = 1.18$.

Data Source	Number of Grid Cells	σ_{AIP}	Error of σ_{AIP}
EXP. [33]	—	0.961	—
Coarse mesh	2.48 million	0.968566	1.039×10^{-3}
Fine mesh	4.20 million	0.967687	1.597×10^{-4}
Dense mesh	5.95 million	0.967527	—

The differences between the total pressure contours and the wall static pressure coefficient [33] predicted by the three meshes are also very small, as shown in Figures 5 and 6, respectively. The five low-total-pressure zones as depicted in our previous study by Ansys Fluent [33] are all captured by the three meshes, as well as the bat-shaped distribution, as shown in Figure 5. The predicted static pressure distributions along the top and bottom surfaces of the serpentine inlet all agree well with the experimental results for all the meshes. In summary, there is no significant difference between the results obtained by the three generated meshes both qualitatively and quantitatively. Considering the computational efficiency and accuracy for the later simulation of gusts, the fine mesh is selected for all the subsequent gust response computations.

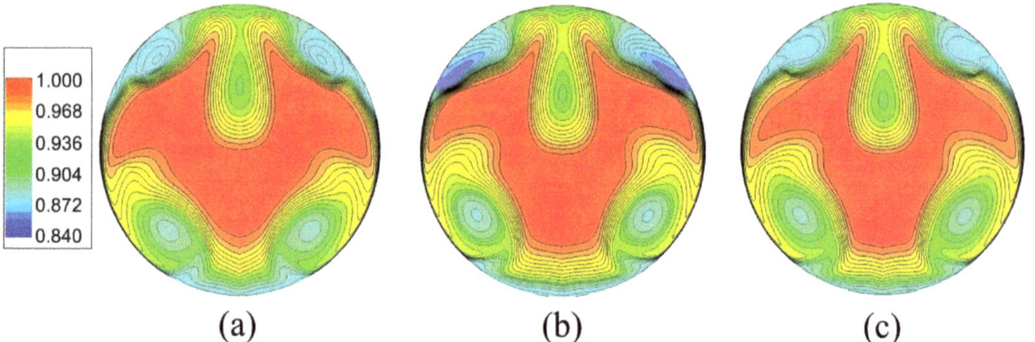

Figure 5. Comparison of the total pressure contours at the AIP calculated with different levels of mesh resolutions at M = 0.7 and $p_{AIP}/p_0 = 1.18$: (**a**) Coarse grid; (**b**) Fine grid; (**c**) Dense grid.

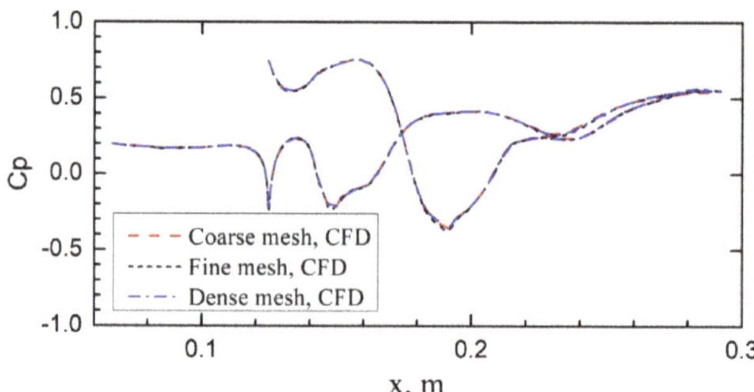

Figure 6. Surface static pressure distributions calculated with different-sized meshes at M = 0.7 and $p_{AIP}/p_0 = 1.18$.

3.2. Unsteady Aerodynamic Solver Validation

To validate the accuracy of the current solver for prediction of the fundamental aerodynamic performance with no gust, an experimental case at the test condition of M = 0.7 and $p_{AIP}/p_0 = 1.18$ was first calculated with the current solver. The time-averaged surface static pressures on both the top and bottom surfaces of the inlet along with the variation distribution of the transient values are shown in Figure 7, where our previous wind-tunnel experimental measurements and CFD results calculated by Fluent [33] are also shown for comparison. The variation bars represent the mean, peak, and valley pressures counted from an efficient period. Generally, the current solver predicts the wall static pressure well with good agreement with the experimental measurements. The adverse pressure gradients are slightly under-predicted by both solvers, which is due to the common deficiency of RANS methods in predicting three-dimensional rotating and separated flows [33,42,43]. The maximum relative errors of the results predicted by OpenFOAM and Fluent relative to the experimental measurements are about 24.5% and 20.1%, respectively. Despite these differences, the trends are consistent with that of the experimental results, and the size of the flow separation region on the lee side of the top surface is correctly predicted.

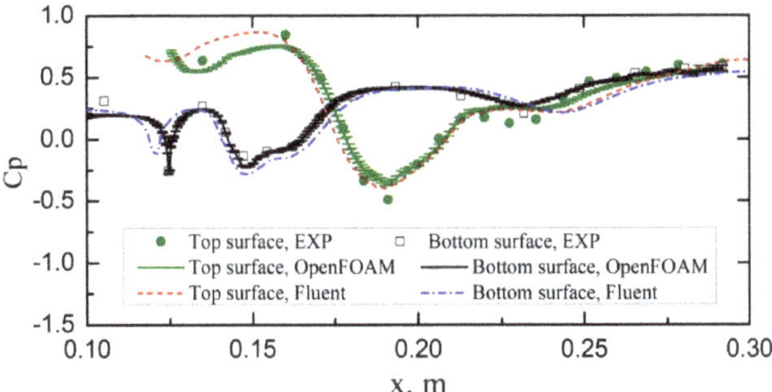

Figure 7. Comparison of the surface static pressure results between the experiment and both CFD methods at M = 0.7 and p_{AIP}/p_0 = 1.18 in the absence of gust.

Figure 8 presents a comparison between the time-averaged total pressure contours at the AIP predicted by the current unsteady solver and the experimental observation. The typical characteristics of the flow structure are successfully simulated by the current solver. The five low-total-pressure zones as depicted in our previous study by Ansys Fluent [33] are also captured by the current solver, as well as the bat-shaped distribution. By changing the pressure at the outlet of the diffuser, the inlet performance at different operating conditions can be obtained. Figure 9 shows the total pressure recovery and circumferential total pressure distortion index [33] calculated by the two solvers. It can be seen that the results are consistent with the experimental data in trend and have a small relative error in magnitude. In conclusion, the current solver is qualified in predicting the flow field characteristics and aerodynamic performance of the serpentine inlet.

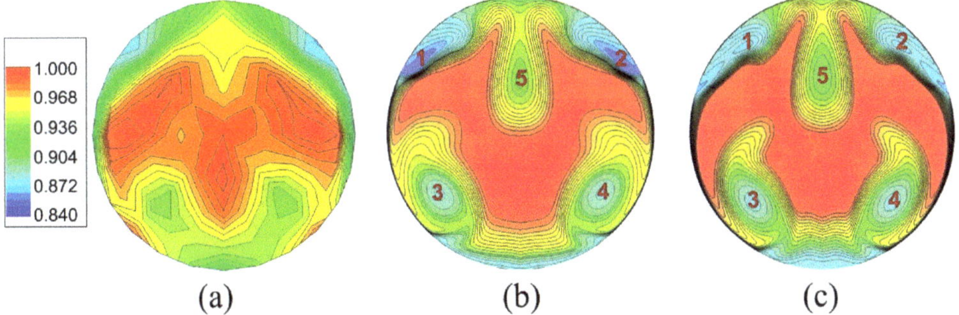

Figure 8. Comparison of the total pressure contours at the AIP between (**a**) the experiment, (**b**) the current OpenFOAM solver, and (**c**) the previous Fluent solver [33] at M = 0.7 and p_{AIP}/p_0 = 1.18.

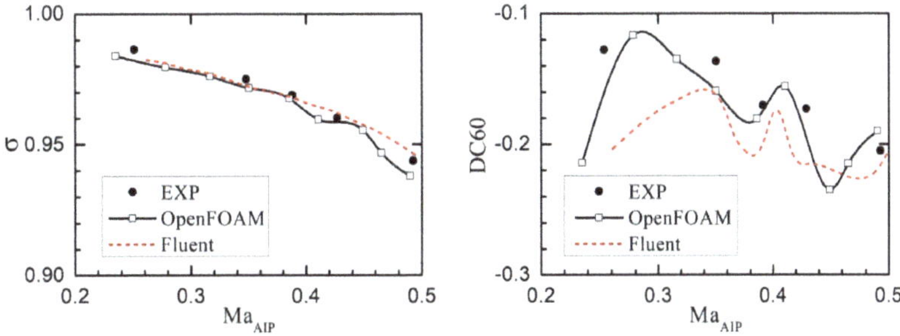

Figure 9. Comparison of the inlet performance results between the experiment and both CFD methods at various AIP Mach numbers (M = 0.7).

3.3. Inlet Low-Speed Performance without Gusts

The following study primarily concerns the aerodynamic performance of the serpentine inlet at low velocities, because gust encounters are more common and adverse at aircraft landing and take-off phases. Thus, for all the following gust response calculations, the freestream velocity is fixed at u_0 = 80 m/s (i.e., M = 0.235), and the backpressure p_{AIP}/p_0 = 0.85, leading to an AIP Mach number of 0.45.

The flow field characteristics at an inlet Mach number M = 0.235 are presented in Figure 10. The vortex tubes are visualized using the Q-criterion. Figure 10a shows the iso-surface of $Q' = 1 \times 10^{-5}$, where $(Q' = Q/(Q_{max} - Q_{min}))$. Similar to the high-speed behavior at M = 0.7 as observed by Sun and Tan [33], the three dominant vortex tubes are also present but more chaotic at M = 0.235 with some more minor vortex tubes accompanying the three major ones (Figure 10a). These vortex tubes are oriented both in the streamwise and lateral directions, indicating a strong three-dimensional effect of the highly curved serpentine inlet geometry on the flow field. There is only a slight flow separation at the lee side of the top surface thanks to the application of the bump (Figure 10b). The bat structure of the total pressure contours becomes less sharp at the low-speed condition (Figure 10c). Besides, regarding the five low-total-pressure regions, the ones referred to as No. 1, 2, and 5 have shrunk significantly, while the two at both sides of the bottom part (No. 3 and 4) are slightly enlarged. The changes agree well with the pattern of the secondary flow at the AIP (Figure 10d).

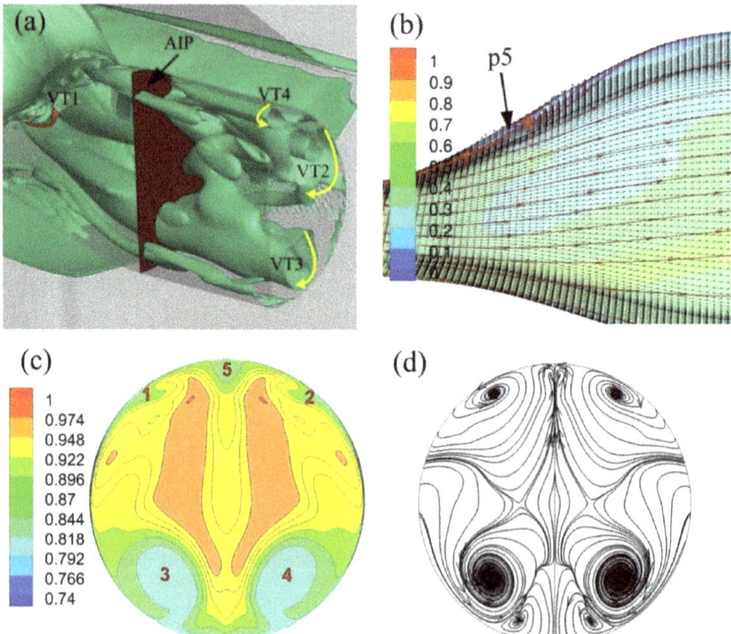

Figure 10. Flow field characteristics of the serpentine inlet at M = 0.235 and p_{AIP}/p_0 = 0.85 under the no-gust condition: (**a**) the vortex tube (VT) structure, (**b**) Flow separation occurring at the lee side of the top surface coupled with the Mach number contour, (**c**) Total pressure contour at the AIP, and (**d**) Pattern of the secondary flow at the AIP.

As there is unsteady flow separation occurring on the top surface, it is necessary to clarify the unsteadiness of flow field characteristics before investigating gust responses. Figure 11 shows the unsteady fluctuation and the spectrum of the wall static pressure probed at the location 'p5' in the separation zone in Figure 10b. It is seen from Figure 11a that both low- and high-frequency fluctuations of the wall static pressure are present due to the shedding of the separation bubbles. Figure 11b indicates that there are two predominant peak zones in the spectrum, one within the range of gust frequency studied in the present study (25–300 Hz) and the other one located at a much higher frequency zone between 2000 and 3000 Hz. Figures 12 and 13 show the corresponding results for total pressure recovery and distortion at the AIP, respectively. Since these are surface-averaged quantities, these curves do not show fluctuations at very high frequencies. Note that the mean values of the total pressure recovery and distortion are about 0.945 and −0.36, respectively. Although the AIP has fully been out of the separation zone and about 1 D away from the vortex core at the separation zone, the unsteady shedding vortices from the top surface still cause both performance parameters to slightly fluctuate at a low peak frequency of around 100 Hz. This is an important finding for analyzing the inlet aerodynamic responses to the gusts in the last subsection.

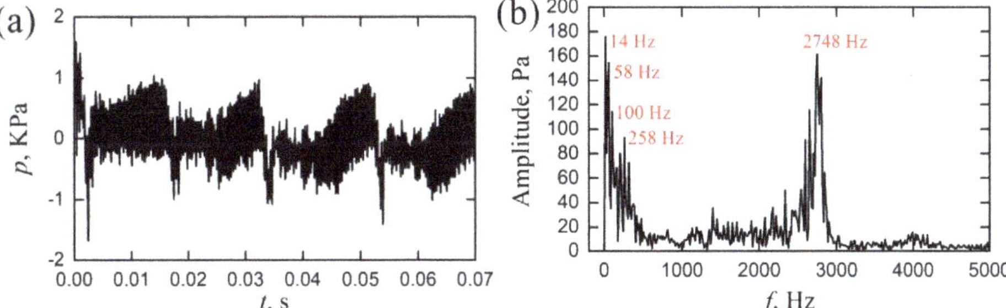

Figure 11. (**a**) The unsteady fluctuation and (**b**) the spectrum of the wall static pressure probed at position p5.

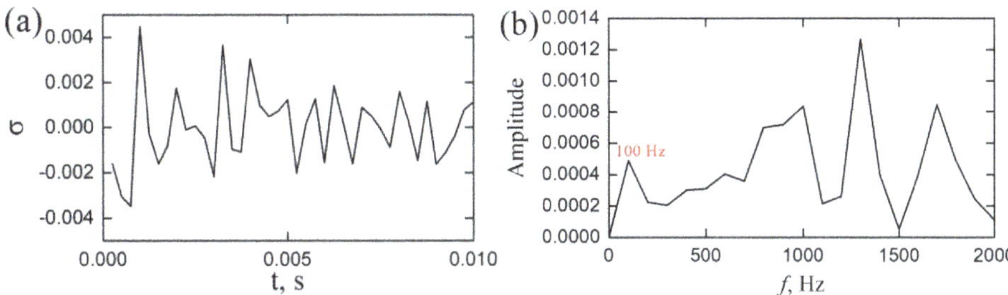

Figure 12. (**a**) The unsteady fluctuation and (**b**) the spectrum of the total pressure recovery at the AIP.

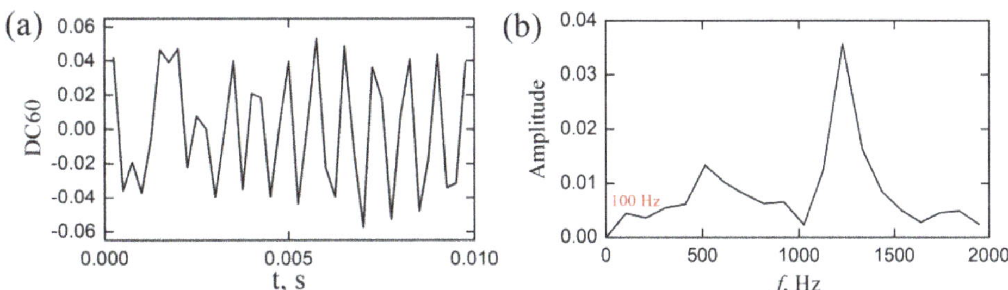

Figure 13. (**a**) The unsteady fluctuation and (**b**) the spectrum of the total pressure distortion at the AIP.

3.4. Gust Model Validation

Before the study of the response of the serpentine inlet to sinusoidal gusts, the accuracy of the current gust modeling method should also be validated, which includes two aspects one concerns producing a gusty flow field and the other is gust response. For the former consideration, a structured grid with a uniform cell size of 0.01 m is generated for a rectangular computational domain where a sinusoidal gust with the frequency f = 10 Hz amplitude $\tilde{u}_g$ = 20 m/s plus a mean velocity u_0 = 80 m/s is introduced at the left boundary as shown in Figure 14. Both the top and bottom boundaries of the domain are set as the "slip" wall. To investigate the gust velocity distribution at different locations, five probe positions are selected to monitor the local airflow velocities. Figure 15 presents a comparison of the gust velocity profile of a full period between the CFD value monitored at the origin of the coordinate, O, and the theoretical value. It is expected that the current numerical result agrees quite well with the theoretical one with no significant numerical dissipation.

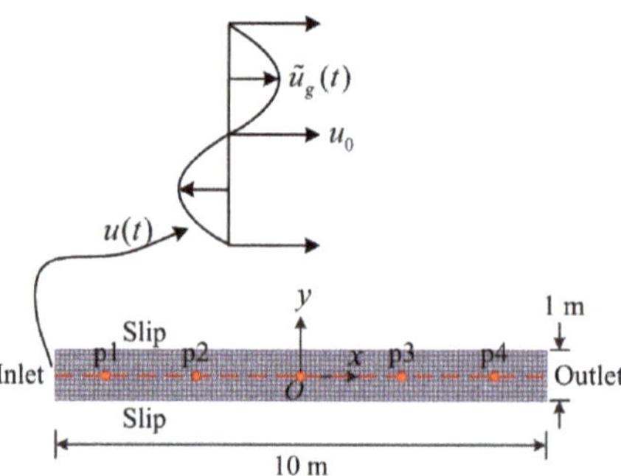

Figure 14. Illustration of the gust model and the probed positions for characterization of the horizontal sinusoidal gusty inflow condition implemented by this study.

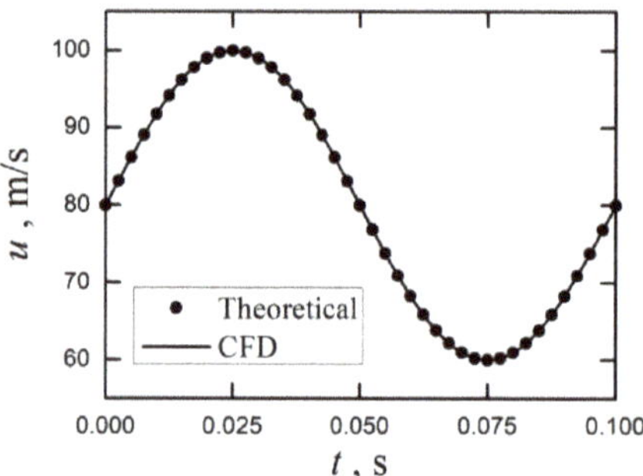

Figure 15. Comparison of the horizontal gust velocity between the theoretical value and the current CFD result.

To further validate the accuracy of the current FVM for gust response predictions, a NACA 0006 airfoil was selected for two-dimensional simulations. The reasons why an external flow case is chosen are twofold. The first is the fact that there is little data about internal flow gust responses like the inlet gust responses of interest in this study. The second is data availability of airfoil gust response for reference and comparison. The NACA 0006 airfoil selected here is a thin symmetric airfoil that is very suitable for reduced order modeling (ROM) such as the Küssner function and convolution integral model [25]. The computational domain and grid are shown in Figure 16. The total numbers of nodes and cells are 637,548 and 377,464, respectively. The height of the first layer adjacent to the walls is 1×10^{-5} c, where c is the chord length of the airfoil and c = 0.02 m. Two kinds of vertical gust shapes are simulated, i.e., 1-cosine and sinusoidal, whose shape definitions are also accessible in [25]. As shown in Figure 17, the gust velocity simulated by the FVM method is highly consistent with the theoretical value for both gust shapes, and the airfoil lift coefficient response also agrees generally well with the theoretical value calculated by the CIM. There is some deviation at the ending part of both lift responses because the CIM results converge to the mean no-gust value after one period, while the CFD results continue with another gust cycle. To be concluded, the currently implemented FVM method is capable of simulating various gust shapes and responses.

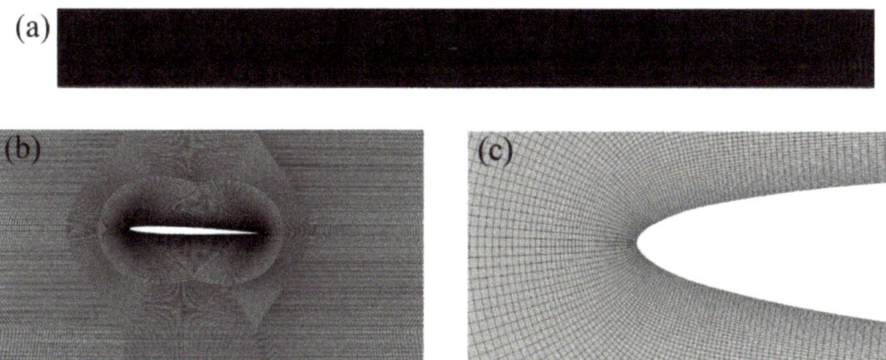

Figure 16. Computational grid for simulation of the gust responses of the NACA 0006 airfoil: (**a**) Global view of the whole computational domain; (**b**) Close view of the mesh near the airfoil; (**c**) Close view of the O-type mesh at the leading edge of the airfoil.

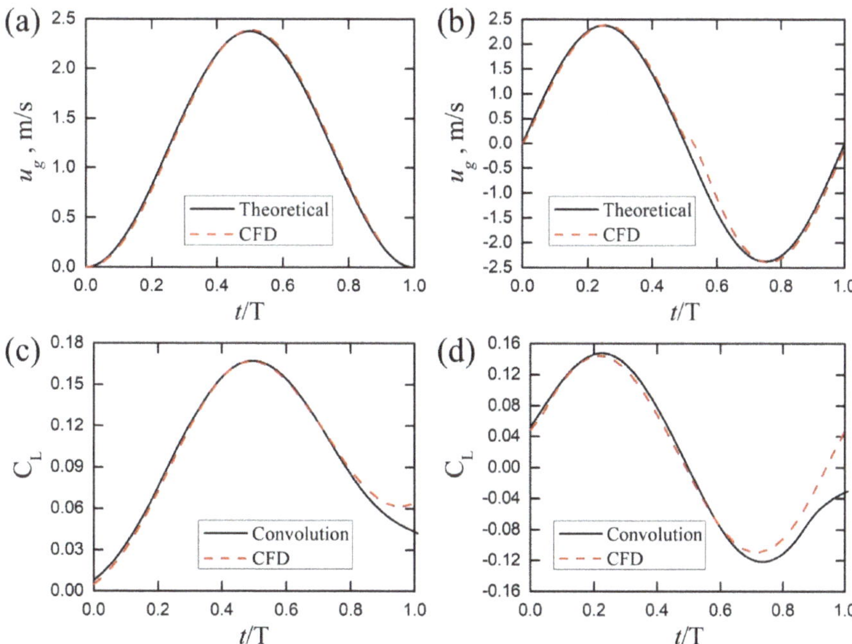

Figure 17. Comparisons of (**a**) the 1-cosine and (**b**) sinusoidal vertical gust velocity profiles between the theoretical and current CFD results, followed by comparisons of lift coefficient (C_L) responses of the NACA 0006 airfoil to the two gusts in (**c**) and (**d**), respectively.

3.5. Gust Discretization Examination

As the internal flow characteristics of the serpentine inlet are highly sensitive to the freestream conditions, the resolution of the gust velocity profile should be guaranteed to minimize the unfavorable effect of gust shape discretization on the accuracy of the calculated results. Therefore, the influence of the number of gust discretization intervals, N, on the aerodynamic responses of the serpentine inlet during a full period of the gust is first investigated. The settings of the material and gust parameters for the following gust tests are tabulated in Table 4.

Table 4. Values of the material parameters for the inlet gust response calculations.

Variable	Value
Ambient pressure, p_0, Pa	101,325
Inlet exit pressure, p_e, Pa	86,126.25
Air density, ρ_0, kg/m^3	1.17
Air dynamic viscosity, μ_0, Pa·s	1.82×10^{-5}
Ambient temperature, T_0, K	300
Inlet exit temperature, T_e, K	300
Freestream Mach number, M	0.235
Freestream velocity, u_0, m/s	80
Gust frequency, f, Hz	25, 50, 75, 100, 125, 150, 175, 200, 225, 250, 275, 300
Gust amplitude, $\tilde{u}_g$, m/s	4, 8, 12, 16, 20

The changes of the horizontal flow velocity probed at three locations in the symmetry plane and the various inlet performances at the AIP with the variation of N are shown in Figure 18. It is seen from Figure 18a to Figure 18c that the influence of N on the flow velocity becomes more and more apparent when approaching the AIP. The gusty flow velocity at locations far from the serpentine inlet coincides with the specified gust profile. However, both the mean velocity and fluctuation amplitude are considerably increased near and inside the serpentine inlet due to the suction effect of the favorable backpressure set at the exit of the serpentine inlet. It can also be seen that the flow velocity profile gradually loses the original sinusoidal shape in the proximity of the serpentine inlet and that an interaction with the intrinsically unsteady flow field within the serpentine inlet takes place, resulting in strong high-frequency oscillations at the AIP, as shown in Figure 18c. In general, the current gust discretization resolutions of interest have obtained negligible differences in the simulation results. To minimize its influence on the prediction accuracy of the gust response of the inlet, the maximum number of gust discretization intervals, i.e., N = 2000, is selected for all the later calculations.

3.6. Gust-Inlet Coupled Flow Field Characteristics

Gust responses of inlet performance are fundamentally contributed by the interaction between the gust and the inlet. Therefore, it is necessary to first have an overall understanding of the characteristics of the gust-inlet interacted flow field. Figure 19 shows the velocity vectors combined with the streamline distributions near the flow separation zone at the top surface for the gust case at four-time instants corresponding to the four extreme points of the total pressure recovery shown later. They are t1/T = 1.375 (acceleration phase), t1/T = 1.825 (deceleration phase), t1/T = 2.325 (acceleration phase) and t1/T = 2.80 (deceleration phase). Compared with the results in the absence of gust as shown in Figure 10c, a remarkable change is the enlarged flow separation zone at all the instants under the interaction of gust, suggesting that both the acceleration and deceleration phases of the gust promote the extent of flow separation in the serpentine inlet. Besides, a couple of vortexes is formed at the two instants of the deceleration phase (Figure 19b,d), which will be discussed later in more detail.

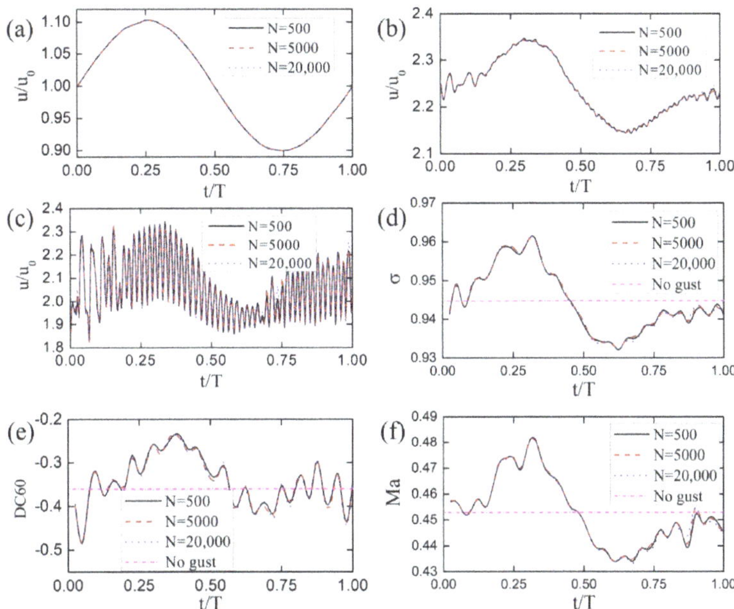

Figure 18. Effects of the number of gust discretization intervals for a full-period gust, N, on the CFD results of the gust velocity at three probe locations in the symmetry plane (**a**) in the farfield, (**b**) at the middle of the entrance of the diffuser and (**c**) at the middle of the AIP position; Effects of N on (**d**) the total pressure recovery, (**e**) circumferential distortion and (**f**) outlet Mach number at the AIP. The simulated gust frequency is $f = 50$ Hz and amplitude $\tilde{u}_g = 8$ m/s.

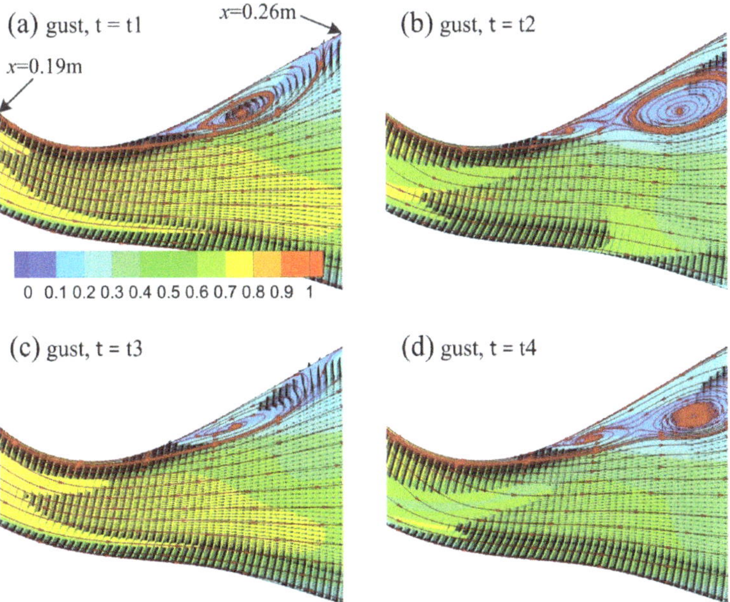

Figure 19. Velocity vector, streamline, and Mach number distribution near the flow separation zone for the gust case of $f = 100$ Hz and $\tilde{u}_g = 8$ m/s at the four instants.

As found from the results above, both the acceleration and deceleration phases of the periodic gust have caused strong flow separations at the lee side of the top surface. However, the dominating mechanisms behind the two cases are not the same. Figure 20 presents a comparison of the static pressure distribution on the top surface in the symmetry plane between the no-gust case and the four instants of the gust case of f = 100 Hz and $\tilde{u}_g$ = 8 m/s. It is seen that the adverse pressure gradient is much larger than that in the absence of gust at the two instances of the acceleration phase (i.e., t1 and t3). This is due to the velocity increase during the acceleration phase and a resulting decreased static pressure at the lee side of the top surface upstream of the separation region (x = 0.19 m), which leads to a stronger adverse pressure gradient further downstream. The main flow characteristic during the deceleration phase (i.e., t2 and t4) is characterized by the development of two separation vortices. The vorticity of these coupling vortices is significantly higher compared to the single vortex in the absence of gust and also to the single vortex developing in the acceleration phase, as is demonstrated in Figure 21. As is shown in Figure 20b, the adverse pressure gradient in the region between x = 0.19 m and x = 0.22 m during the deceleration phase (i.e., t2 and t4) is essentially equal to that of the no-gust case. However, the kinetic energy of the flow is significantly lower, which leads to earlier and much stronger flow separation on the lee side. In conclusion, this analysis demonstrates that flow separation is significantly enhanced during both the acceleration and the deceleration phase, but the flow is more strongly affected during the deceleration phase.

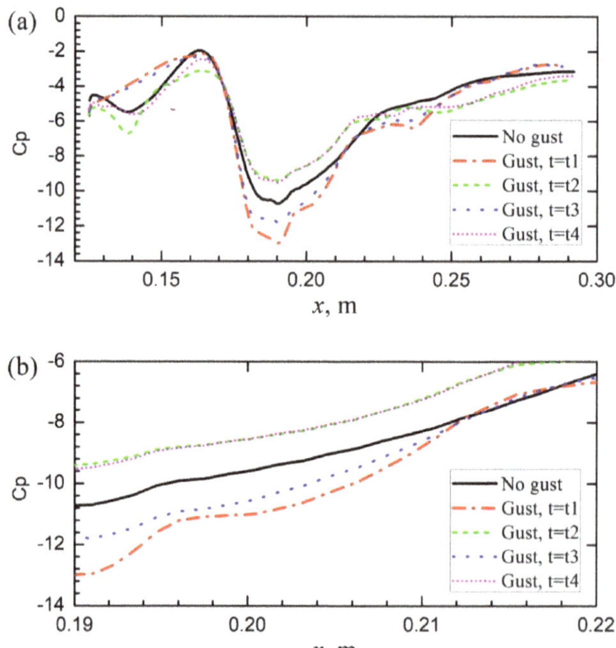

Figure 20. Static pressure distributions on the top surface of the inlet under the no-gust condition and at different phases of the gust (f = 100 Hz and $\tilde{u}_g$ = 8 m/s) at M = 0.235 and p_{AIP}/p_0 = 0.85: (**a**) the overall view and (**b**) amplified view in the adverse-pressure-gradient zone.

Figure 22 presents a comparison of the total pressure contours at the AIP between the four instances of the gust. It is observed that both the structure and magnitude of the total pressure distribution are changed due to the interaction between the gust and the inlet. All the five low-total-pressure zones marked in Figure 10c have been enlarged by the gust, indicating greater total pressure losses within these zones. The secondary vortices are even larger at times t2 and t4, and this aspect can be detrimental to the engine further downstream. However, the average total pressure losses remain within 4% of the total pressure recovery in the absence of gust, as will be discussed in more detail later in this paper. Moreover, the difference in the total pressure distribution seems not to have changed the structure of the secondary vortices at the AIP other than the scope of the vortex pairs, as shown in Figure 23.

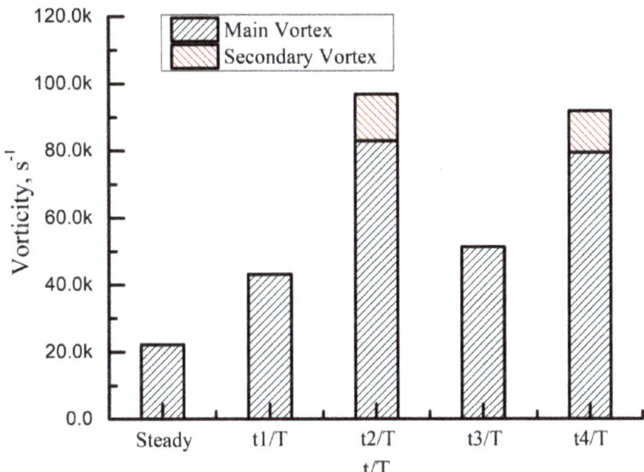

Figure 21. Vorticity of the vortex cores shown in Figure 19. For both the cases with two vortex cores, the upstream one has the larger vorticity and is thus named as 'main vortex'.

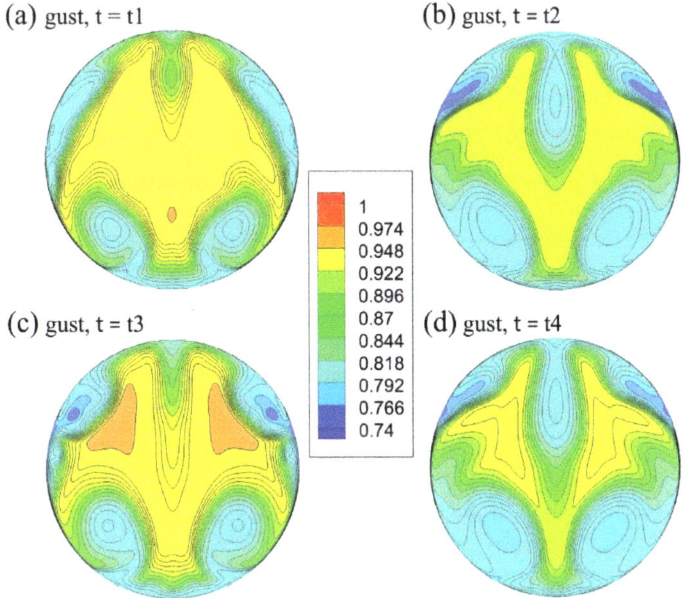

Figure 22. Total pressure contours at the AIP at four instances of the gust case of f = 100 Hz and $\tilde{u}_g$ = 8 m/s.

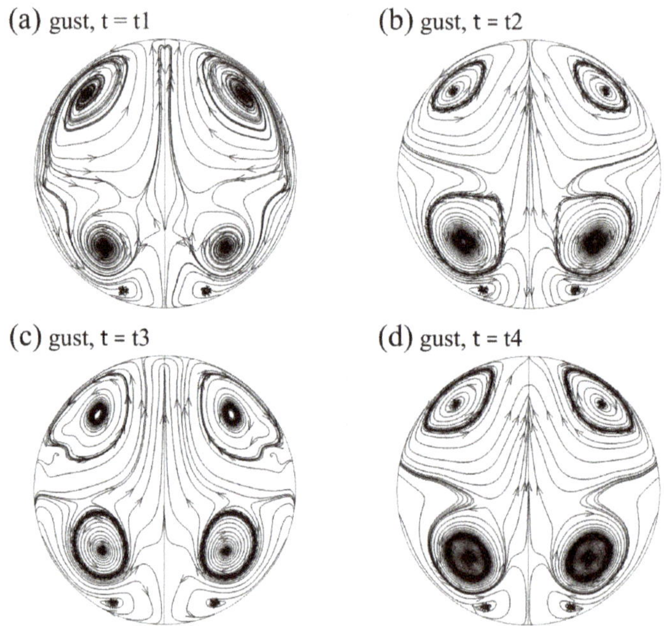

Figure 23. Streamline distributed at the AIP at four instances of the gust case of f = 100 Hz and $\tilde{u}_g = 8$ m/s.

3.7. Inlet Aerodynamic Performance Responses to Gusts

This subsection deals with the influences of sinusoidal gusts on the inlet aerodynamic performance for both the gust frequency and amplitude. For the gust frequency influence study, the gust amplitude is fixed at $\tilde{u}_g = 10\%u_0 = 8$ m/s while the frequency varies from 25 to 300 Hz. To study the influence of gust amplitude, the gust frequency is fixed at f = 50 Hz while the amplitude varies from 4 to 20 m/s, considering the higher convergence rate of the performance response at this frequency.

Figure 24 compares the inlet performance between various gust frequencies. Both performance parameters start to respond to the sinusoidal gusts immediately from the initiation of gusts and finally converge to the mean value (not shown for frequencies larger than 50 Hz). Due to the intrinsic unsteadiness of the inlet flow field without gusts, as has been discussed in Section 3.3, several sinusoidal peaks of the total pressure at the AIP are observed when the gusts are present in the flow field. However, the existence of gusts has significantly intensified the fluctuation of the total pressure, compared with that solely due to the three-dimensional effect and boundary-layer separation as well as flow separation on the top surface of the inlet in the absence of gusts. Here, we define a parameter called response time to account for the period during which gusts impart significantly large fluctuations in the inlet performance. It can be seen that the response time of both performance parameters is nearly identical to the gust period at the relatively low frequencies of f = 25 Hz and 50 Hz. However, both the amplitude and time of the two performance responses increase rapidly as the gust frequency is increased to 75 Hz. This is considered to be due to the intrinsic flow field characteristics in the absence of gusts, which have a fluctuating frequency of around 100 Hz as revealed in Section 3.3. Therefore, the interaction between the gust and the inlet may have caused a resonance in the flow field characteristics beyond 75 Hz, which magnifies the unsteady fluctuations in the flow field, as will be verified quantitatively later in Table 5.

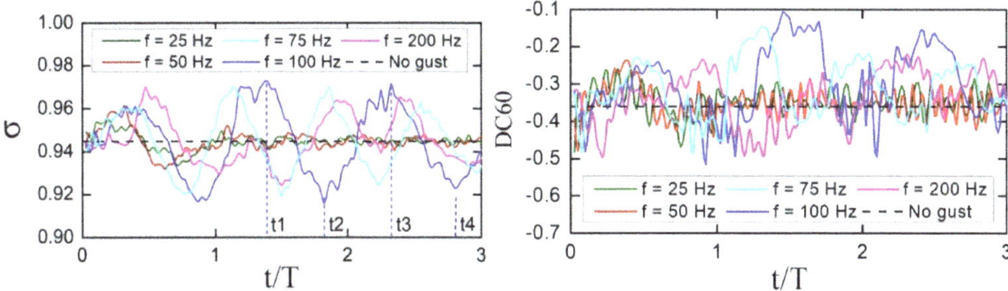

Figure 24. Time histories of the inlet performance at various gust frequencies combined with the amplitude of $\tilde{u}_g = 8$ m/s.

Table 5. Comparison of the maximum amplitudes of total pressure recovery and distortion fluctuations between the cases without/with gusts.

Variable	Without Gust	With Gust
σ_m/σ_0	0.48%	3.83%
$DC60_m/DC60_0$	16.00%	56.39%

Taking the same practice of extracting the amplitude of gust response in external flows [44], Figure 25 presents the minimum, maximum, and average values of the inlet performance responses over a wide range of frequencies, accompanied by the results of the no-gust case. Note that the minimum and maximum values represent the valley and crest of the unsteady responses shown in Figure 24. It is clear to observe that the minimum- and maximum-value curves are approximately symmetrical to the mean values which are close to the results with no gust. Despite this result, the impacts of the gust-inlet interaction are different from the two performance parameters. To the total pressure recovery, its reduction by the deceleration phase of gust yet can be remedied by the following counterpart phase. To the total pressure distortion, however, its reduction means an increasing non-uniformity of the flow field, which will directly be transferred to the engine face further downstream and may cause a sudden and instantaneous occurrence of stall and surge of the compressor in the engine. To be concrete, for σ, the most severe effect of the gusts occurs at the gust frequency $f = 100$ Hz, at which σ reaches the lowest, as shown in Figure 12a. This should be due to the resonance caused by the interaction between the gust and the inlet flow field possessing a natural fluctuation frequency of 100 Hz, as shown in Figure 12b. For DC_{60}, it is interesting to observe a slight enhancement of the average value between $f = 75$ Hz and 150 Hz. However, as analyzed above, the minimum value of the distortion has the most important significance to the inlet, which remains at a low level within the same range of frequency and generally decreases with the increasing gust frequency. More quantitative evaluations of the gust influence on the variations of the total pressure recovery and distortion will be presented later.

Figure 26 compares the time histories of inlet total pressure recovery and distortion index at the AIP for various gust amplitudes. Generally, the total pressure recovery responses show a similar sinusoidal pattern as that of the gust shape with progressive growth in amplitude with the increasing gust amplitude. In contrast, the majority of the distortion response happens in the first half period, during which the gust plays a positive role in increasing the streamwise flow velocity as well as decreasing the distortion.

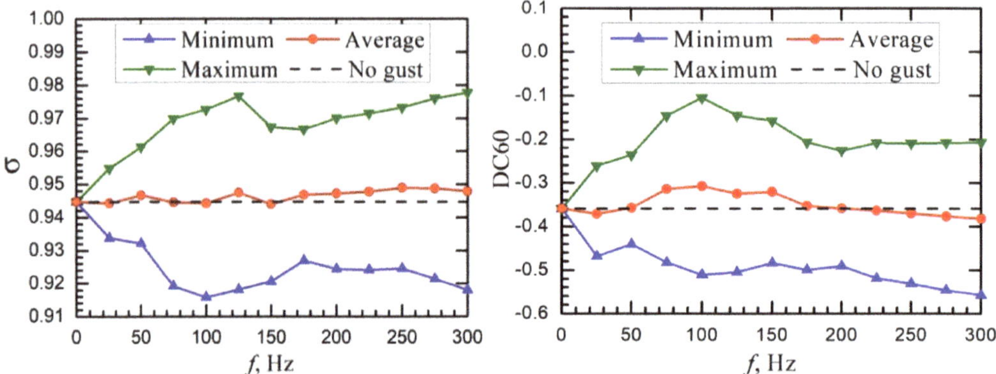

Figure 25. Performance responses of the inlet to the gusts over a wide range of frequency at the constant amplitude of $\tilde{u}_g = 8$ m/s. The mean results of the no-gust case are also shown for comparison.

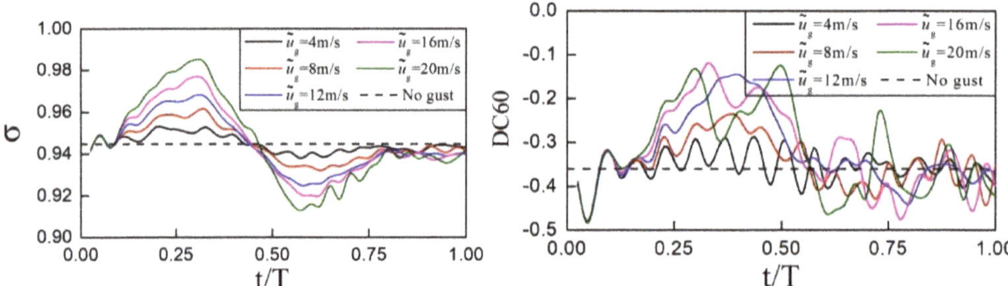

Figure 26. Time histories of the inlet performance responses to the gusts with various amplitudes at a frequency of $f = 50$ Hz.

Figure 27 presents the dependence of the minimum, maximum, and average values of the inlet performance responses on gust amplitude, as well as the unsteady calculation results of the no-gust case. Compared with the case of gust frequency, the dependence on gust amplitude is relatively simpler. In general, the minimum and maximum responses of total pressure recovery assume approximately symmetrical to the mean values, causing the average response to be lying in the vicinity of the mean values. Moreover, the average response of the distortion index lies above the mean value beyond $\tilde{u}_g = 8$ m/s, indicating a positive role of the gusts in decreasing the flow distortion at the AIP of the serpentine inlet. However, it should be clarified again that the gust-induced deterioration of the flow field homogeneity is a more important response index.

The above results have shown that sinusoidal gusts induce significant oscillations in inlet aerodynamic performance. Therefore, the oscillation amplitude is also an important index to weigh the influence extent of gusts. Like the practice adopted in wing-gust response studies [9,44], two non-dimensional index parameters are defined here, i.e., total pressure recovery amplitude

$$\sigma_m / \sigma_0 = \frac{\sigma_{max} - \sigma_{min}}{2\sigma_0} \times 100\% \tag{12}$$

and total pressure distortion amplitude

$$DC60_m / DC60_0 = \frac{DC60_{max} - DC60_{min}}{2DC60_0} \times 100\% \tag{13}$$

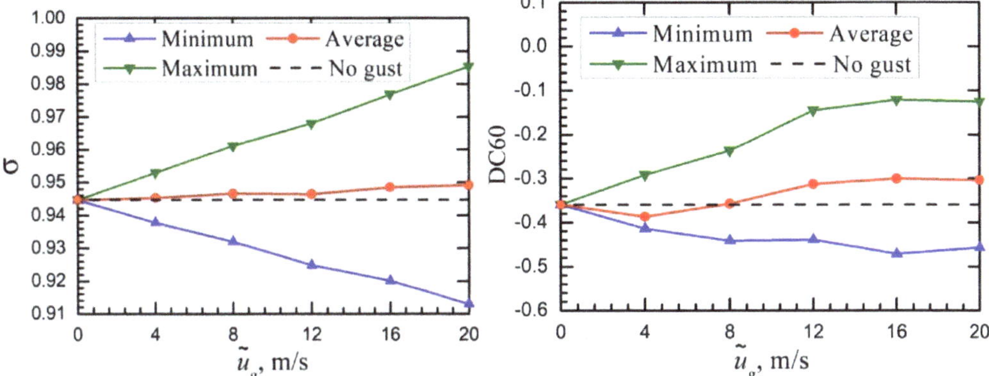

Figure 27. The inlet performance responses to the gusts with various amplitudes at a frequency of $f = 50$ Hz. The results of the no-gust case are also presented for comparison purpose.

The results of both parameters are presented in Figure 28. Both the total pressure recovery and distortion amplitudes experience an ascending, descending, and re-ascending process as gust frequency increases, with the maximum at $fD/u_0 = 0.102$ and 0.081 (i.e, $f = 125$ Hz and 100 Hz, respectively). On the other hand, the total pressure recovery amplitude increases linearly with the increasing gust amplitude over the full range, while the distortion amplitude increases monotonously till $\tilde{u}_g/u_0 = 0.2$ (i.e, $\tilde{u}_g = 16$ m/s) and drops beyond that. These results indicate that the dependence of the inlet performance response on the gust frequency is much more complicated than that on the gust amplitude.

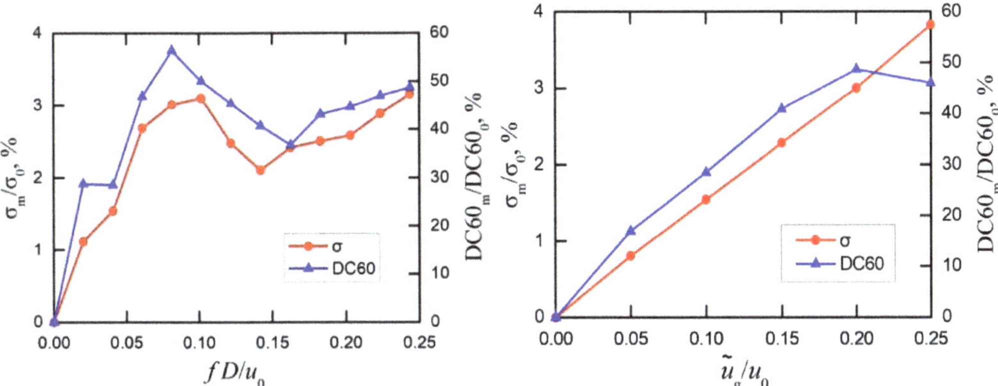

Figure 28. Non-dimensional amplitudes of the total pressure recovery and distortion responses for the non-dimensional gust frequency fD/u_0 and non-dimensional gust amplitude $\tilde{u}_g/u_0$, respectively.

To compare the effect of gusts on increasing the fluctuations of the inlet performance with the intrinsic unsteadiness of the flow field in the absence of gusts, Table 5 lists the maximum amplitudes of total pressure recovery and distortion fluctuations for the gust cases investigated in this study as well as for the no-gust case. This demonstrates that the gusts have amplified the total pressure recovery and distortion fluctuations by several times the fluctuations caused by the intrinsic unsteadiness of the flow field. This may validate the above hypothesis of the production of resonance in the flow field by the interaction between the gusts and the inlet.

4. Conclusions and Outlook

In conclusion, the effects of horizontal sinusoidal gusts on the flow field and aerodynamic performance of a serpentine inlet were investigated based on a field velocity method implemented in OpenFOAM. Before starting the gust response simulations, preliminary work was done to guarantee the numerical solution accuracy. First, the accuracy of the baseline CFD solver was validated via a thorough comparison between the numerical results and wind-tunnel experimental data for the flow field and aerodynamic characteristics of the inlet at a high flight Mach number M = 0.7 in the absence of gusts. Second, the accuracy of the gust modeling method and the dependence of the modeling results on the model parameters were carefully examined to guarantee accurate inputs of the gust shapes of interest. Third, because gusts are more often to occur and more severe at aircraft take-off and landing phases, the flow field of the inlet at a relatively lower flight Mach number M = 0.235 was calculated to serve as a baseline for all the following gust response simulations. Finally, some gust response computations were run to investigate both the flow field and aerodynamic characteristics of the serpentine inlet under horizontal sinusoidal gust conditions.

It was found that both the flow field and aerodynamic performance of the inlet are significantly affected by the horizontal sinusoidal gusts. As for the flow field responses to the gusts, an increased area of the low-momentum fluid zone is formed by the gust, which includes not only the amplification of the flow separation zone in the absence of the gusts but also new additions in the other locations in the flow field. The deceleration phase of the sinusoidal gusts is more detrimental to the flow field characteristics of the serpentine inlet than the acceleration phase. More total pressure losses have occurred at typical locations of the AIP under gust conditions compared with the no-gust case. However, the secondary flow pattern at the AIP did not experience significant influence from the gusts, but the scopes of the vortex pairs were altered by the gust at different moments. The aerodynamic performance of the serpentine inlet has also significantly been affected by the horizontal sinusoidal gusts. Increased oscillations in both the total pressure recovery and distortion are observed, with both the amplitude and time increased rapidly beyond the gust frequency of 75 Hz. The minimum and maximum responses of both performance responses assume approximately symmetrical to the mean values, causing the average response to be lying in the vicinity of the mean value. Both the total pressure recovery and distortion amplitudes show a complex changing trend with the increasing gust frequency but an approximately linear relationship with the gust amplitude in most cases. It is worthy of special attention that the deterioration of the total pressure distortion caused by sinusoidal gusts cannot be counteracted and directly affects the quality of the air ingested by the engine, although the average responses are close and sometimes even above the mean values in the no-gust case.

From the results of this paper, it is clear that gusts can influence significantly the internal flow characteristics of the engine inlets. Therefore, it is suggested that more attention should be paid to the adverse effects of gusts on inlet performance which directly affects the quality of the air entering the gas generator of gas turbine engines. To study the aerodynamic performance of inlets under gusty conditions, considering the velocity and scale of atmospheric gusts, engine inlets possessing complex geometries and operating at subsonic flight speeds deserve special attention. It should be pointed out that the present study only provides a preliminary insight into the influence of gusts on aircraft engines. Further studies may be conducted by combing the compressor with the inlet to have a more comprehensive understanding of the inlet-compressor coupled responses to horizontal sinusoidal gusts. Last but not the least, experimental tests are necessary to characterize the gusty inflow condition and visualize the gust-affected internal flow characteristics in wind tunnels. A major disadvantage of the current study is that all the gust response results in the current study are based on pure CFD simulations, although the gust modeling method has carefully been examined before the massive calculation. The good news is that we have designed and manufactured a novel double-vane gust generator apparatus recently, which will be mounted in our low-speed wind tunnel for tests soon.

Author Contributions: Conceptualization, S.S.; methodology, Z.W.; software, Z.W.; validation, S.S., Z.W. and H.H.; formal analysis, S.S.; resources, H.T.; data curation, Z.W.; writing—original draft preparation, S.S. and Z.W.; writing—review and editing, G.B.; supervision, Z.W.; project administration, H.T.; funding acquisition, H.T. All authors have read and agreed to the published version of the manuscript.

Funding: This research was funded by [the National Natural Science Foundation of China] grant number [12172174] and [the start-up fund of Nanjing University of Aeronautics and Astronautics] grant number [90YQR21012].

Data Availability Statement: The data within the article will be available upon request.

Conflicts of Interest: The authors declare that they have no known competing financial interests or personal relationships that could have appeared to influence the work reported in this paper.

Nomenclature

c	airfoil chord length
C_L	lift coefficient
dS	surface element
D	exit diameter of the diffuser, 65 mm
DC_{60}	circumferential total pressure distortion index
$\vec{F}_c$	vector of convective fluxes
$\vec{F}_v$	vector of viscous fluxes
f	gust frequency
k	turbulent kinetic energy
L	length of the serpentine inlet, mm
M	Mach number
N	number of gust discretization intervals
p	static pressure, Pa
p_0	ambient pressure
p_e	inlet exit pressure
Q	Q-criterion based vorticity
$\vec{Q}$	source term
t	time
T	gust period
T	temperature
T_0	ambient temperature
T_e	inlet exit temperature
$\tilde{u}_g$	gust amplitude
u	velocity component along the x-coordinate direction
v	velocity component along the y-coordinate direction
w	velocity component along the z-coordinate direction
$\vec{V}$	velocity vector
W	width of the serpentine inlet, mm
σ	mass flow averaged total pressure recovery coefficient
Ω	control volume
ω	specific dissipation rate
ρ_0	Ambient air density
μ_0	Ambient air dynamic viscosity
Superscripts *	total
Subscripts	state
AIP	aerodynamic interface plane
0	freestream condition
g	gust condition

τ	grid time metrics
avg	average value
min	minimum value
max	maximum value

References

1. Proctor, F.H.; Hamilton, D.W.; Rutishauser, D.K.; Switzer, G.F. *Meteorology and Wake Vortex Influence on American Airlines FL-587 Accident*; NASA/TM-2004-213018; NASA: Washington, DC, USA, 2004.
2. David, R.H. 15 November 1993. Available online: https://www.ntsb.gov/safety/safety-recs/RecLetters/A93_136_141.pdf (accessed on 17 October 2022).
3. Richards, M. 8 July 2015. Available online: https://data.ntsb.gov/Docket/?NTSBNumber=GAA15CA172 (accessed on 17 October 2022).
4. Wu, Z.; Cao, Y.; Ismail, M. Gust loads on aircraft. *Aeronaut. J.* **2019**, *123*, 1216–1274. [CrossRef]
5. Bertolin, R.; Chaves Barbosa, G.; Cunis, T.; Kolmanovsky, I.V.; Cesnik, C.E. Gust Rejection of a Supersonic Aircraft during Final Approach. In Proceedings of the AIAA Scitech 2022 Forum, San Diego, CA, USA, 3–7 January 2022; p. 2174.
6. Halwas, H.K. Side Gust Effects on the Performance of Supersonic Inlet with and without Bleed System Using RANS and URANS. Ph.D. Thesis, University of Illinois, Chicago, IL, USA, 2022.
7. Ullah, A.H.; Rostad, B.L.; Estevadeordal, J. Three-cylinder rotating system flows and their effects on a downstream dimpled airfoil. *Exp. Therm. Fluid Sci.* **2021**, *124*, 110343. [CrossRef]
8. Engin, K.; Aydin, E.; Zaloglu, B.; Fenercioglu, I.; Cetiner, O. Large Scale Spanwise Periodic Vortex Gusts or Single Spanwise Vortex Impinging on a Rectangular Wing. In Proceedings of the 2018 Fluid Dynamics Conference, Atlanta, GA, USA, 25–29 June 2018; p. 3086.
9. Wei, N.J.; Kissing, J.; Westner, T.T.; Wegt, S.; Schiffmann, K.; Jakirlic, S.; Hölling, M.; Peinke, J.; Tropea, C. Insights into the periodic gust response of airfoils. *J. Fluid Mech.* **2019**, *876*, 237–263. [CrossRef]
10. Cordes, U.; Kampers, G.; Meißner, T.; Tropea, C.; Peinke, J.; Hölling, M. Note on the limitations of the Theodorsen and Sears functions. *J. Fluid Mech.* **2017**, *811*. [CrossRef]
11. Boulbrachene, K.; De Nayer, G.; Breuer, M. Assessment of two wind gust injection methods: Field velocity vs. split velocity method. *J. Wind. Eng. Ind. Aerodyn.* **2021**, *218*, 104790. [CrossRef]
12. Li, Y.; Qin, N. Gust load alleviation by normal microjet. *Aerosp. Sci. Technol.* **2021**, *117*, 106919. [CrossRef]
13. Wales, C.; Jones, D.; Gaitonde, A. Prescribed Velocity Method for Simulation of Aerofoil Gust Responses. *J. Aircr.* **2015**, *52*, 64–76. [CrossRef]
14. Badrya, C.; Jones, A.R.; Baeder, J.D. Unsteady aerodynamic response of a flat plate encountering large-amplitude sharp-edged gust. *AIAA J.* **2022**, *60*, 1549–1564. [CrossRef]
15. Badrya, C.; Baeder, J.D.; Jones, A.R. Application of prescribed velocity methods to a large-amplitude flat-plate gust encounter. *AIAA J.* **2019**, *57*, 3261–3273. [CrossRef]
16. Badrya, C.; Biler, H.; Jones, A.R.; Baeder, J.D. Effect of gust width on flat-plate response in large transverse gust. *AIAA J.* **2021**, *59*, 49–64. [CrossRef]
17. Heinrich, R.; Reimer, L. Comparison of Different Approaches for Gust Modeling in the CFD Code TAU. In Proceedings of the International Forum on Aeroelasticity & Structural Dynamics, Bristol, UK, 24–27 July 2013.
18. Reimer, L.; Ritter, M.; Heinrich, R.; Krüger, W. CFD-Based Gust Load Analysis for a Free-Flying Flexible Passenger Aircraft in Comparison to a DLM-Based Approach. In Proceedings of the AIAA Computational Fluid Dynamics Conference, Dallas, TX, USA, 22–26 June 2015.
19. Robert, B. Developing an Accurate CFD Based Gust Model for the Truss Braced Wing Aircraft. In Proceedings of the 31st AIAA Applied Aerodynamics Conference, San Diego, CA, USA, 24–27 June 2013; p. 3044.
20. Bekemeyer, P.; Thormann, R.; Rimme, S. Rapid gust response simulation of large civil aircraft using computational fluid dynamics. *Aeronaut. J.* **2017**, *121*, 1–13. [CrossRef]
21. Bekemeyer, P.; Timme, S. Flexible aircraft gust encounter simulation using subspace projection model reduction. *Aerosp. Sci. Technol.* **2019**, *86*, 805–817. [CrossRef]
22. Liu, H.; Huang, R.; Zhao, Y.; Hu, H. Reduced-order modeling of unsteady aerodynamics for an elastic wing with control surfaces. *J. Aerosp. Eng.* **2017**, *30*, 04016083. [CrossRef]
23. Halder, R.; Damodaran, M.; Khoo, B. Deep learning based reduced order model for airfoil-gust and aeroelastic interaction. *AIAA J.* **2020**, *58*, 4304–4321. [CrossRef]
24. Bekemeyer, P.; Ripepi, M.; Heinrich, R.; Görtz, S. Nonlinear unsteady reduced-order modeling for gust-load predictions. *AIAA J.* **2019**, *57*, 1839–1850. [CrossRef]
25. Wu, Z.; Wang, Q.; Huang, H. A methodological exploration for efficient prediction of airfoil response to gusts in wind engineering. *Proc. Inst. Mech. Eng. Part A J. Power Energy* **2019**, *233*, 738–750. [CrossRef]
26. Wasserbauer, J.F. *Dynamic Response of a Mach 2.5 Axisymmetric Inlet with Engine or Cold Pipe and Utilizing 60 Percent Supersonic Internal Area Contractio*; NACA TN D-5338; NASA: Washington, DC, USA, 1969.

27. Grenson, P.; Beneddine, S. Analysis of shock oscillations of an external compression supersonic inlet through unsteady numerical simulations. In Proceedings of the 2018 Applied Aerodynamics Conference, Atlanta, GA, USA, 25–29 June 2018; p. 3011.
28. Halwas, H.K.; Aggarwal, S. Effect of Side Gust on Performance of External Compression Supersonic Inlet. *J. Aircr.* **2019**, *56*, 569–582. [CrossRef]
29. Halwas, H.K.; Aggarwal, S. Side Gust Effects on the Performance of a Supersonic Inlet with Bleed. *J. Aircr.* **2019**, *56*, 2357–2370. [CrossRef]
30. Kozakiewicz, A.; Frant, M. Analysis of the gust impact on inlet vortex formation of the fuselage-shielded inlet of an jet engine powered aircraft. *J. Theor. Appl. Mech.* **2013**, *51*, 993–1002.
31. Kozakiewicz, A.; Frant, M. Numerical Analysis of the Intake Vortex Formation in the Case of a Double Fuselage Shielded Inlet. *J. Theor. Appl. Mech.* **2014**, *52*, 757–766.
32. Übelacker, S.; Hain, R.; Kähler, C. Experimental Investigation of the Flow in a Stalling Engine Inlet. In Proceedings of the AIAA Applied Aerodynamics Conference, Atlanta, GA, USA, 16–20 June 2014.
33. Sun, S.; Tan, H. Flow Characteristics of an Ultracompact Serpentine Inlet with an Internal Bump. *J. Aerosp. Eng.* **2018**, *31*, 04017089. [CrossRef]
34. Sun, S.; Guo, R.W. Serpentine Inlet Performance Enhancement Using Vortex Generator Based Flow Control. *Chin. J. Aeronaut.* **2006**, *19*, 10–17. [CrossRef]
35. Menter, F.R. Two-equation eddy-viscosity turbulence models for engineering applications. *AIAA J.* **1994**, *32*, 1598–1605. [CrossRef]
36. Blazek, J. *Computational Fluid Dynamics: Principles and Applications*; Butterworth-Heinemann: Oxford, UK, 2015.
37. Wu, Z.; Bangga, G.; Cao, Y. Effects of lateral wind gusts on vertical axis wind turbines. *Energy* **2019**, *167*, 1212–1223. [CrossRef]
38. Wu, Z. Rotor power performance and flow physics in lateral sinusoidal gusts. *Energy* **2019**, *176*, 917–928. [CrossRef]
39. Greenshields, C. *OpenFOAM User Guide Version 4.0*; OpenFOAM Foundation Ltd.: London, UK, 2016.
40. Kraposhin, M.V.; Ryazanov, D.A.; Elizarova, T.G. Numerical algorithm based on regularized equations for incompressible flow modeling and its implementation in OpenFOAM. *Comput. Phys. Commun.* **2022**, *271*, 108216. [CrossRef]
41. *Ansys ICEM CFD Tutorial Manual*; Release 2022 R1; ANSYS, Inc.: Canonsburg, PA, USA, 2022.
42. Tajnesaie, M.; Jafari Nodoushan, E.; Barati, R.; Azhdary Moghadam, M. Performance comparison of four turbulence models for modeling of secondary flow cells in simple trapezoidal channels. *ISH J. Hydraul. Eng.* **2020**, *26*, 187–197. [CrossRef]
43. Ye, C.; Wang, F.; Wang, C.; van Esch, B.P. Assessment of turbulence models for the boundary layer transition flow simulation around a hydrofoil. *Ocean. Eng.* **2020**, *217*, 108124. [CrossRef]
44. Wu, Z.; Bangga, G.; Lutz, T.; Kampers, G.; Hölling, M. Insights into airfoil response to sinusoidal gusty inflow by oscillating vanes. *Phys. Fluids* **2020**, *32*, 125107. [CrossRef]

Article

On the Self-Similarity in an Annular Isolator under Rotating Feedback Pressure Perturbations

Zhongqi Luo [1], Hexia Huang [1,2,*], Huijun Tan [1,*], Gang Liang [1], Jinghao Lv [1], Yuwen Wu [2] and Liugang Li [3]

[1] Jiangsu Province Key Laboratory of Aerospace Power System, College of Energy and Power Engineering, Nanjing University of Aeronautics and Astronautics, Nanjing 210016, China
[2] National Key Laboratory of Transient Physics, Nanjing University of Science and Technology, Nanjing 210094, China
[3] Science and Technology on Space Physics Laboratory, Beijing 100076, China
*   Correspondence: huanghexia@nuaa.edu.cn (H.H.); thj@263.net (H.T.)

Abstract: In this paper, the transient flow simulation in an annular isolator under rotating feedback pressure perturbations simplified from the rotating denotation wave (RDW) is performed. The instantaneous flow characteristics and the self-similarity of the isolator flow-field are investigated in detail. It is found that a helical moving shock wave (MSW) and a quasi-toroidal terminal shock wave (TSW) are induced in the isolator. Hence, the flow-fields on the meridian planes could be classified into three zones, i.e., the undisturbed zone, the terminal shock wave/moving shock wave/boundary layer interaction (TSW/MSW/BLI) zone and the moving shock wave/boundary layer interaction (MSW/BLI) zone. The TSW/MSW/BLI zone is characterized by the coupling of the TSW/BLI and the MSW/BLI due to their small axial distance, which intensifies the adverse pressure gradient on the meridian planes, thus rolling up large separation bubbles developing along the MSW driven by the circular pressure gradient. In the MSW/BLI zone, the shock induces the boundary layer to separate, forming a helical vortex located at the foot of the MSW. During the upstream propagation process, the pattern of the MSWs transforms from a moving normal shock wave to a moving oblique shock wave with decreased strength. Moreover, after the collision with the MSWs, P, T_{emp} and S of the flow elevate with the prompt decrease of v_a, while v_θ increases to a higher level. Despite the deflection effect of the MSWs on the streamlines, the flow direction of the air still maintains an almost axial position at the exit, except in the adjacent region of the MSW. Likewise, three types of zones can be determined in the flow pattern at the exit: the rotating detonation wave/boundary layer interaction (RDW/BLI) zone, the expansion zone, and the vortices discharge zone. Comparing the transient flow patterns at different moments in one cycle and between adjacent cycles, an interesting discovery is that the self-similarity property is observed in the flow-field of the annular isolator under rotating feedback pressure perturbations. The global flow structure of the isolator at different moments shows good agreement despite its rotation with the RDW, and the surface pressure profiles of the corresponding meridian planes all match perfectly. Such a characteristic indicates that the rotation angular velocity of the TSW and the MSW are equal and hold invariant, and the isolator flow could be regarded as a quasi-steady flow. On this basis, the theoretical model of the inclination angles of the MSW by the coordinate transformation and velocity decomposition is developed and validated. The relative errors of the inclination angles between the predicted and measured results are below 3%, which offers a rapid method to predict the shape of the MSW, along with a perspective to better understand the physical meaning of the shape of the MSW.

Keywords: terminal shock wave/moving shock wave/boundary layer interaction; rotating denotation wave; helical vortex; isolator

Citation: Luo, Z.; Huang, H.; Tan, H.; Liang, G.; Lv, J.; Wu, Y.; Li, L. On the Self-Similarity in an Annular Isolator under Rotating Feedback Pressure Perturbations. *Aerospace* **2023**, *10*, 188. https://doi.org/10.3390/aerospace10020188

Academic Editor: Kung-Ming Chung

Received: 31 December 2022
Revised: 12 February 2023
Accepted: 14 February 2023
Published: 16 February 2023

1. Introduction

Due to the potential benefits of high thermal efficiency, rapid heat release and compact combustor, the rotating detonation engine (RDE) has attracted growing attention in recent years [1–3]. In a typical air-breathing rotating detonation combustor (RDC), a detonation wave propagates circumferentially along the annular combustor, ensuring the stability and continuity of detonative combustion [4]. As a critical component of the rotating detonation ramjet/scramjet engine (RDRE/RDSE), the isolator plays a key role in isolating the aerodynamic and thermal perturbations propagating upstream from the RDC and preventing the inlet from unstart [5,6]. Unlike the conventional deflagration combustion, the feedback pressure perturbations induced by the rotating detonation wave (RDW) propagates upstream in the axial and circumferential directions periodically, with a frequency of 1–10 kHz and an ultra-high amplitude [7]. These feedback pressure perturbations induce a helical moving shock wave (MSW) in the isolator, which interacts with the boundary layer and the terminal shock wave (TSW), thus directly affecting the aerodynamic performance of the inlet/isolator and the operating characteristics of the RDE, bringing great challenges to the design of the isolator. Hence, it is of great importance to investigate in depth the flow characteristics in the isolator under rotating feedback pressure perturbations with high frequency and amplitude.

Since the concept of rotating detonation proposed by Voitsekhovskii in the 1960s, [8] a series of studies have been carried out on flow characteristics in air-breathing RDCs. Zhdan [9] and Zhu [10] numerically studied the two-dimensional flow structure of hydrogen-oxygen RDC and carbon-air RDC, respectively. Two-dimensional and three-dimensional simulations on the RDC of H_2/O_2 and H_2/air were conducted by Uemura, and the detonation mechanism and related dynamics of the RDC were analyzed in a subtle way [11]. Rui tracked the paths of flow particles in a two-dimensional RDC, and evaluated the effect of the wave system of the detonation on the paths [12]. Smirnov [13,14] reconstructed the flow-field of the RDC, and the effect of the different equivalence ratios of the mixture on the modes of the RDW was analyzed. A common finding in experimental studies is that the variation of the operating conditions seems to result in the transition of propagation modes of the RDW, which could be mainly classified as the single wave mode, the multiple waves mode, the contrarotating waves mode, and the colliding waves mode, etc. [15–19]. Moreover, the RDW instability is another critical issue in the RDC flow-field investigations, a various types of which have been confirmed and analyzed in both experiments and simulations [20–24].

As a matter of common knowledge, the rotating feedback pressure perturbations propagating upstream from the RDC is one of the key factors affecting the flow structure and performance of the isolator, which have been observed by both simulations and experiments [16,25–27]. For the rotating detonation based aero-turbine engine, the isolator [28–30] is utilized and studied to dampen the effect of the feedback pressure on the compressor, the inflow of which is determined both by the inlet and compressor. In the isolator of a ramjet/scramjet-type RDE, where no compressor or turbine exists, the feedback pressure propagation from the RDC induces the upstream oblique shock wave (OSW), i.e., the helical MSW in the isolator, fairly affecting the characteristics of the isolator flow-field. The three-dimensional simulations conducted by Liu showed that the location of the upstream OSW would be affected by the total temperature and velocity of the inflow [31]. Furthermore, they validated the existence of the pressure feedback of the upstream OSW in the experiments [32]. In the numerical study by Dubrovskii, [33] due to the supersonic condition of the inflow, the pressure feedback could not propagate upstream to the isolator, two OSWs cycled along with the RDW in the combustor instead. Zhao [34] numerically explored the upstream influence of the RDW on the supersonic inlet, proposing that the TSW appears to adjust its location and strength with time in the diffuser. Wu performed the simulations of the RDE with a direct-connect Laval nozzle [35]. Their study indicated that the unsteady upstream OSW may destroy the internal flow of the RDE, ultimately affecting the fuel intake. On this basis, they further applied the particle trajectory method to investigate the influence of the OSWs on particle behavior and evaluated the pressure gain performance in detail [36].

From the literature, it is understood that the previous studies on feedback pressure and the upstream OSW are mainly focused on their impacts on the flow-field and operation of the RDC. The flow-field structure in the isolator under rotating feedback pressure perturbations and the mechanism of the interaction between the helical MSW and the boundary layer have not been thoroughly understood. Nevertheless, the propagation of the rotating feedback pressure perturbations and the moving shock wave/boundary layer interaction (MSW/BLI) would break the original axial symmetry of the isolator flow, enhancing the three-dimensionality and unsteadiness of the flow-field, making it extremely complex. This undoubtedly could have great influence on the thermal and aerodynamic performance of the RDE, thus requiring further studies. In addition, the issue on the self-similarity property of the isolator flow-field under such conditions has not been addressed and discussed. Furthermore, The theoretical model of the inclination angles of the MSW has never been developed. Therefore, in this paper, the above undisclosed flow characteristics in an annular isolator under rotating feedback pressure perturbations are investigated in depth, aiming to fill such a gap in the existing knowledge, and provide a theoretical basis for the isolator design of the RDE.

This paper is built on previous studies and is hereby, organized as follows. First, Sections 2 and 3 introduce the physical model of the isolator studied and the relevant details of the numerical setup in sequence. On this basis, the transient isolator flow characteristics under simplified rotating feedback pressure perturbations is discussed in detail in Section 4, which also details the terminal shock wave/helical moving shock wave/boundary layer interaction (TSW/MSW/BLI), along with the impacts of the terminal shock wave and the moving shock wave on the main flow and the flow patterns at the exit. Then, the spatial-temporal self-similarity property in the isolator flow-field at such a condition is discovered and discussed in Section 5. Then, in Section 6, a theoretical prediction method of the inclination angle of the helical MSW is put forward, based on the velocity decomposition method, and compared with the numerical result. Ultimately, the important results and findings obtained from the current study are discussed in Section 7.

2. Physical Model of the Isolator

The curved meridian plane of the co-axial annular isolator model studied in this paper is schematically presented in Figure 1, of which the length L_{iso} is 228.43 mm, the height of the entrance H_{ent} and the exit H_{exi} is 12 mm and 14 mm, respectively. In addition, the inner radius of the entrance $R_{\mathrm{ent\text{-}in}}$ and the exit $R_{\mathrm{exi\text{-}in}}$ of the isolator is 152.28 mm and 143 mm in sequence. Table 1 summarizes the specific geometric parameters of the isolator.

The isolator comprises two sections, that is, the S-shaped section and the straight terminal section with constant cross-section. The quadruplicate polynomial is utilized as the centerline function of the S-shaped section shown in Equation (1), where x varies from 0 to $L_{\mathrm{S\text{-}shaped}}$ (the length of the S-shaped section, which is 208.43 mm) and $R_{\mathrm{off,iso}}$ refers to the radial offset of the isolator equivalent to that of the S-shaped section:

$$\frac{r - (R_{\mathrm{ent-in}} + 0.5 H_{\mathrm{ent}})}{R_{\mathrm{off,iso}}} = -4\left(\frac{x}{L_{\mathrm{S-shaped}}}\right)^3 + 3\left(\frac{x}{L_{\mathrm{S-shaped}}}\right)^4 \tag{1}$$

In addition, the duct height H (which is defined as the radial difference between the outer and inner wall) distribution function of the S-shaped section along the axis are a cubic polynomial, as shown in Equation (2):

$$H = \left[3\left(\frac{x}{L_{\mathrm{S-shaped}}}\right)^2 - 2\left(\frac{x}{L_{\mathrm{S-shaped}}}\right)^3\right](H_{exi} - H_{\mathrm{ent}}) + H_{\mathrm{ent}}, \tag{2}$$

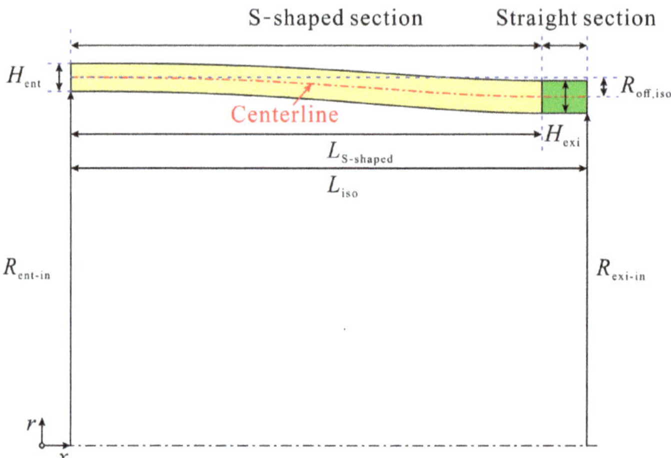

Figure 1. Sketch of the isolator model.

Table 1. Geometric parameters of the isolator.

Parameter	Value
L_{iso}, mm	228.43
$L_{S-shaped}$, mm	208.43
R_{ent-in}, mm	152.28
H_{ent}, mm	12.00
R_{exi-in}, mm	143.00
H_{exi}, mm	14.00
$R_{off,iso}$, mm	8.28

3. Methodology

3.1. Numerical Methods

Since the unsteady shock wave system/boundary layer interaction plays a vital role in the isolator study, a reliable numerical method that can capture the unsteady motion of the shock waves accurately is quite essential. In this paper, the three-dimensional unsteady compressible RANS equations are solved via the computational fluid dynamic software ANSYS FLUENT, which is based on the finite volume method. The turbulence is modeled by the κ–ω SST model, [37] while the piecewise polynomial method and the Sutherland model are selected to compute the specific heat and the viscosity coefficient, separately. All of the flow equations are discretized both spatially and temporally with second-order accuracy and the Green–Gauss cell based gradient approximation is adopted. In addition, unsteady flow solutions are sought at a fine time step of $\Delta t = 2 \times 10^{-7}$ s with 45 iterations per time step, which is 1×10^{-3} times of the cycle T (equaling to 2×10^{-4} s) of rotating feedback pressure perturbations. During calculation, the mass flow, the Mach number and the total pressure at the exit are monitored, along with the residuals. An absolute convergence criterion of 10^{-5} is satisfied at every time step in the continuity equation's scaled-residuals.

The computational domain shown in Figure 2a is all filled with hexahedral cells, and the near-wall grids are all encrypted with the spacing of 0.01 mm, ensuring the most of the y^+ and z^+ values are kept below 1. The details in the mesh sensitivity study are presented in Section 3.3. In addition, three primary metrics to evaluate the quality of the grids generated are paid special attention here, which are aspect ratio, equiangle skewness, and orthogonal quality. The aspect ratio is a measure of the stretching of the elements, which is referred to as the ratio of the lengths of the largest and the smallest edge, for the three-dimensional cell utilized here. The minimum value of the aspect ratio is 1.64 while the maximum is 179.18. The average value is 27.77, which is quite below the threshold of 100. The equiangle skewness is one of the primary measures determining how close a face or a cell is to the equilateral, affecting the accuracy and stability of the calculation. The maximum value of the grid is 0.049, the minimum is 0.003, and the average is 0.024, which is quite close to 0. The orthogonal quality ranges from 0 to1. When the orthogonal quality is closer to 1, the best quality of the grid is defined. The minimum orthogonal quality for all cells is 0.988, which is extremely close to 1. Hence, the quality of the grid used in this paper is rather good.

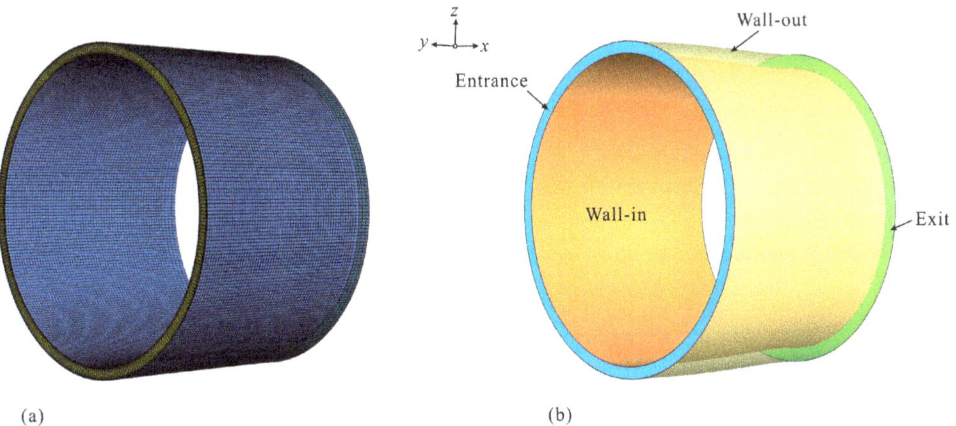

Figure 2. Surface grid and boundary conditions. (**a**) Surface grid. (**b**) Computational domain and boundary conditions.

As shown is Figure 2b, there are three types of boundary conditions utilized in this paper. The computational entrance is set as the pressure far field, and the exit of the isolator is set as the pressure outlet. Moreover, the inner and outer walls are set as an adiabatic non-slip wall. In this study, the incoming flow conditions are listed in Table 2, and the distribution of which is given by a profile calculated from an axisymmetric inlet with the design Mach number of 5 at the flight height of 24 km. Accordingly, the freestream pressure P_0 of the inlet/isolator is 2971.75 Pa. The inflow static pressure P_{ent} in Table 2 is normalized against P_0.

Table 2. Incoming flow conditions at the entrance.

Parameter	Value
Mean Mach number M_{ent}	2.67
Total pressure $P_{t,ent}$, kPa	1423.07
Static pressure P_{ent}	18.94 P_0
Static temperature T_{ent}, K	554.45
Nominal boundary layer thickness δ_{ent}, mm	1.09

3.2. Implementation of Rotating Feedback Pressure Perturbations

As mentioned in Section 1, the RDW alters the flow-field in the isolator essentially via the upstream propagation of its rotating feedback pressure perturbations. Hereby, imposing the unsteady pressure with the key properties of the RDW on the exit of the isolator would be a reasonable simplified simulation method, [38,39] instead of a real RDW, as the emphasis of this paper is placed on the effects of the rotating feedback pressure propagation on the isolator flow-field. Primarily, the radial discrepancy of the feedback pressure is ignored as $H_{\text{exi}} << R_{\text{exi-in}}$ and $H_{\text{ent}} << R_{\text{ent-in}}$. Moreover, the instabilities of the RDW are neglected [40], which renders the simplified RDW periodical.

As the RDW sweeps over the pressure transducer at the exit, the measured value increases steeply to a peak value and then falls back to the valley value at a relatively gentle rate [41]. In addition, since only the single RDW mode is considered here, the spatial distribution of the static pressure around the exit is essentially the same as the temporal distribution of the static pressure in a certain radial column of cell faces at the exit, despite the difference on the independent variable and the corresponding phase difference. Thus, a simplified model of the RDW is developed [38,39,42].

The column of cell faces with the azimuthal angle θ equaling $180°$ and named as the 1st column is chosen as an example to introduce the temporal function of the simplified model. The rotating frequency of the RDW is set to 5 kHz, a linear function is selected to model the ascent stage of the feedback pressure in one cycle given by Equation (3):

$$\frac{P_{b,1}}{p_0} = 45.92\left(100\frac{t}{T}\right) + 104.6 \tag{3}$$

where $P_{b,1}$ denotes the feedback pressure for the 1st column of cell faces at the exit, and t/T ranges from 0 to 0.0382. The subsequent steep drop of $P_{b,1}$ is also given by a linear function, as follows:

$$\frac{P_{b,1}}{p_0} = -13.09 \times \left(100\frac{t}{T}\right) + 330 \tag{4}$$

where t/T is from 0.0382 to 0.1146. Then, the final slight descending of $P_{b,1}$ is given by a trigonometric function with an exponential function as its independent variable listed in Equation (5), where the corresponding t/T is from 0.1146 to 1:

$$\frac{P_{b,1}}{p_0} = 156.64\sin\left(\pi \cdot e^{-0.02k}\right) + 104.6 \tag{5}$$

where

$$k = m \cdot \frac{t}{T} \tag{6}$$

In Equation (6), m is defined as the magnification factor, equivalent to 800 in this study.

Figure 3 presents the spatial distribution and the time-varying curve of the feedback pressure computed by Equations (3)–(5). The main parameters of the feedback pressure perturbations at the exit P_b are presented in Table 3, among which the pressure parameters are all non-dimensionalized by P_0.

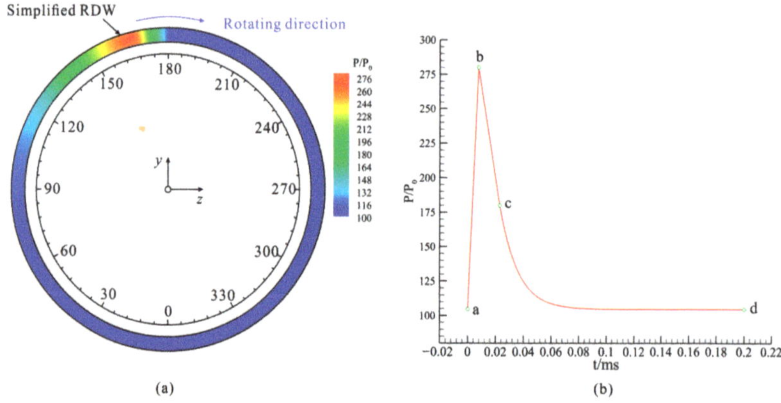

Figure 3. Spatial and temporal distribution of the feedback pressure. (**a**) Static pressure contour at the exit when t satisfies the equation floor $(t, T) = 0$. (**b**) The curve of the feedback pressure with time in a cycle, where P_b in section a-b, b-c and c-d is calculated by Equations (3)–(5), successively.

Table 3. Main parameters of P_b.

Parameter	Value
Peak value	$280\,P_0$
Valley value	$104.6\,P_0$
Variation period T, s	2×10^{-4}
Variation frequency f, Hz	5000

3.3. Validation of the Numerical Method and Grid Sensitivity Verification

The capability of describing the unsteady shock wave propagation of the numerical approach adopted hereinafter is inspected with the experimental results in [43], where a planar shock wave is transmitted in a 90° branched duct with a rectangular cross section (20 mm in height, and 40 mm in width). The propagation Mach number of the original shock wave M_s is 2.4 before its diffraction around the 90° sharp corner, when the pre-shock static pressure and temperature are 100 kPa and 288.15 K, respectively. Hence, the pressure ratio of the moving shock is about 6.55. The comparison of the shock wave structure between the experimental and numerical schlieren images are exhibited in Figure 4, the former of which was obtained with a 24 frame Cranz-Schardin spark camera, that clearly shows that there are no movable parts in the light ray's path. The experimental shock tube possesses a test section equipped with plane, parallel windows of high optical quality glass. The optical field of view is 200 mm·110 mm with the depth of 40mm [44]. As can be noted from the schlieren images, there is no significant disparity between the experiment and simulation from the perspectives of the shock wave structure. Consequently, the numerical method introduced above has a good accuracy in capturing the moving shock wave in a restricted duct.

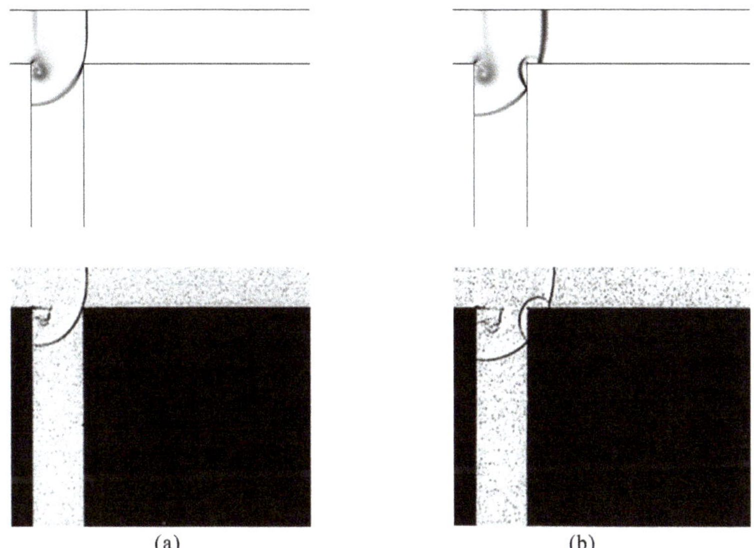

(a)   (b)

Figure 4. Comparison of between the numerical (**top**) and experimental (**bottom**) schlieren images. (**a**) $t = 28$ μs. (**b**) $t = 40.5$ μs.

For the purpose of verifying the grid convergence, three-dimensional structured grids in the isolator designed in Section 2 are generated with three different grid densities: the coarse grid (3 million cells), the fine grid (6 million cells), and the dense grid (9 million cells). Figure 5 compares the computed outer-wall pressure distribution of the isolator along the x-direction under an unthrottled state, the incoming flow condition of which is listed in Table 2. As it can be seen, the pressure curve of the fine grid is basically consistent with that of the dense grid, while the coarse case shows a tiny inequality in the region where the internal shock impinges on the walls. Hereby, the fine grid is utilized for the rest of the analysis in this paper.

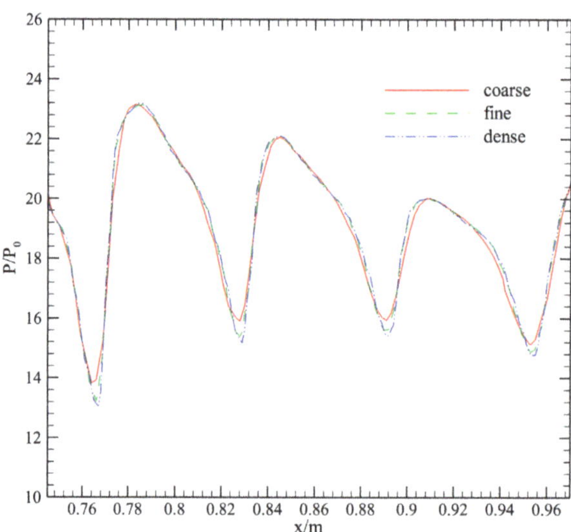

Figure 5. Comparison of the computed outer-wall pressure distribution for the three grid densities.

4. Transient Flow Characteristics in the Isolator Affected by the Rotating Feedback Pressure

Since the isolator is an axisymmetric duct with steady and circumferentially uniform inflow conditions at the entrance and cyclical conditions of the rotating feedback pressure at the exit, it is rational to presume that the flow behaviors in the isolator during the stable operating stage would present a high periodicity as well, which will be discussed minutely in Section 5. Therefore, the instantaneous flow-field at the initial time of a certain cycle (i.e., $t = nT$, or floor $(t, T) = 0$) during the stable operating stage is firstly selected and analyzed in this section.

4.1. Terminal Shock Wave/Moving Shock Wave/Boundary Layer Interaction

Figure 6a,b present the pressure contours transformed into rectangular planes on the outer and inner wall of the isolator, respectively. The rotating feedback pressure perturbations induce a forward and circumferential motion of the helical moving shock wave (MSW) in the isolator, as indicated in Figure 6c. As the high-pressure perturbations propagate forward, the strength of the MSW attenuates gradually, the screw-pitch of the MSW reduces slightly, and the inclination angle between the MSW and the x-axis increases gently. In other words, the MSW shows more "flat" patterns along the negative direction of the x-axis. For facilitating the analysis, the helical MSWs on the walls are converted into multiple segments of curved MSWs on the x-θ plane, as indicated by blue arrows in Figure 6a,b. It can be observed that the MSWs on the outer wall appear stronger than those on the inner wall. This is mainly determined by the curvature effect of the walls [45], since the outer wall of the annular duct compresses the inflow and enhances the MSWs; while the flow near the inner wall tends to expand, thus weakening the MSWs. Moreover, the time-average feedback pressure equaling 114.95 P_0 at the exit also induces a quasi-toroidal terminal shock wave (TSW) marked by red arrows in Figure 6a,b. Due to the rotation of the feedback pressure, the TSW oscillates in the x direction as the θ varies, which differs from that in the axisymmetric isolator under steady and uniform backpressure. [46] Though a series of expansion waves are generated downstream of the MSWs, the pressure along the axis-direction in the isolator still presents an overall ascending trend.

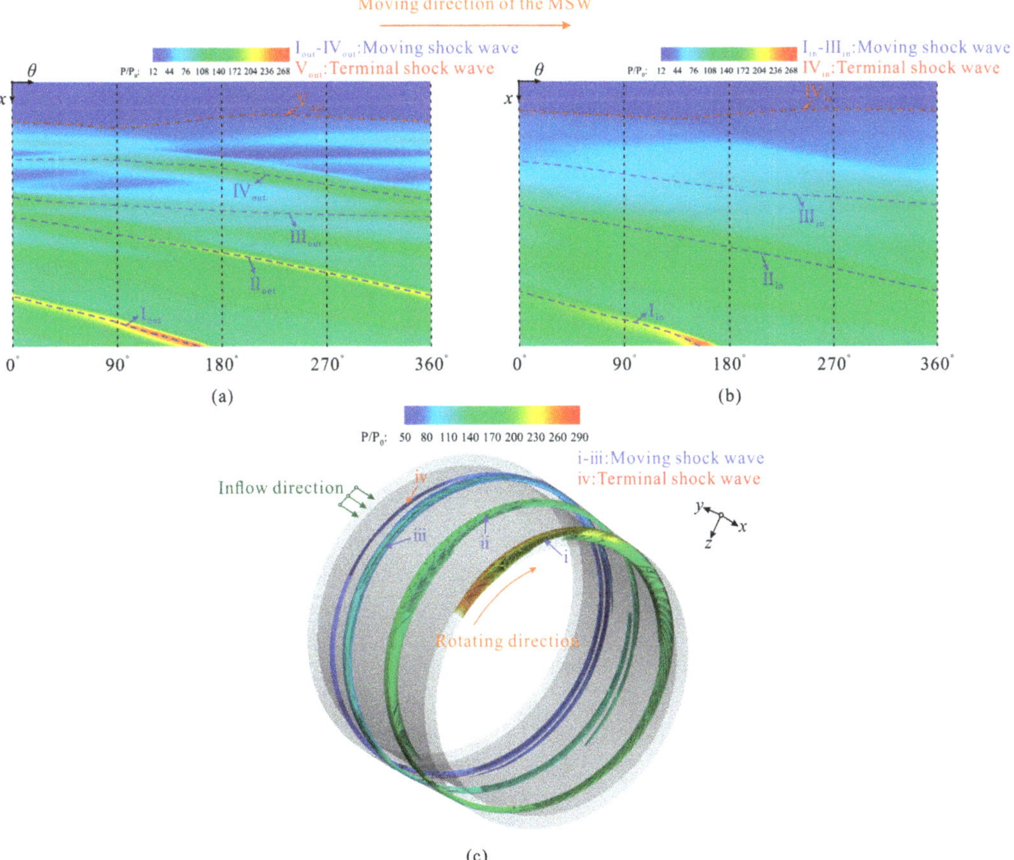

Figure 6. Shock wave structures. (**a**) Pressure contour on the outer wall after the coordinate change. (**b**) Pressure contour on the inner wall after the coordinate change. (**c**) Three-dimensional schematic diagram of the moving shock wave and the terminal shock wave.

In an effort to visualize the terminal shock wave/moving shock wave/boundary layer interaction (TSW/MSW/BLI) intuitively, the flow-fields on eight meridian planes with equidifferent angles are rotated and shifted to the corresponding positions on the same x-y plane, as shown in Figure 7. It should be noted that the dashed lines in Figure 7 represents the curve of the shock waves with the angle, instead of the actual shock patterns in the meridian planes. Due to the circumstantial and axial curvature effects, the pressure on the outer wall is higher than that on the inner wall in Figure 7b, as observed in Figure 6a,b, and the low-momentum flow accumulates to the inner wall eventually after the first bend. As shown in Figure 7, the flow-fields could be classified into three zones, i.e., the undisturbed zone, the TSW/MSW/BLI zone dominated by the TSW indicated by blue dashed line iv, and the MSW indicated by the red dashed line iii, and the MSW/BLI zone, where only the MSWs indicated by red dashed lines i and ii exist. In the undisturbed zone, the streamlines are not affected by the TSW and the MSW downstream, and basically parallel to the x-axis. A coupling of the terminal shock wave/boundary layer interaction (TSW/BLI) and the moving shock wave/boundary layer interaction (MSW/BLI) takes place in the TSW/MSW/BLI zone. On the meridian planes with azimuthal angle θ ranging from 45°

to 180°, the axial distance between the TSW and MSW iii is rather small, resulting in the enhanced coupling of the adverse pressure gradient generated from two shock waves, thus rolling up large separation bubbles developing along MSW iii driven by the circular pressure gradient. This migration of separation bubbles contributes to the maximum separation zone scale on the meridian plane at $\theta = 180°$. While on the meridian planes with θ ranging from 225° to 360°, the axial distance between the two shock waves augments as the azimuthal angle increases, and the strength of the pressure gradient coupling drops, along with the shrinking of the separation bubbles. As for the MSW/BLI zone affected by MSW ii, with the counterflow in the vicinity of the inner wall, the separation bubbles and the streamline deflections occur successively in the region with θ ranging from 45° to 360°. Else, MSW i with the maximal strength turns the entire inflow in the reverse direction at $\theta = 135°$, which is quite different from the steady shock/boundary layer interaction. An interesting discovery is that notwithstanding the deflecting effect the MSWs have on the flow, the flow direction of air still remains axial upstream of the exit, except in the adjacent region of MSW i. This undoubtedly will provide a favorable impact on the detonation combustion organization in the RDC.

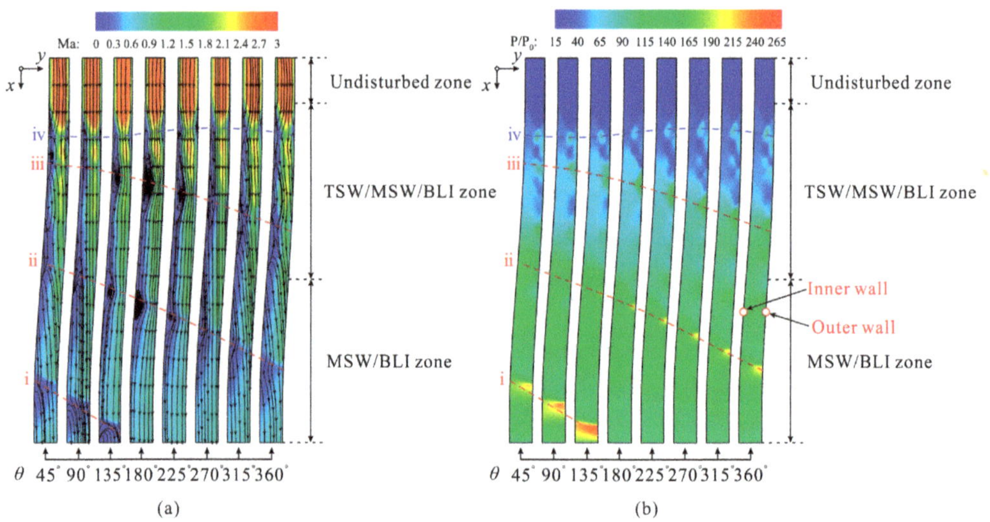

Figure 7. Flow patterns on eight meridian planes of different angles. (**a**) Mach number contours. (**b**) Pressure contours.

The surface pressure distribution curves and the secondary flow patterns on the meridian plane at $\theta = 135°$ are extracted to obtain more details about the TSW/MSW/BLI, as depicted in Figure 8. During the process of propagation, the MSWs are weakened from the moving normal shock wave to the moving oblique shock wave, and eventually merge with the TSW. It can be seen that the pressure on the inner wall varies more smoothly than that on the outer wall, and a phase difference between the inner and outer surface pressure appears at the peak generated by MSW i. Whereas, no phase difference arising at the peak of MSW ii and iii can be noticed, which means that the velocity difference of the MSW's upstream propagation and the flow on two sides of the walls counteract each other. Pressure oscillations on the outer wall can be found, which correspond to the collision between the airflow and the outer wall, as exhibited in Figure 8. This flow migration is caused by the vortices rolled up by the TSW/MSW/BLI and the MSW/BLI located around MSW ii, while the MSW/BLI caused by MSW i forms a separation line resembling a semi-circle and a reattachment line shaped as a straight line.

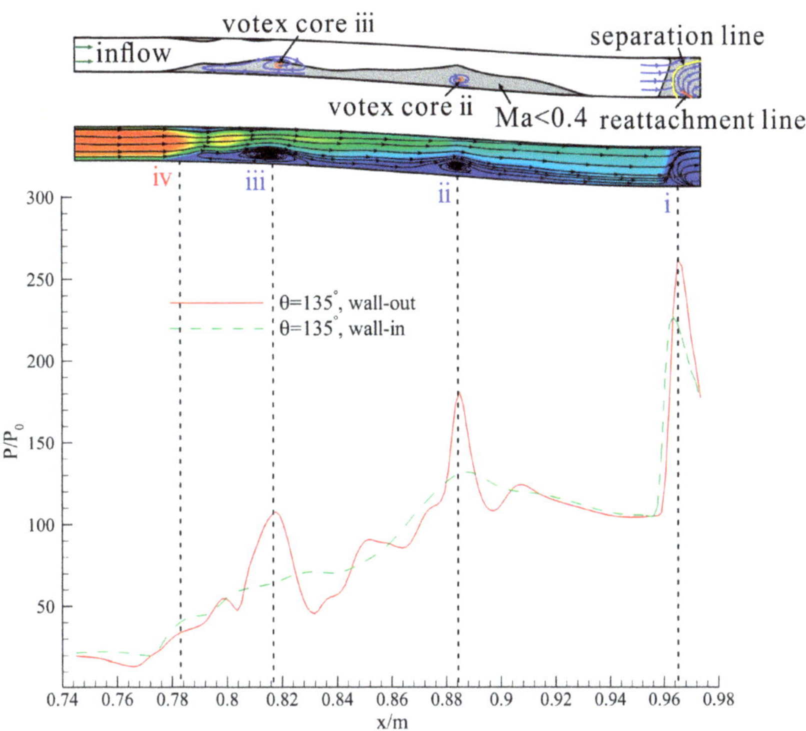

Figure 8. Surface pressure distributions and secondary flow structure on the meridian plane at $\theta = 135°$.

The spatial streamlines originated from the points at eight azimuthal angles ($\theta = 45°$, $90°$, $135°$, $180°$, $225°$, $270°$, $315°$, and $360°$) at the entrance are traced forward, the target of which is to portray and investigate the vortex structure induced by the TSW/MSW/BLI. The tracked streamlines can be divided into three types: the main flow streamlines emphasized by red lines in Figure 9a,b, the low-momentum flow streamlines near the outer wall emphasized by blue lines in Figure 9a, and the low-momentum flow streamlines near the inner wall emphasized by blue lines in Figure 9b. As shown in Figure 9, the streamlines of the main flow with a high momentum present a negligible lateral motion in the circumferential direction but slightly lifts and dives in the radial direction, due to the separation bubbles. It is interesting to note that the vortices generated from the boundary layer near both sides of walls all develop along the TSW and the MSWs, as shown in Figure 9 (the vortices shown in Figure 9c are identified by the Liutex-Ω_R criterion proposed in [47,48] with the value of $\Omega_R = 0.6$ in yellow). The motivation of the circular motion of the vortices are the circular pressure gradient contributed by the inclination of the shock waves, which is a typical glancing shock wave/boundary layer interaction (GSW/BLI) [49,50]. In addition, the scales of the vortices show some disparity between the vicinity of the outer and inner walls, which corresponds to the flow-fields in Figure 7a. Driven by the high-momentum flow, the inner vortices depart from the isolator at the azimuthal angle θ, slightly less than 315°, as illustrated by the orange circle in Figure 9b.

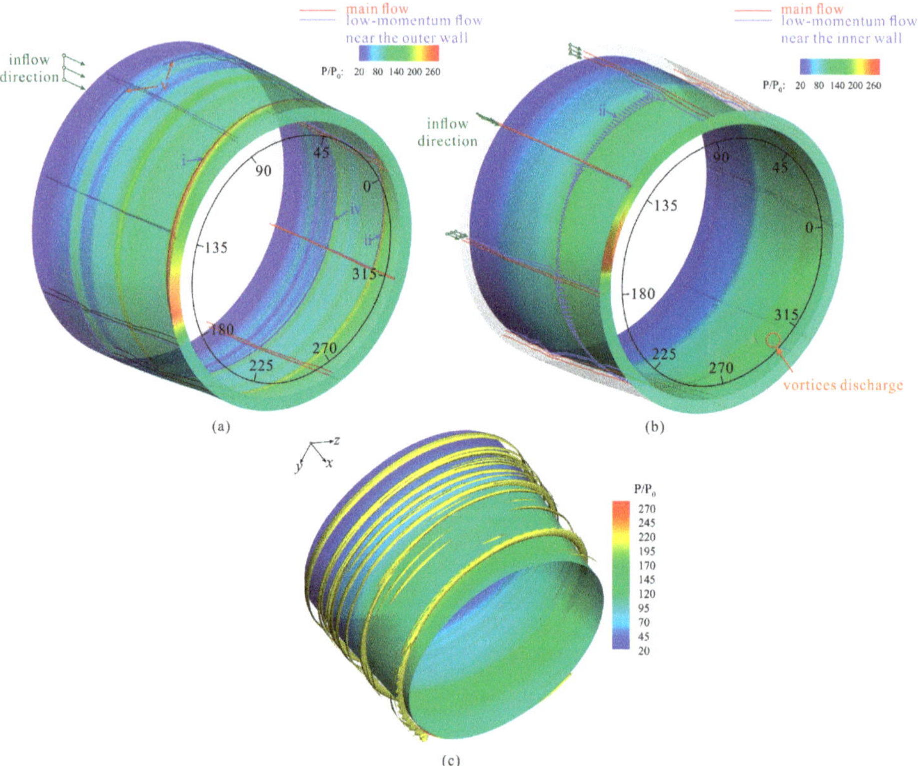

Figure 9. Vortex structures originate. (**a**) Three-dimensional streamlines released from the entrance at eight angles without the vicinity of the inner surface. (**b**) Three-dimensional streamlines released from the entrance at eight angles without the vicinity of the outer surface. (**c**) Three-dimensional vortices and the inner-wall pressure contour.

4.2. Impact of the Shock Wave System on the Main Flow

The aerodynamic parameter profiles of the centerlines of two chosen meridian planes located on both sides of the simplified RDW are employed to explore the effects of the TSW illustrated by the green dashed line iv and the MSWs denoted by the green dashed lines i–iii on the high-momentum flow behaviors ulteriorly, as exhibited in Figures 10 and 11. The oscillation of the parameters upstream of the TSW is the common product of the shock and expansion waves. In general, the trends of the parameters on two centerlines are the same despite some differences caused by MSW i when encountering the TSW and the MSWs. Take the centerline of the meridian plane at $\theta = 135°$ presented in Figures 10a and 11a for example. First, the static pressure P, temperature T_{emp}, and entropy S elevate promptly when the flow on the centerline meets with TSW iv, while the velocity in three directions (v_a, v_θ, v_r) all drop off, which is in accordance with the variation rules of the flow parameters when passing by the stationary shock wave. It means that TSW iv generated from the rotating feedback pressure still belongs to the class of the stationary shock wave in one sense. Whereafter, P, T_{emp}, and S decreases, yet the velocity increases due to the expansion waves. Then, the flow encounters MSW iii and ii in sequence. The variation trends of P, T_{emp}, S, and velocity in the axial and radial directions are consistent with that after TSW iv, despite some differences in values. However, the circumstantial velocity v_θ varies in an opposite way, i.e., v_θ increases instantly during the collision, then decreases gradually to even lower than 0 under the expansion effect. In other words, first the air flows along the MSWs just downstream, then eventually it turns to deviate from the MSWs. Ultimately, the flow encounters the intense MSW i, and P, T_{emp}, S, and v_θ rise immediately to a quite higher level before dropping off rapidly, while v_a shows a reverse trend. In addition, v_r increases slightly upstream of MSW i due to the elevation effect of the backflow zone. Yet, it is noteworthy that the absolute change value and the absolute terminal value of v_θ and v_r are rather small and ignorable, compared with v_a, corresponding to the main-flow streamlines in Figures 7a and 9.

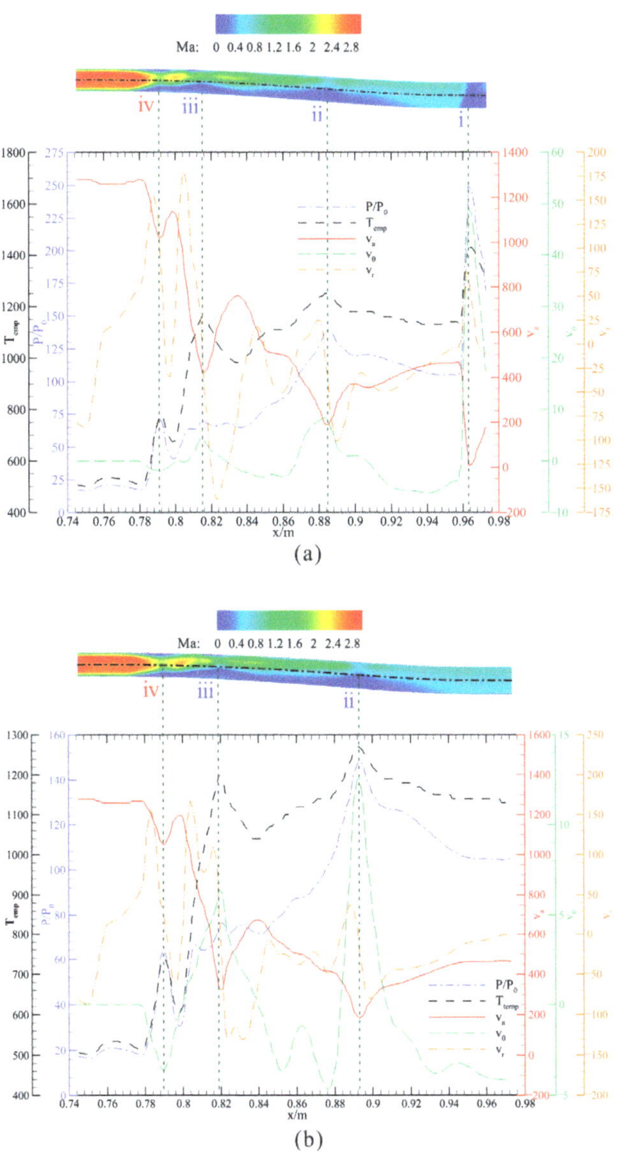

Figure 10. Aerodynamic parameter distribution curves of the sectional centerlines along the axial direction. (**a**) $\theta = 135°$. (**b**) $\theta = 180°$.

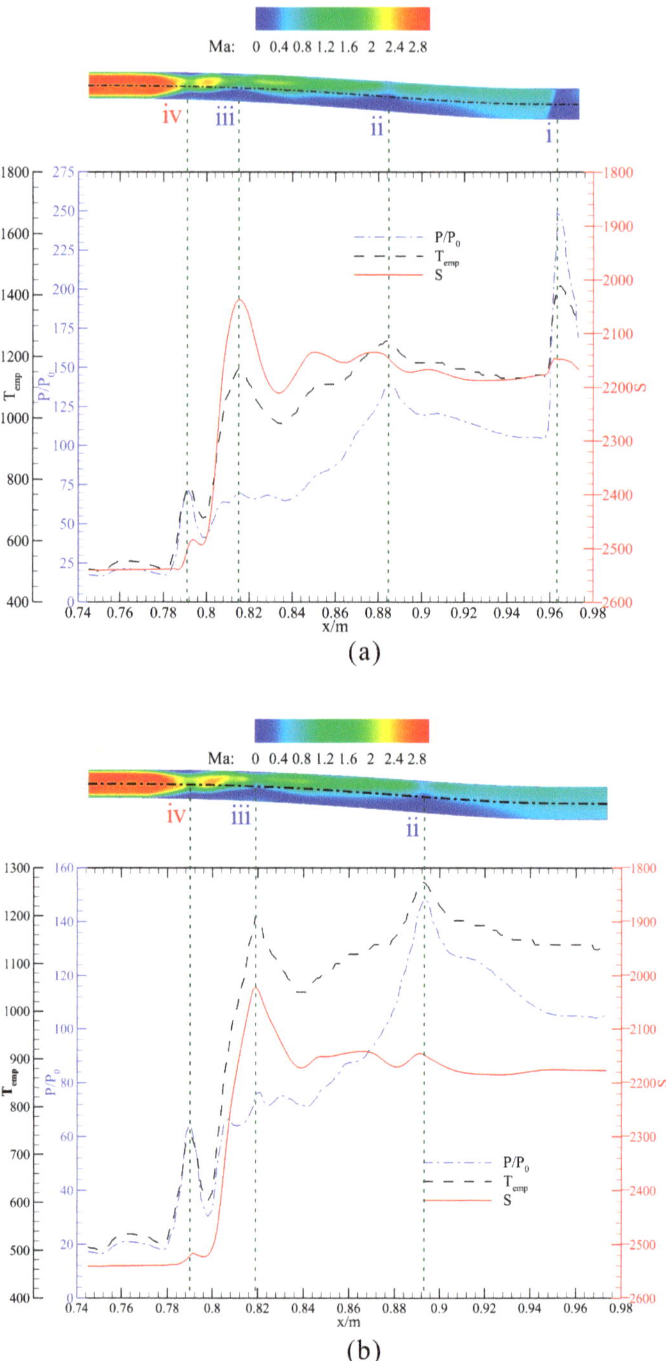

Figure 11. Entropy profile of the sectional centerlines along the axial direction. (**a**) $\theta = 135°$. (**b**) $\theta = 180°$.

4.3. Flow Patterns at the Exit

Another important issue that should be discussed is the flow pattern at the exit, as presented in Figure 12. As can be seen, three types of zones are determined in Figure 12b, based on the morphology of the streamlines: zone A is referred to as the rotating detonation wave/boundary layer interaction (RDW/BLI) zone, zone B, consisting of B_1 and B_2, is dominated by the expansion waves, and zone C is where the helical vortices are exhausted from the isolator. Still, there are some generalities in the three zones. The radial velocity v_r maintains a negative at every azimuthal angle θ, owing to the radial pressure gradient induced at the first bend of the "S-shaped" section. As a result of the radial pressure gradient, the low-speed flow accumulates in the minor-radius region, driving the high-speed flow to migrate to the major-radius one. In zone A, disturbed by the simplified RDW, the positive circumferential velocity v_θ of the flow is induced after the RDW, that is, the streamlines point towards the RDW. Thus, a separation line S_1 that is almost normal to the walls is formed under the strong adverse pressure gradient that originated from the RDW, then a reattachment line R_1 that is nearly parallel to the outer wall could be noted in the angle range of 120°–150°, due to the expansion effects. A quasi-triangular gap in the low-speed region can be found near the inner wall, which is caused by the high-speed streamlines released from line R_1. As a result, the low-speed region is shaped like a "Λ". As for zone B, due to the expansion effect behind the MSW, v_θ decreases to below 0 gradually, causing the flow to rotate counter-clockwise (the view direction is along the negative x-axis). It can be found that the twist of the streamlines in zone B_2 is more violent than that in zone B_1. In zone C, the vortices generated from the boundary layer near the inner wall exhaust into the RDC, as shown in Figure 9b, hence the streamlines here are perturbated, which contribute to the formation of the separation line S_2 and the reattachment line R_2.

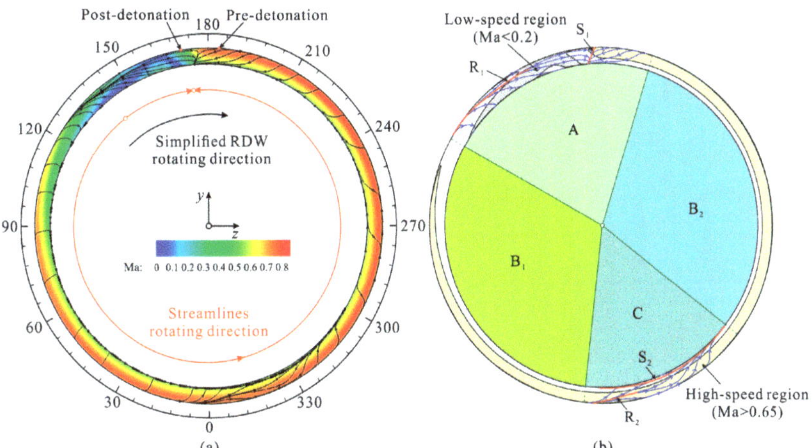

Figure 12. Flow topology on the exit. (**a**) Absolute Mach number contour and the streamlines. (**b**) Sketch of the distilled flow structure.

5. Self-Similarity in the Isolator Affected by the Rotating Feedback Pressure

5.1. Similarity of the Flow Patterns between Adjacent Cycles

To demonstrate the correlations of the flow-field between two adjacent cycles, Figure 13 compares the wall pressure and the sectional Mach number contours in two neighbor cycles. An interesting discovery is that the pressure contour on the outer wall at $t = nT$ and $t = (n + 1)T$ are extremely similar, both consisting of four curved oblique MSWs with analogous patterns and comparative strength. As mentioned above, the flow-field of the meridian planes in Figure 13a,b could both be split into three zones, i.e., the undisturbed zone, the TSW/MSW/BLI zone, and the MSW/BLI zone. The distribution and shape of the low-momentum flow regions resemble each other, as well. Further, the flow patterns in the meridian planes situated on both sides of the RDW ($\theta = 135°$, and $\theta = 225°$) at $t = nT$ and $t = (n + 1)T$ are contrasted in Figure 14, which corresponds to points A_1, B_1, A_2, B_2 emphasized in Figure 15. As can be seen, the Mach number and wall pressure profiles of two meridian planes at $t = nT$ are basically consistent with those at $t = (n + 1)T$, correspondingly. Thus, it can be deduced that the flow-field in the adjacent cycles are self-similar.

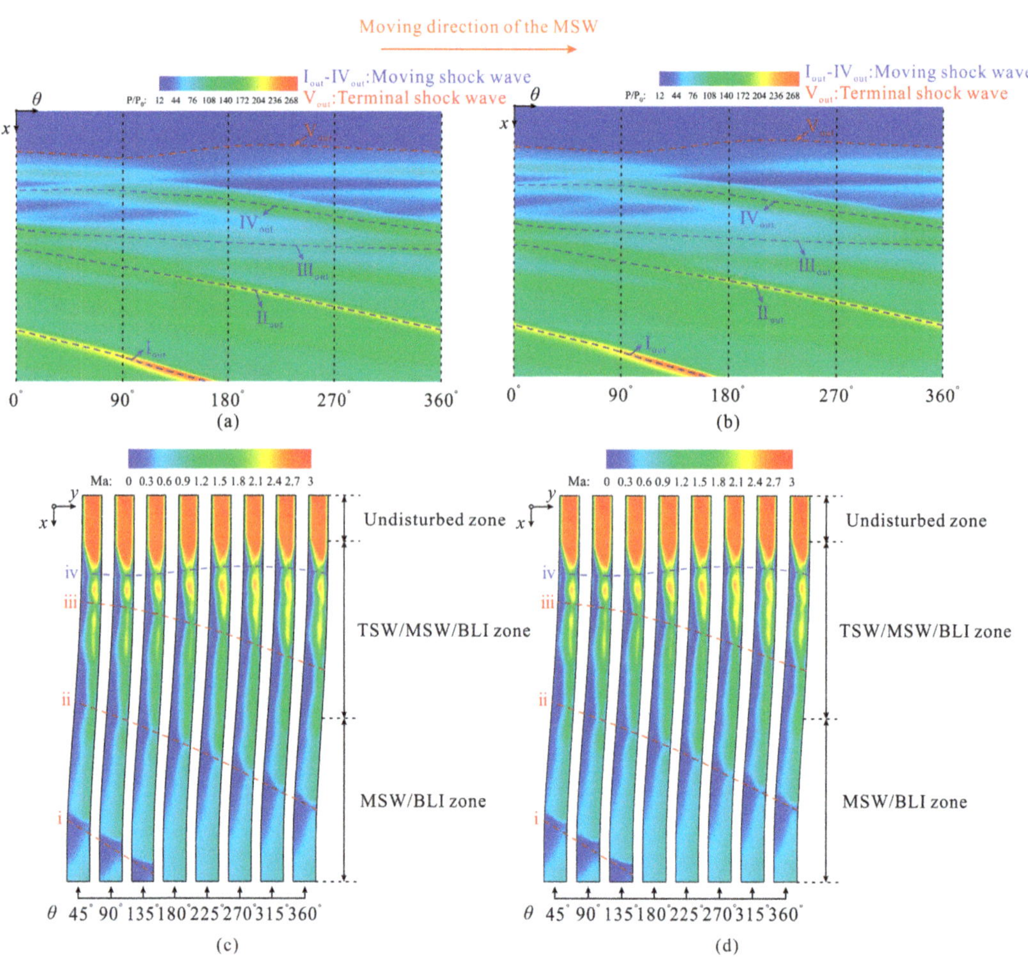

Figure 13. Critical flow structure in the adjacent cycles. (**a**) Pressure contour on the outer wall, $t = nT$. (**b**) Pressure contour on the outer wall, $t = (n + 1)T$. (**c**) Mach number contour of eight meridian planes, $t = nT$. (**d**) Mach number contour of eight meridian planes, $t = (n + 1)T$.

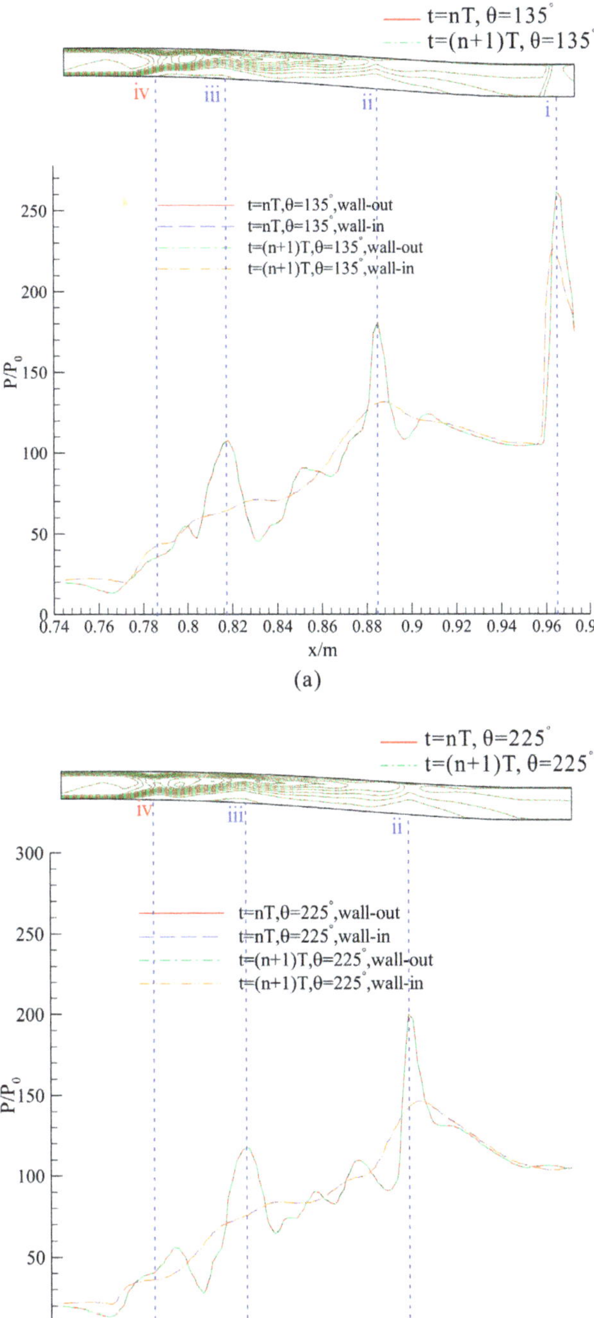

Figure 14. Comparison of the Mach number and pressure profiles of the meridian planes in adjacent cycles. (**a**) $\theta = 135°$. (**b**) $\theta = 225°$.

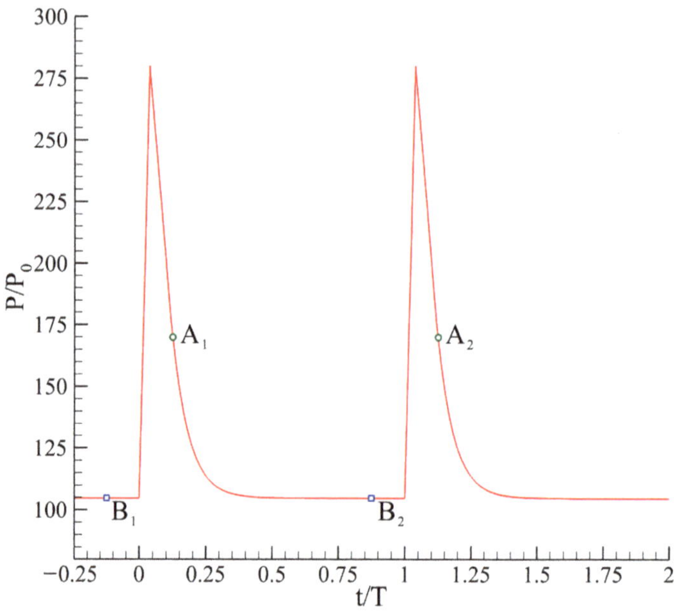

Figure 15. Feedback pressure change curve with time, where points A_1 and A_2 denote the meridian plane with $\theta = 135°$ at $t = nT$ and $t = (n + 1)T$, respectively, and points B_1 and B_2 denote the meridian plane with $\theta = 225°$ at $t = nT$ and $t = (n + 1)T$, respectively.

5.2. Comparison of the Flow Patterns at Different Moments in One Cycle

Then, the overall flow-fields at different moments in one cycle are compared in Figures 16 and 17. It should be noted that the time difference among the selected four moments equals $T/4$, that is, the feedback pressure at the exit rotates a corresponding angle in sequence, which equals $360°/4 = 90°$ in this case. Hence, it is valid to assume that the isolator flow-field should also rotate $90°$. Take the meridian plane with $\theta = 135°$ at $t = nT$ corresponding to point A_1 in Figure 18 as a reference plane, the comparison among the sectional flow patterns of A_1 and that with $\theta = 135° + 90° = 225°$ at $t = (n + 1/4)T$ (B_2 in Figure 18), that with $\theta = 135° + 2 \cdot 90° = 315°$ at $t = (n + 2/4)T$ (C_2 in Figure 18) and that with $\theta = 135° + 3 \cdot 90° = 45°$ at $t = (n + 3/4)T$ (D_2 in Figure 18) is rather reasonable for the validation of flow-field similarity at different moments in one cycle.

Likewise, it can be discovered that the global flow structures of the isolator at different moments in one cycle all show good agreement, though the flow-field rotates $90°$. The structure of the shock waves, the scale and distribution of the low-momentum flow, even the surface pressure change curves and the Mach number contours of the corresponding meridian planes exhibited in Figure 19 all match perfectly, which confirms the self-similarity of the isolator flow in one cycle under rotating P_b.

Based on the preceding discussions, we can draw the conclusion that the flow-field of the isolator under rotating feedback pressure perturbations is similar during the stable operating stage, which indicates that the rotation angular velocity of the TSW and the MSWs are equal and hold invariant, building the key foundation for the theoretical model of the inclination angle of the MSWs through velocity decomposition, previously mentioned in Section 6.

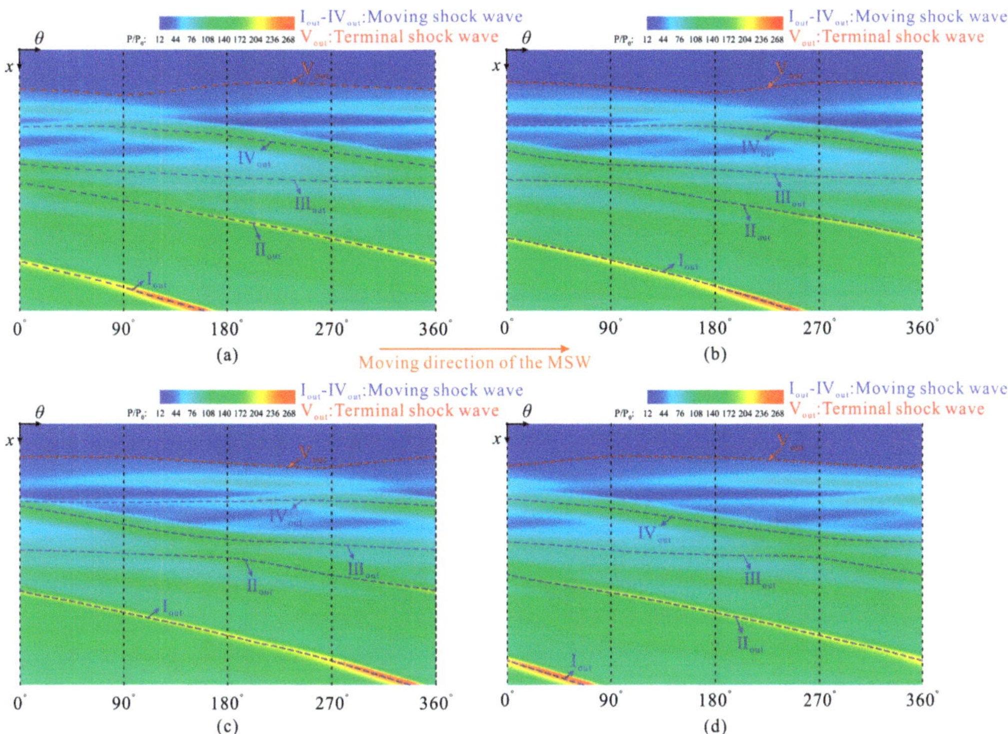

Figure 16. Pressure contours on the outer wall at different moments in one period. (**a**) $t = nT$. (**b**) $t = (n + 1/4)T$. (**c**) $t = (n + 2/4)T$. (**d**) $t = (n + 3/4)T$.

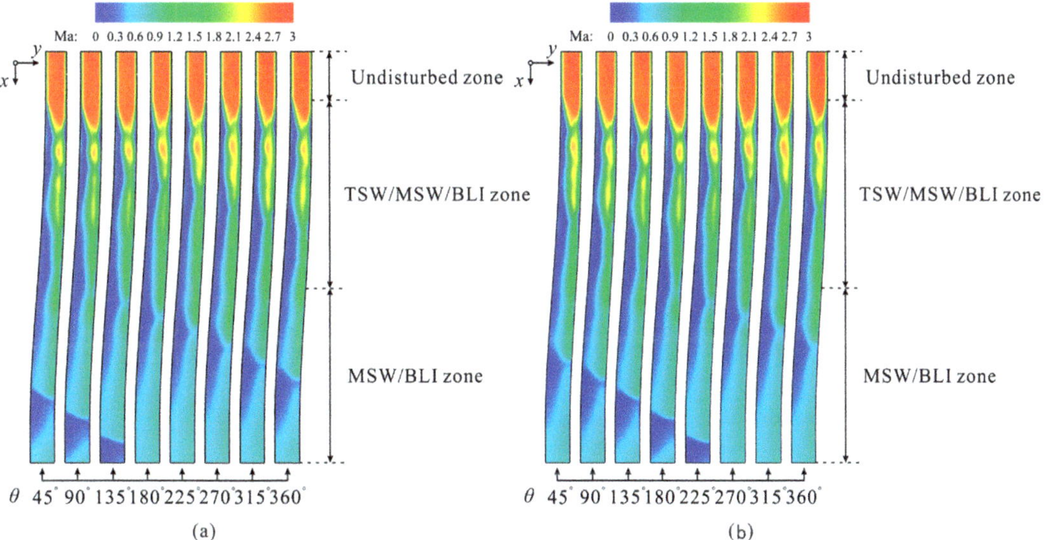

Figure 17. *Cont.*

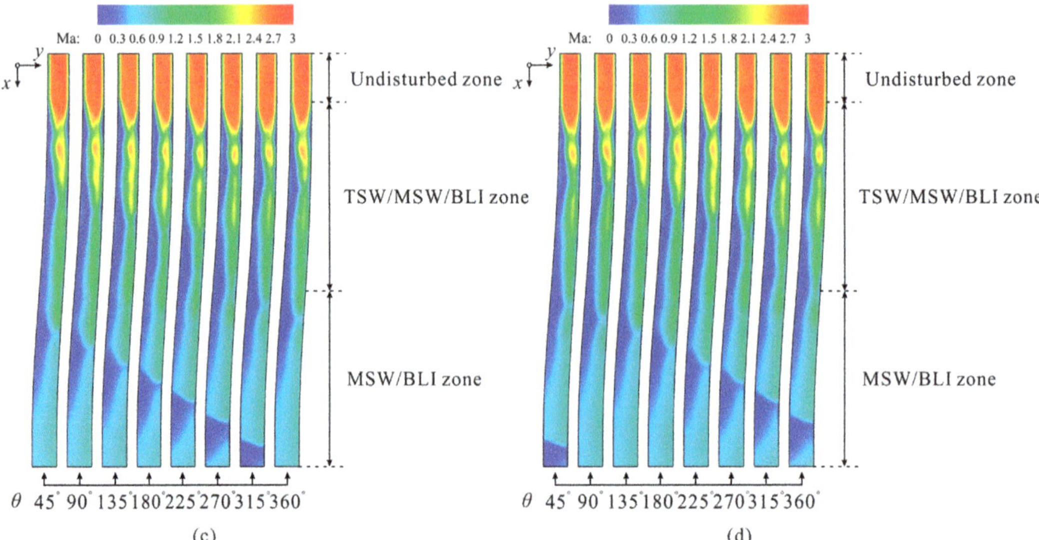

Figure 17. Mach number contours of eight meridian planes at different moments in one period. (**a**) $t = nT$. (**b**) $t = (n + 1/4)T$. (**c**) $t = (n + 2/4)T$. (**d**) $t = (n + 3/4)T$.

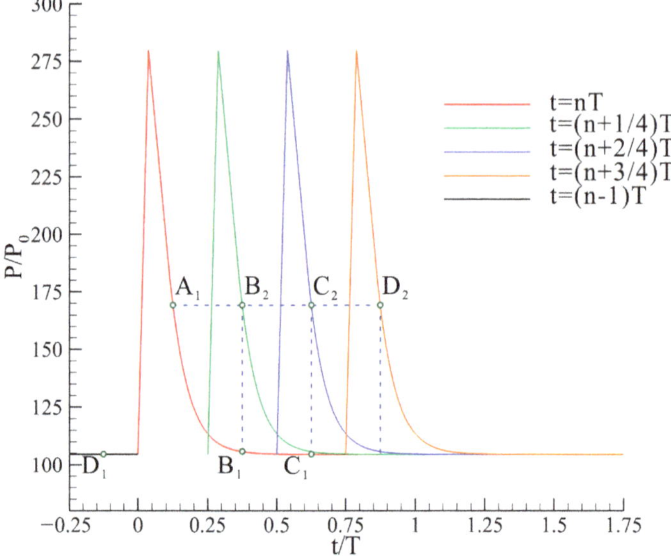

Figure 18. Feedback pressure change curve with time, where points A_1, B_1, C_1, and D_1 denote the meridian plane with $\theta = 135°$, $225°$, $315°$, and $45°$ at $t = nT$ successively, point B_2 denotes that with $\theta = 225°$ at $t = (n + 0.25)T$, point C_2 denotes that with $\theta = 315°$ at $t = (n + 0.5)T$, and point D_2 denotes that with $\theta = 45°$ at $t = (n + 0.75)T$.

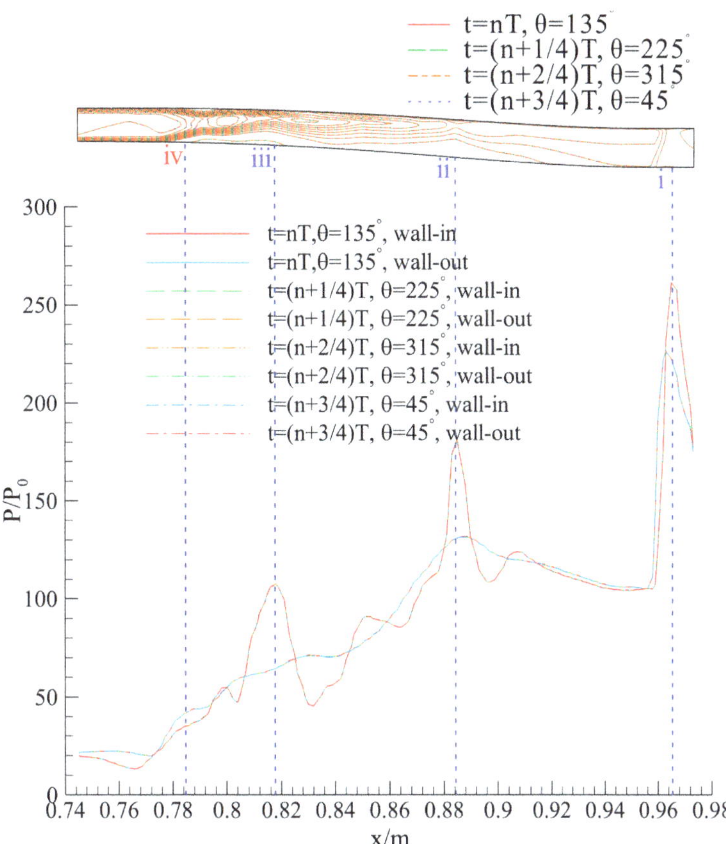

Figure 19. Mach number and pressure profiles of corresponding meridian planes at different moments in one period.

6. Theoretical Model of the Inclination Angles of the Moving Shock Wave, Based on Velocity Decomposition

The shock wave structure on the central plane of the isolator at $t = nT$, as shown in Figure 20, is chosen to calculate the inclination angle α of the MSWs. It can be discovered that the MSWs appear approximately as straight lines, indicating that the MSW angle α varies with little discrepancy in one round. Hence, the average parameters could be utilized to compute the inclination angle of each MSW separately. MSW i is taken as an example to introduce the theoretical model, the schematic of which is exhibited in Figure 21. The inclination angle of MSW i α_i is defined as the inclination angle between MSW i and the axis. Since the absolute values of the radial and circumferential velocities of the pre-shock air are ignorable, compared with the pre-shock axial velocity according to Figure 10, it is valid to assume that the pre-shock air flows along the axis. So α_i also refers to the inclination angle between MSW i and the pre-shock flow direction. In addition, there is a negative correlation between α_i and β_i, which is given as:

$$\alpha_i + \beta_i = 90°, \tag{7}$$

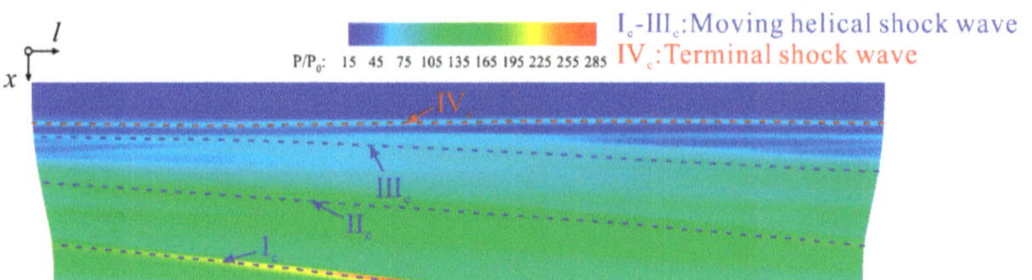

Figure 20. Shock wave structure on the central plane at $t = nT$ in x-l plane, where l denotes the arc length, and the dependent ratio of x and l equals 1.

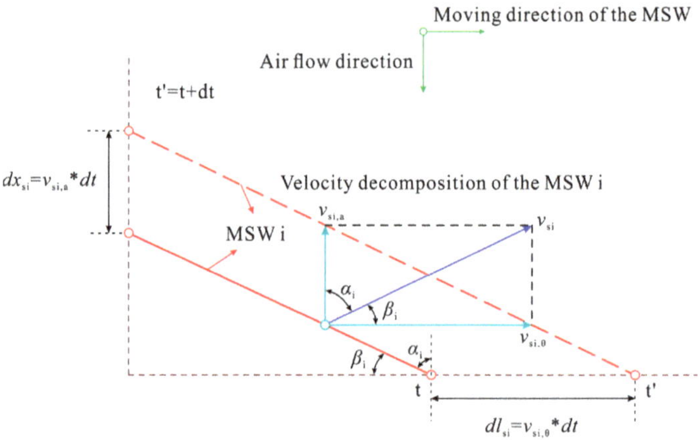

Figure 21. Sketch of MSW i.

The movement of MSW i between $t = nT$ illustrated by the solid red line and $t' = nT + dt$ illustrated by the dashed red line could be deconstructed into two parts:

$$dx_{si} = v_{si,a} \cdot dt, \tag{8}$$

$$dl_{si} = v_{si,\theta} \cdot dt, \tag{9}$$

where dx_{si} and dl_{si} denote the movement distance of MSW i in the axis direction and circumferential direction, respectively; $v_{si,a}$ and $v_{si,\theta}$ denote the absolute velocity of MSW i in the axis direction and circumferential direction, separately.

In addition, MSW i at $t = nT$ is parallel to that at $t' = nT + dt$ ($dt \to 0$) due to the self-similarity property of the flow-field demonstrated in Section 5. Thus, α_i also refers to the inclination angle between the axial velocity $v_{si,a}$ and the resultant velocity v_{si} of MSW i, computed by:

$$v_{si} = \sqrt{v_{si,\theta} \cdot v_{si,\theta} + v_{si,a} \cdot v_{si,a}} \tag{10}$$

$$\alpha_i = \mathrm{atan}(v_{si,\theta}/v_{si,a}), \tag{11}$$

It is noteworthy that v_{si} refers to the absolute velocity of MSW i, and it is known that the angular velocity of the MSWs equals that of the simplified RDW, which is defined as:

$$\omega = 2\pi f, \tag{12}$$

Hence,

$$v_{si,\theta} = \omega \cdot r_i, \tag{13}$$

$$v_{si,a} = v_{pi,a} - v_{pre-i,a}, \tag{14}$$

where $v_{\text{pre-i,a}}$ denotes the absolute velocity of the pre-shock air in the axial direction, which is defined by temperature $T_{\text{pre-i}}$ and Mach number $M_{\text{pre-i}}$ of the air before MSW i:

$$v_{\text{pre}-i,a} = \sqrt{\gamma R T_{\text{pre}-i}} \cdot M_{\text{pre}-i} \qquad (15)$$

and $v_{\text{pi,a}}$ denotes the axial velocity of MSW i relative to the pre-shock flow, i.e., the axial propagation velocity of MSW i, which is the next key factor to be calculated.

The calculation method of $v_{\text{pi,a}}$ is illustrated in Figure 22. The sketch of MSW i and the absolute velocity of MSW i and the pre-shock air at $t = nT$ is presented in Figure 22a. Firstly, a microelement of MSW i ds_{si} is extracted, and the MSW i coordinate system is established, as shown in Figure 22b. On this basis, the relative velocity of the pre-shock air $v_{\text{pre-i,rel}}$ could be deconstructed into $-v_{\text{si},\theta}$ in the circumferential direction and $-v_{\text{pi,a}}$ in the axial direction, and MSW i microelement ds_{si} consists of dx_{si} and dl_{si}. Hence, α_i could also be obtained by:

$$\alpha_i = \text{atan}(dl_{\text{si}}/dx_{\text{si}}), \qquad (16)$$

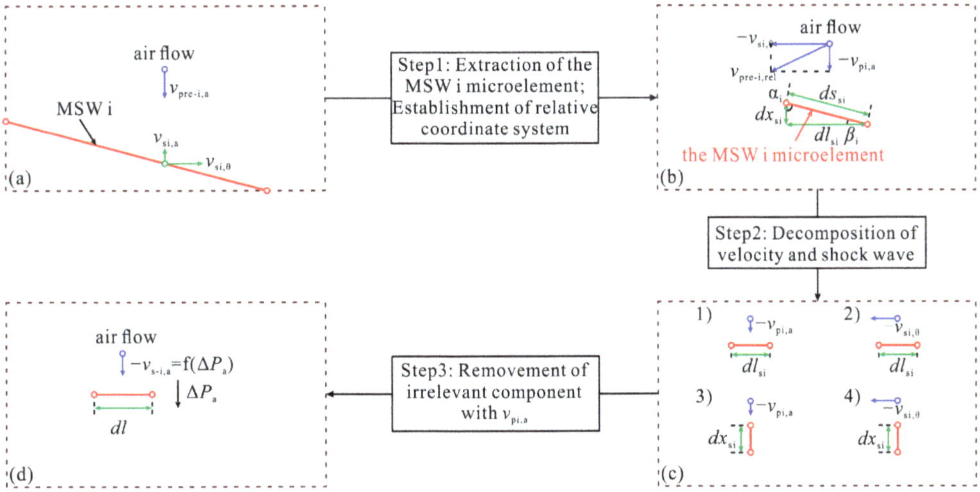

Figure 22. Sketch of the calculation method. (**a**) MSW i and the absolute velocity of air and MSW i. (**b**) the MSW i microelement and the relative velocity of air. (**c**) Components of the relative velocity and MSW i. (**d**) Relevant component with $v_{\text{pi,a}}$.

Then, the relative velocity of the pre-shock air $v_{\text{pre-i,rel}}$ and MSW i microelement ds_{si} are both axially and circumferentially deconstructed into four types shown in Figure 22c. Type (1) refers to the collision between the air with an axial velocity $-v_{\text{pi,a}}$ and the circumferential microelement of MSW i dl_{si}, type (2) exhibits the encounter between the air with a circumferential velocity $-v_{\text{si},\theta}$ and MSW i circumferential microelement dl_{si}, while type (3) and (4) indicate the meeting of MSW i axial microelement dx_{si} and the air with the axial ($-v_{\text{pi,a}}$) and circumferential ($-v_{\text{si},\theta}$) velocity, successively.

It can be discovered that only type (1) is relevant with the calculation of $v_{\text{pi,a}}$, since the components of the pre-shock relative velocity are both parallel to the components of MSW i microelement in type (2) and (3), which will not induce a pressure rise; though the circumferential component of the pre-shock relative velocity $-v_{\text{si},\theta}$ is perpendicular to the axial component of MSW i microelement dx_{si} in type (4), there is no correlation between the circumferential pressure rise induced and $v_{\text{pi,a}}$. Hence, the functions of $v_{\text{pi,a}}$ and ΔP_a are as follows:

$$M_{\text{pi,a}} = \frac{\gamma+1}{2\gamma} \cdot \left(\frac{P_{\text{post}-i}}{P_{\text{pre}-i}} \right)_a + \frac{\gamma-1}{2\gamma} \qquad (17)$$

$$\left(\frac{P_{\text{post}-i}}{P_{\text{pre}-i}}\right)_a = \frac{P_{\text{pre}-i} + \Delta P_a}{P_{\text{pre}-i}} \tag{18}$$

$$v_{\text{pi,a}} = \sqrt{\gamma R T_{\text{pre}-i}} * M_{\text{pi,a}} \tag{19}$$

where ΔP_a denotes the axial component of the pressure gradient originating from MSW i, and $M_{\text{pi,a}}$ denotes the axial propagation Mach number of MSW i.

Last, the integral average calculation along MSW i is performed, and MSW i angle α_i computed by Equations (11) and (16) are compared, as shown in Table 4, along with those of MSW ii and iii. To distinguish from each other, the $\bar{\alpha}$ calculated by Equation (11) is referred to as the predicted angle $\bar{\alpha}_{\text{aero}}$, and the inclination angle obtained via Equation (16) is defined as the measured angle $\bar{\alpha}_{\text{geom}}$. The relative error δ_α is defined as:

$$\delta_\alpha = \frac{\left(\bar{\alpha}_{\text{aero}} - \bar{\alpha}_{\text{geom}}\right)}{\bar{\alpha}_{\text{geom}}} \times 100\% \tag{20}$$

Table 4. Integral average results of the MSWs.

Parameters	Value		
	MSW i	**MSW ii**	**MSWb iii**
$\bar{v}_{\text{si,a}}$, m/s	657.50	485.44	−3.75
$\bar{v}_{\text{si,θ}}$, m/s	4716.84	4797.93	4932.15
$\bar{\alpha}_{\text{aero}}$, °	82.07	84.22	90.04
$\bar{\alpha}_{\text{geom}}$, °	83.94	85.99	87.77
δ_α, %	−2.23	−2.06	2.58

As can be seen, the absolute value of the relative errors of the inclination angles are below 3%, indicating that the theoretical method based on the velocity decomposition on the MSW angles α is reasonable, with a proper accuracy and feasibility. Moreover, the predicted results give a clear and sensible explanation of the physical meaning of the MSW angles and the propagation of the MSWs, which provides some guidance to further investigations on the isolator with the simplified RDW under complex boundary conditions and with the real RDW.

7. Conclusions

Herein, the three-dimensional simulations on an annular isolator under rotating feedback pressure perturbations simplified from the single-mode rotating denotation wave (RDW), which is modeled as a periodical function consists of two linear variation zones and one curve change zone given by a transformed trigonometric function, are performed. The transient flow characteristics dominated by the moving shock wave/boundary layer interaction (MSW/BLI) is investigated thoroughly, and the self-similarity property of the flow-field is evaluated. Furthermore, the theoretical model of the inclination angle of the moving shock wave (MSW) by velocity decomposition is developed and validated.

It is found that a helical MSW and a terminal shock wave (TSW) are generated in the isolator due to the upstream and circumferential propagation of the feedback pressure perturbations. As a result, the flow-fields could be divided into three regions, i.e., the undisturbed zone, the terminal shock wave/moving shock wave/boundary layer interaction (TSW/MSW/BLI) zone, and the MSW/BLI zone. In the TSW/MSW/BLI zone, the rather small axial distance between the TSW and the MSW connects the two adverse-pressure gradients induced by the shocks, intensifying the flow separation on the meridian planes with smaller azimuthal angles, which rolls up vortices in the vicinity of the outer and inner walls. These vortices all develop along the TSW and the MSWs, presenting a typical glancing shock wave/boundary layer interaction (GSW/BLI). As a result, the migration of the separation bubbles makes the separation scale on the meridian plane at $\theta = 180°$ the maximum. In the MSW/BLI zone, the shock induces the boundary layer to separate, forming a helical vortex located at the foot of the MSW. During the upstream propagation process, the pattern of the MSWs transforms from a moving normal shock wave to a moving oblique shock wave with decreased strength, the screw-pitch of the MSW reduces slightly, and the inclination angle of the MSW increases gently. In other words, the MSW shows more "flat" patterns along the negative direction of the x-axis. Furthermore, the TSW shows some characteristics of a standing shock wave, causing the static pressure, temperature, and entropy of the flow to increase instantly and the velocity to drop in three directions. Whereafter, following the collisions with MSW, the static pressure, temperature, entropy, and circumferential velocity of the air rise instantly, while the axial velocity drops to a lower level. It is interesting to find that though the MSW do deflect the streamlines, the air still maintains an axial flow at the exit except in the adjacent region of the MSW with the neglected radial velocity and circumferential velocity. Likewise, three types of zones can be determined in the flow pattern at the exit: the rotating detonation wave/boundary layer interaction (RDW/BLI) zone, the expansion zone, and the vortices discharge zone.

Then, the self-similarity property is observed in the flow-field of the annular isolator under rotating feedback pressure perturbations, which connects the transient flow patterns at different moments with each other. The global flow structure of the isolator at different moments all show good agreement despite its rotation with the RDW, and the surface pressure profiles of the corresponding meridian plane all match perfectly. Such a characteristic indicates that the rotation angular velocity of the TSW and the MSW are equal and hold invariant, and the isolator flow could be regarded as a quasi-steady flow, i.e., independence of time. On this basis, the theoretical model of the MSW angles by coordinate transformation and velocity decomposition is introduced and validated. This model establishes contacts between the geometric form and the velocity triangle of the MSW for the first time, the latter of which is determined by its variation period and the original axial pressure gradient. The relative errors between the predicted and the geometrical results are below 3%, which confirms the reasonability and preciseness of the theoretical model, offering a rapid method to predict the shape of the MSW, and a perspective to better understand the physical meaning of the shape of the MSW.

Author Contributions: Conceptualization, Z.L.; methodology, H.H. and H.T.; validation, Z.L., H.H. and H.T.; formal analysis, Z.L.; investigation, Z.L.; resources, H.H.; data curation, H.H. and H.T.; writing—original draft preparation, Z.L.; writing—review and editing, H.H., H.T., G.L.; visualization, J.L.; supervision, H.H. and H.T.; project administration, Y.W.; funding acquisition, L.L. All authors have read and agreed to the published version of the manuscript.

Funding: This work was funded by the National Natural Science Foundation of China (Grant Nos. 12272177, 51906104, and U20A2070), the National Science and Technology Major Project (No. J2019-II-0014-0035), 1912 Project (No. 2019-JCJQ-DA001-164), the Foundation of National Key Laboratory of Transient Physics (No.6142604200212), and the priority academic program development of Jiangsu higher education institutions.

Institutional Review Board Statement: Not applicable.

Data Availability Statement: The data within the article will be available upon request.

Conflicts of Interest: The authors declare that they have no known competing financial interest or personal relationships that could have appeared to influence the work reported in this paper.

References

1. Lu, F.K.; Braun, E.M. Rotating detonation wave propulsion: Experimental challenges, modeling, and engine concepts. *J. Propul. Power* **2014**, *30*, 1125–1142. [CrossRef]
2. Wolanski, P. Detonative propulsion. *Proc. Combust. Inst.* **2013**, *34*, 125–158. [CrossRef]
3. Anand, V.; Gutmark, E. Rotating detonation combustors and their similarities to rocket instabilities. *Prog. Energy Combust. Sci.* **2019**, *73*, 182–234. [CrossRef]
4. Rong, G.; Cheng, M.; Sheng, Z.-H.; Liu, X.-Y.; Zhang, Y.-Z.; Wang, J.-P. Investigation of counter-rotating shock wave and wave direction control of hollow rotating detonation engine with Laval nozzle. *Phys. Fluids* **2022**, *34*, 056104. [CrossRef]

5. Hexia, H.; Huijun, T.; Wang, J.; Le Ning, S.S. A fluidic control method of shock train in hypersonic inlet/isolator. In Proceedings of the 50th AIAA/ASME/SAE/ASEE Joint Propulsion Conference, Cleveland, OH, USA, 28–30 July 2014.

6. Matsuo, K.; Miyazato, Y.; Kim, H.D. Shock train and pseudo-shock phenomena in internal gas flows. *Prog. Aerosp. Sci.* **1999**, *35*, 33–100. [CrossRef]

7. Bykovskii, F.A.; Zhdan, S.A.; Vedernikov, E.F. Continuous spin detonations. *J. Propul. Power* **2006**, *22*, 1204–1216. [CrossRef]

8. Voitsekhovskii, B. Stationary spin detonation. *Sov. J. Appl. Mech. Tech. Phys.* **1960**, *3*, 157–164.

9. Zhdan, S.A. Mathematical model of continuous detonation in an annular combustor with a supersonic flow velocity. *Combust. Explos. Shock Waves* **2008**, *44*, 690–697. [CrossRef]

10. Zhu, W.; Wang, Y.; Wang, J. Flow field of a rotating detonation engine fueled by carbon. *Phys. Fluids* **2022**, *34*, 073311. [CrossRef]

11. Uemura, Y.; Hayashi, A.K.; Asahara, M.; Tsuboi, N.; Yamada, E. Transverse wave generation mechanism in rotating detonation. *Proc. Combust. Inst.* **2013**, *34*, 1981–1989. [CrossRef]

12. Zhou, R.; Wang, J.P. Numerical investigation of flow particle paths and thermodynamic performance of continuously rotating detonation engines. *Combust. Flame* **2012**, *159*, 3632–3645. [CrossRef]

13. Smirnov, N.; Nikitin, V.; Stamov, L.; Mikhalchenko, E.; Tyurenkova, V. Rotating detonation in a ramjet engine three-dimensional modeling. *Aero. Sci. Technol.* **2018**, *81*, 213–224. [CrossRef]

14. Smirnov, N.; Nikitin, V.; Stamov, L.; Mikhalchenko, E.; Tyurenkova, V. Three-dimensional modeling of rotating detonation in a ramjet engine. *Acta Astronaut.* **2019**, *163*, 168–176. [CrossRef]

15. Liu, S.-J.; Lin, Z.-Y.; Liu, W.-D.; Lin, W.; Zhuang, F.-C. Experimental realization of H2/air continuous rotating detonation in a cylindrical combustor. *Combust. Sci. Technol.* **2012**, *184*, 1302–1317. [CrossRef]

16. Lin, W.; Zhou, J.; Liu, S.; Lin, Z.; Zhuang, F. Experimental study on propagation mode of H2/Air continuously rotating detonation wave. *Int. J. Hydrogen Energy* **2015**, *40*, 1980–1993. [CrossRef]

17. Wang, C.; Liu, W.; Liu, S.; Jiang, L.; Lin, Z. Experimental investigation on detonation combustion patterns of hydrogen/vitiated air with annular combustor. *Exp. Therm Fluid Sci.* **2015**, *66*, 269–278. [CrossRef]

18. Bluemner, R.; Bohon, M.; Paschereit, C.; Gutmark, E. Counter-rotating wave mode transition dynamics in an RDC. *Int. J. Hydrogen Energy* **2019**, *44*, 7628–7641. [CrossRef]

19. Han, J.; Bai, Q.; Zhang, S.; Weng, C. Experimental study on propagation mode of rotating detonation wave with cracked kerosene gas and ambient temperature air. *Phys. Fluids* **2022**, *34*, 075127. [CrossRef]

20. Naples, J.; Hoke, J.; Karnesky; Schauer, F. Flowfield Characterization of a Rotating Detonation Engine. In Proceedings of the 51st AIAA Aerospace Sciences Meeting Including the New Horizons Forum and Aerospace Exposition, Dallas, TX, USA, 7–10 January 2013.

21. Anand, V.; George, A.S.; Driscoll, R.; Gutmark, E. Characterization of instabilities in a rotating detonation combustor. *Int. J. Hydrogen Energy* **2015**, *40*, 16649–16659. [CrossRef]

22. Liu, Y.S.; Wang, Y.H.; Li, Y.S.; Li, Y.; Wang, J.P. Spectral analysis and self-adjusting mechanism for oscillation phenomenon in H_2/O_2 continuously rotating detonation engine. *Chin. J. Aeronaut.* **2015**, *28*, 669–675. [CrossRef]

23. Wu, Y.X.; Ma, H.; Peng, L.; Gao, J. Experimental research on initiation characteristics of a rotating detonation engine. *Exp. Therm. Fluid Sci.* **2016**, *71*, 154–163.

24. Liu, Y.; Zhou, W.; Yang, Y.; Liu, Z.; Wang, J. Numerical study on the instabilities in H2-air rotating detonation engines. *Phys. Fluids* **2018**, *30*, 046106. [CrossRef]

25. Schwer, D.; Kailasanath, K. Effect of Inlet on Fill Region and Performance of Rotating Detonation Engines. In Proceedings of the 47th AIAA/ASME/SAE/ASEE Joint Propulsion Conference & Exhibit, San Diego, CA, USA, 31 July–3 August 2011.

26. Anand, V.; St George, A.; Driscoll, R.; Gutmark, E. Analysis of air inlet and fuel plenum behavior in a rotating detonation combustor. *Exp. Therm. Fluid Sci.* **2016**, *70*, 408–416. [CrossRef]

27. Schwer, D.A.; Kailasanath, K.; Kaemming, T. Pressure characteristics of a ram-RDE diffuser. *Aerosp. Sci. Technol.* **2019**, *85*, 187–198. [CrossRef]

28. Frolov, S.M.; Dubrovskii, A.V.; Ivanov, V.S. Three-dimensional numerical simulation of operation process in rotating detonation engine. *Prog. Propuls. Phys.* **2013**, *4*, 467–488.

29. Frolov, S.M.; Dubrovsky, A.V.; Ivanov, V.S. Three-Dimensional Numerical Simulation of a Continuously Rotating Detonation in the Annular Combustion Chamber with a Wide Gap and Separate Delivery of Fuel and Oxidizer. In Proceedings of the 5th EUCASS, Munich, Germany, 1–6 July 2013.

30. Dyson, R.W. Flow Diode and Method for Controlling Fluid Flow Origin of the Invention. US Patent 9169855 B1, 4 November 2015.

31. Liu, S.; Liu, W.; Jiang, L.; Lin, Z. Numerical investigation on the airbreathing continuous rotating detonation engine. In Proceedings of the 25th ICDERS, Leeds (ICDERS, 2015), Leeds, UK, 2–7 August 2015; Volume 157.

32. Wang, G.; Liu, W.; Liu, S.; Zhang, H.; Peng, H.; Zhou, Y. Experimental verification of cylindrical air-breathing continuous rotating detonation engine fueled by non-premixed ethylene. *Acta Astronaut.* **2021**, *189*, 722–732. [CrossRef]

33. Dubrovskii, V.; Ivanov, A.; Zangiev; Frolov, S. Three-dimensional numerical simulation of the characteristics of a ramjet power plant with a continuous-detonation combustor in supersonic flight. *Russ. J. Phys. Chem. B* **2016**, *10*, 469–482. [CrossRef]

34. Mengmeng, Z.; Buttsworth, D.; Gollan, R.; Jacobs, P. Simulation of a Rotating Detonation Ramjet Model in Mach 4 Flow. *Simulation* **2018**, *10*, 1321.

35. Wu, K.; Zhang, S.; Luan, M.; Wang, J. Effects of flow-field structures on the stability of rotating detonation ramjet engine. *Acta Astronaut.* **2020**, *168*, 174–181. [CrossRef]
36. Wu, K.; Zhang, S.; Shen, D.; Wang, J. Analysis of flow-field characteristics and pressure gain in air-breathing rotating detonation combustor. *Phys. Fluids* **2021**, *33*, 126112. [CrossRef]
37. Menter, F.R. Two-equation eddy-viscosity turbulence models for engineering applications. *AIAA J.* **1994**, *32*, 1598–1605. [CrossRef]
38. Jianhua, C. *Backpressure Characteristics of Annular Isolator in CRDR*; National University of Defense Technology: Changsha, China, 2017.
39. Yue, S. *Numerical Research on the Effects of Ramjet Rotating Detonation on the Incoming Flow*; National University of Defense Technology: Changsha, China, 2019.
40. Peng, H.Y.; Liu, W.D.; Liu, S.J.; Zhang, H.L. Experimental investigations on ethylene-air continuous rotating detonation wave in the hollow chamber with laval nozzle. *Acta Astronaut.* **2018**, *151*, 137–145. [CrossRef]
41. Peng, H.-Y.; Liu, W.-D.; Liu, S.-J.; Zhang, H.-L.; Jiang, L.-X. Flowfield analysis and reconstruction of ethylene–air continuous rotating detonation wave. *AIAA J.* **2020**, *58*, 5036–5045. [CrossRef]
42. Rentao, Z. *Study of Flow Characteristics of Rotating Detonation Engine Inlet*; Nanjing University of Aeronautics and Astronautics: Changsha, China, 2019.
43. Igra, O.; Wang, L.; Palcovitz, J.; Heilig, W. Shock wave propagation in a branched duct. *Shock Waves* **1998**, *8*, 375–381. [CrossRef]
44. Mazor, G.; Igra, O.; Ben-Dor, G.; Mond, M.; Reichenbach, H.; Smith, F.T. Head-on collision of normal shock waves with a rubber-supported wall. *Philos. Trans. R. Soc. Lond. Ser. A Phys. Eng. Sci.* **1992**, *338*, 237–269. [CrossRef]
45. Chao, W. *Self-Sustaining and Propagation Mechanism of Airbreathing Continuous Rotating Detonation Wave*; National University of Defense Technology: Changsha, China, 2016.
46. Yuan, H.; Liu, F.; Wang, X.; Zhou, Z. Design and analysis of a supersonic axisymmetric inlet based on controllable bleed slots. *Aerosp. Sci. Technol.* **2021**, *118*, 107008. [CrossRef]
47. Liu, J.; Liu, C. Modified normalized Rortex/vortex identification method. *Phys. Fluids* **2019**, *31*, 061704. [CrossRef]
48. Dong, X.; Gao, Y.; Liu, C. New normalized Rortex/vortex identification method. *Phys. Fluids* **2019**, *31*, 011701. [CrossRef]
49. Kalkhoran, M.; Smart, M.K. Aspects of shock wave-induced vortex breakdown. *Prog. Aerosp. Sci.* **2000**, *36*, 63–95. [CrossRef]
50. Liang, G.; Huang, H.; Tan, H.; Luo, Z.; Tang, X.; Li, C.; Cai, J. Shock train/glancing shock/boundary layer interaction in a curved isolator with sidewall contraction. *Phys. Fluids* **2022**, *34*, 116106. [CrossRef]

Article

Mechanism Underlying the Effect of Self-Circulating Casings with Different Circumferential Coverage Ratios on the Aerodynamic Performance of a Transonic Centrifugal Compressor

Haoguang Zhang [1], Hao Wang [1,*], Qi Li [2], Fengyu Jing [1] and Wuli Chu [1]

[1] School of Power and Energy, Northwestern Polytechnical University, Xi'an 710072, China
[2] Xi'an Aerospace Propulsion Institute, Xi'an 710199, China
* Correspondence: hao1999@mail.nwpu.edu.cn

Abstract: The aim of this research was to explore the mechanisms underlying the effect of self-circulating casing treatment with different circumferential coverage ratios on the aerodynamic performance of a transonic centrifugal compressor. A three-dimensional unsteady numerical simulation was carried out on a Krain impeller. The circumferential coverage ratios of the self-circulating casings were set to 36%, 54%, 72% and 90%, respectively. The numerical results showed that the Stall Margin Improvement (*SMI*) increased with the increase in circumferential coverage ratios. The self-circulating casing with a 90% circumferential coverage ratio exhibited the highest *SMI* at 20.22%. Internal flow field analysis showed that the self-circulating casing treatment improved the compressor stability by sucking the low-speed flow in the blade tip passage and restraining the leakage vortexes breaking, which caused flow blockage. The compressor performance was improved at most of the operating points, and the improvement increased with increase in circumferential coverage ratio. The improvement in compressor performance was mainly attributed to reduction in the area of the high relative total pressure loss in the blade tip passage and significant decrease in the flow loss by the self-circulating casings.

Keywords: transonic centrifugal compressor; self-circulating casing treatment; circumferential coverage ratio; aerodynamic performance; Krain impeller

Citation: Zhang, H.; Wang, H.; Li, Q.; Jing, F.; Chu, W. Mechanism Underlying the Effect of Self-Circulating Casings with Different Circumferential Coverage Ratios on the Aerodynamic Performance of a Transonic Centrifugal Compressor. *Aerospace* **2023**, *10*, 312. https://doi.org/10.3390/aerospace10030312

Academic Editors: Dan Zhao, Chenzhen Ji and Hexia Huang

Received: 16 January 2023
Revised: 28 February 2023
Accepted: 8 March 2023
Published: 22 March 2023

1. Introduction

Centrifugal compressors have several advantages such as compact structure, high single-stage total pressure ratio, low cost and long service life, thus they are widely used in technical fields such as aerospace, ship and energy power [1,2]. The internal flow of centrifugal compressors is a complex process. An unstable flow occurs under low mass flow rate conditions and the compressor performance is reduced by the unstable flows. Therefore, expanding the stable working range of the centrifugal compressor is imperative in improving impeller technology [3].

Casing treatment is an effective way to improve compressor performance and stability. Casing treatment mainly involves slot casing treatment, blade tip injection and self-circulating casing treatment [4]. Previous findings show that slot casing treatment can effectively extend the compressor stable working range, but it reduces the compressor efficiency. Self-circulating casing treatment effectively improves the compressor stability, reduces the compressor efficiency loss and even improves the compressor efficiency. Therefore, self-circulating casing treatment has been widely explored in improving compressor efficiency and stability [5].

Fisher [6] conducted an experimental study on a centrifugal compressor and the results showed that the self-circulating casing treatment expanded the compressor's stable working range and slightly increased the compressor's adiabatic efficiency. Hunziker et al. [7] conducted a numerical study and the findings indicated that the self-circulating casing treatment sucked the clearance leakage flow near the suction port, and airflows discharged from the inject port reduced the Mach number and incidence angle of the incoming flow at the blade tip, thus widening the compressor's stable working range. Several numerical and experimental studies have been conducted on a high-speed centrifugal compressor [8–10]. The results of these studies showed that the unsteady flows inside the compressor were caused by the vortexes which were attributed to the tip leakage flow and the backflow, and the self-circulating casing eliminated the backflows near the throat of the blade tip passage. Tamaki [11] installed small blades in the self-circulating casing to generate vortexes at the compressor inlet, and the self-circulating casing treatment significantly improved the compressor stability margin. Jung [12] reported that the position and width of suction port of the self-circulating casing were key variables that affected the ability of the self-circulating casing to improve the compressor stability. The compressor's stable working range was gradually extended by the backward movement of the suction port's position and increase in the port width. Semlitsch [13] performed numerical studies using the LES method. The results showed that the self-circulating casing increased the compressor's stable working range because part of flows in the compressor returned to the compressor upstream along the self-circulating casing passage under small mass flow rate conditions.

Zheng [14] conducted numerical simulations on a centrifugal compressor and verified the findings through experiments. The results showed that the trajectory of the clearance leakage flow in the blade tip passage was blocked and the flow angle was reduced at the compressor inlet due to the suction effect of the self-circulating casing. Xu [15,16] performed a numerical study on a semi-open centrifugal compressor and the findings indicated that the gas discharged from the inject port increased the local mass flow rate and flow angle, thus the self-circulating casing treatment delayed occurrence of the stall. Kang [17] carried out a numerical study on a single-stage subsonic centrifugal compressor and the results showed that the self-circulating casing sucked the leakage flow and low-energy flows at the blade tip and improved the flow blockage in the blade tip passage.

Cao (2017) [18] conducted a numerical study on a transonic centrifugal compressor and reported that the self-circulating casing bled the partial airflows in the impeller passage and returned them to the compressor inlet, which delayed the change in the compressor inlet incidence angle and the development of shock. Gan [19] performed a numerical simulation on a high-speed centrifugal compressor and reported that the self-circulating casing treatment improved the stall margin of the centrifugal compressor by 6% and decreased the compressor efficiency by 0.5% under the design's mass flow rate conditions. Shang [20] reported that the axial distance, angle and diameter of the suction/inject port determined the flow velocity through the suction/inject port, the impact of the backflow on the main flow and the mass flow of the backflow.

Several studies have been conducted on centrifugal compressors with the self-circulating casing treatment. The effect of the self-circulating casing treatment in improving the internal flow of centrifugal compressors, the mechanisms of improving the compressor stability and the size and position of the self-circulating casing suction and inject port have been extensively explored. However, only a few studies have been conducted on the effect of the circumferential coverage ratio of self-circulating casing on the aerodynamic performance of centrifugal compressors. In addition, most previous studies have only been conducted with the self-circulating casing inject sections placed vertical to the casing line [14,15,17,18,20]. In this study, the design of the self-circulating casing inject section was based on a previous study by our research team [21] and the "Coanda Lines" were used to ensure that the gas flowed along the wall of the self-circulating casing while reducing the flow losses. This study sought to explore the effect of the circumferential coverage ratio of the self-circulating casing on the centrifugal compressor's aerodynamic performance. Four research schemes on the self-circulating casing treatment with circumferential coverage ratios of 36%, 54%, 72% and 90% were designed with a Krain impeller as the research object, and unsteady numerical simulations were performed. The unsteady calculation results on the performance curves and internal flow details were compared to reveal the flow mechanisms underlying the effect of the circumferential coverage ratio of the self-circulating casing on the compressor's aerodynamic performance.

2. Model Design and Numerical Simulation

2.1. Model Design

The research object in this study was a transonic centrifugal impeller SRV2-O with splitter blades. A schematic representation of the SRV2-O impeller is presented in Figure 1. The design parameters of the impeller were derived from a previous study [22]. Table 1 shows the basic design parameters of the SRV2-O. Krain (1995) conducted experimental studies on this type of impeller, and then performed numerical simulations in 2002. The numerical calculation results were in agreement with the experimental results [23]. The shape of the impeller was disclosed, which provided a reference for subsequent studies on this centrifugal impeller.

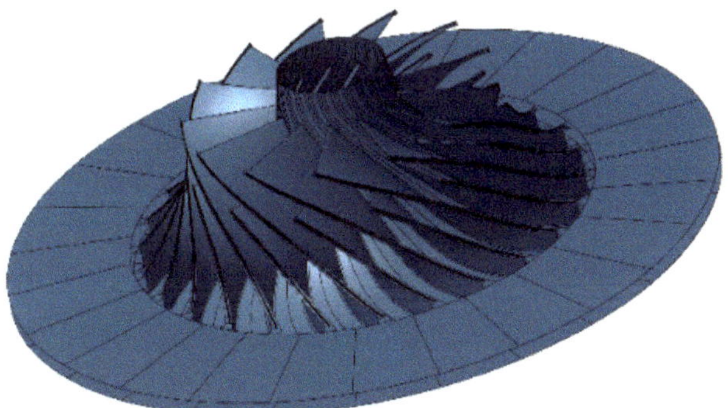

Figure 1. A schematic representation of the SRV2-O impeller.

Table 1. The basic design parameters of the SRV2-O impeller.

Parameters	Value	Parameters	Value
Mass flow (kg/s)	2.55	The number of main/splitter blade	13/13
Rotation speed (rpm)	50,000	Blade tip inlet relative Mach number	1.3
Inlet total pressure (Pa)	101,325	Inlet total temperature (K)	288.15
Inlet blade tip/root radius (mm)	30/78	Outlet impeller radius (mm)	112
Total pressure ratio	6.1	Adiabatic efficiency	0.84
Inlet/outlet blade angle (°)	26.5/52	Blade leading/trailing tip clearance (mm)	0.3/0.5

Thirteen self-circulating casings were uniformly arranged in the compressor along the circumferential direction. The number of self-circulating casings was the same as the number of mainstream passages. The geometric structure of the self-circulating casing is shown in Figure 2, where "m" represents the axial distance of the inject port, "k" represents the axial distance between the inject port trailing edge and the mainstream blade leading edge, "z" represents the axial distance between the suction port leading edge and the mainstream blade leading edge, "x" represents the axial distance of the suction port and "α" and "β" represent the air injection and suction angle of self-circulating casing, respectively. Geometric parameters of the self-circulating casing are presented in Table 2.

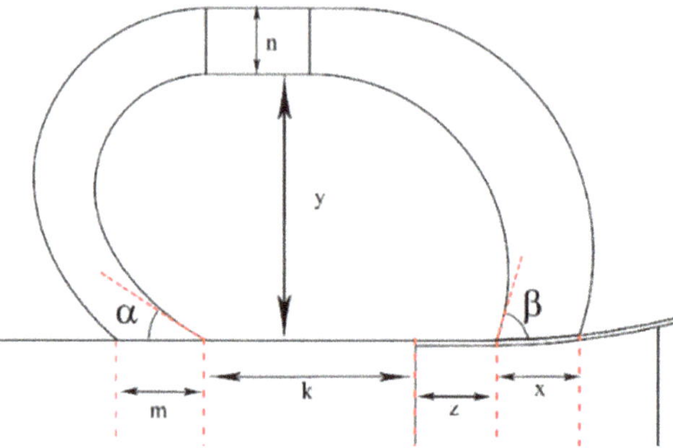

Figure 2. The geometric structure of the self-circulating casing.

Table 2. The geometric parameters of the self-circulating casing.

Parameter	Value	Parameter	Value
A (°)	30	k (mm)	18.57
β (°)	70	z (mm)	3.7
m (mm)	8.56	x (mm)	8
n (mm)	6.23	y (mm)	25

Four schemes of self-circulating casing treatment were designed with different circumferential coverage ratios of 36%, 54%, 72% and 90%, whereas the other geometric parameters of the self-circulating casing were the same. The circumferential coverage ratio is defined as the ratio of the circumferential angle covered by the self-circulating casing to 360°. The three-dimensional structures of the self-circulating casings with the circumferential coverage ratios of 36% and 72% are presented in Figure 3. "SRC" means the self-circulating casing, "C" represents the circumferential coverage ratio and the corresponding number represents the specific values of the circumferential coverage ratio (for example, "SRCC0.36" denotes the self-circulating casing with a circumferential coverage ratio of 36%).

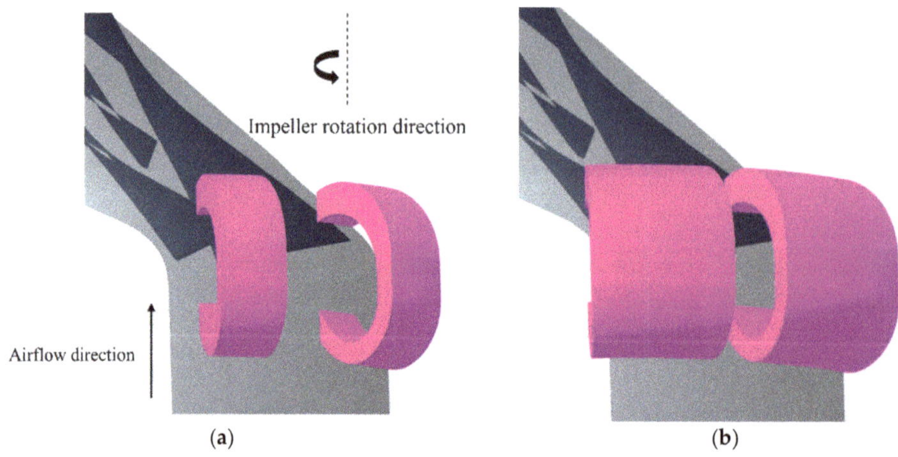

Figure 3. The three-dimensional structure of self-circulating casings. (a) SRCC0.36. (b) SRCC0.72.

2.2. Numerical Simulations

The grids were generated using the IGG/Autogrid5 module in NUMECA-16.1 software. The impeller passages adopted the O4H grid topology, and the butterfly grid was utilized for the blade tip clearance. The H-grid topology was adopted for the inlet and outlet extension sections and the self-circulating casing. The parameter "y^+" was maintained at a value less than 10, whereby "y^+" was related to the distance between the grid first layer and the solid wall. In the numerical simulation, the inlet extension section and the self-circulating casing were configured as the static blocks, and the rotor and diffuser passages were set as the rotating blocks. Two "sliding-block" structures with H-type grid topology were added between the rotor blade tip and the self-circulating structure, and the full non-matching method was applied at the R-S interface to achieve high accuracy of the flow field data transfer between the rotating blocks and the stationary blocks where the self-circulating casing was located. The diagram of computed domain grid is shown in Figure 4.

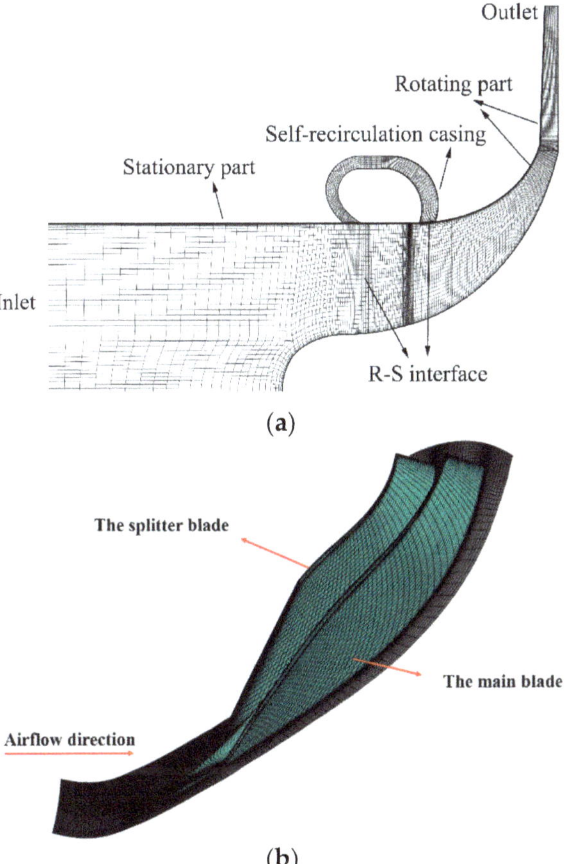

Figure 4. The diagram of computed domain grid. (**a**) A representation of the meridional plane of the computational grid. (**b**) Three-dimensional diagram of the blade grids.

The single-passage unsteady numerical simulations were carried out using the Fine-Turbo in NUMECA software. Full three-dimensional Reynolds time-averaged Navier–Stokes equations were numerically computed. Spalart–Allmaras was selected as the turbulence model. The spatial discretization adopted an upwind Total Variation Diminishing (TVD) scheme with second-order accuracy, and the CFL number was set as 3. The implicit double time-stepping method was used for time discretization in the unsteady numerical simulations. A total of 30 physical time steps were configured for each rotor passage, and the virtual time step in each physical time step was set to 20. The inlet boundary conditions were configured as follows: the total pressure was set to 101,325 Pa, the total temperature was 288.15 K, the air was sucked axially and the outward annular wall and the blade surface were adiabatic and un-sliding. The uniform initial field was adopted with an initial temperature of 288.15 K and a pressure of 300,000 Pa. The average static pressure was determined at the outlet, and the compressor performance curve was obtained by continuously increasing the outlet pressure.

2.3. Numerical Validation

Validation of the grid-independence and the turbulence model were carried out to verify the accuracy of the numerical simulation method. The steady numerical calculation results, the experimental data [22] and numerical simulation results [23] reported by Krain were evaluated and compared. Previous findings indicate that the blockage mass flow rate obtained in the experiment was approximately 2.864 kg/s [24]. The blockage mass flow rate obtained by Krain in the numerical calculation was 5–10% higher than that obtained through the experiments. In this study, the blockage mass flow rate in the numerical calculation was 3.03 kg/s, and the error with the experimental data was 5.7%, which is within the allowable error range [24].

The compressor performance curves were compared with the experimental data by generating 930,000, 1.3 million and 2.2 million grids. The central spatial discretization format and the Second-Order Upwind format were used in the validation. The Spalart–Allmaras and the k-ε turbulence model were selected for comparison with the experimental data. The abscissa of the performance curves' graphs represents the ratio of calculated mass flow rate to blockage mass flow rate, and the ordinate represents the compressor's total pressure ratio and isentropic efficiency, respectively. The terms "Central" and "Upwind" used in the curves represent the central spatial discretization format and the Second-Order Upwind differential format, respectively, "0.93 m", "1.3 m" and "2.2 m" represent the 930,000, 1.3 million and 2.2 million grids, respectively, and "S-A" and "K-E" represent the Spalart–Allmaras and the k-ε turbulence model, respectively. In addition, "Exp" represents the experimental results [22] and "Krain cal" represents the numerical calculation results by Krain et al. [23].

The comparison of the compressor's performance curves under different turbulence models is presented in Figure 5. The results showed that the difference in compressor total pressure curves between the S-A and the k-ε turbulence model was not significant (Figure 5). The compressor efficiency curves of the k-ε turbulence model were highly consistent with the experimental data compared with that of the S-A model. The maximum relative error in efficiency was only 1.9%, but the stalling mass flow rate was markedly bigger relative to that of the S-A model.

The compressor performance curves generated under different differential formats are shown in Figure 6. The total pressure ratio curves for the central spatial discretization format and the Second-Order Upwind format were similar under the same 930,000 grids and S-A turbulence model conditions. However, the efficiency curves for the Second-Order Upwind format were consistent with the experimental data, with a maximum relative error of approximately 3.4%. The calculated stalling mass flow rate for the Second-Order Upwind format was consistent with the experimental data.

The compressor performance curves generated under different grid numbers are shown in Figure 7. The findings indicated that there was no significant difference in the total pressure ratio curves obtained under the different grid numbers (Figure 7). However, under the high mass flow rate conditions, the compressor efficiency curves with the 1.3 million grids number were more consistent with the experimental data compared with the curves generated under the other two grids. Therefore, 1.3 million was selected as the number of grids for the subsequent numerical simulations. A difference was observed between the experimental results and the simulation results obtained with 1.3 million grids. The total pressure for the simulation was higher, and the isentropic efficiency was lower than that obtained from the experiment. The difference was observed mainly because the average total pressure obtained at the outlet was applied as the outlet total pressure in the numerical simulations, whereas the total outlet pressure in the experiment was calculated from the determined outlet total temperature, mass flow rate, wall static pressure and an assumed blockage coefficient. As a result, the compressor performance obtained for the experiment was mainly dependent on the measurement accuracy of the average static pressure and the estimated blockage coefficient. However, the significant variation in static pressure between the hub and the casing lowers the measurement accuracy of the average static pressure, resulting in a discrepancy in the experimental and numerical simulation outlet total pressure.

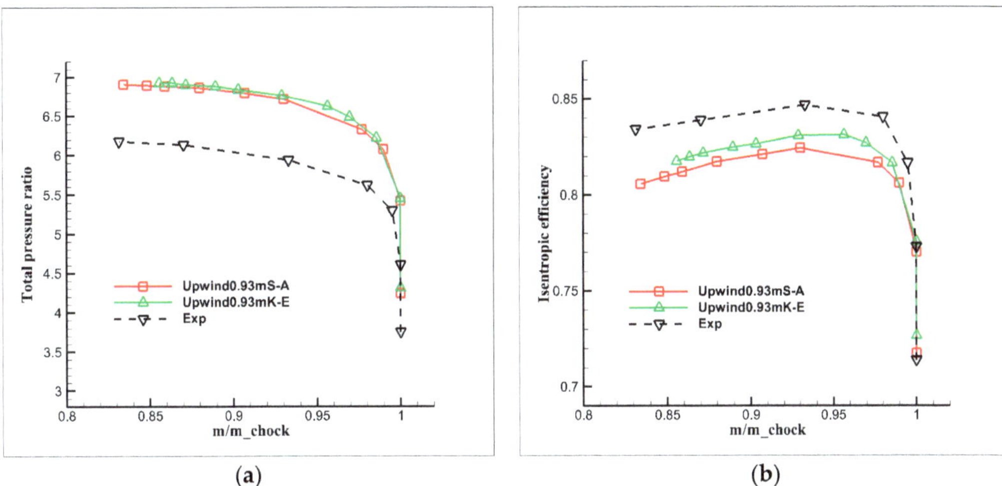

(a)   (b)

Figure 5. Compressor performance curves generated under different turbulence models. (**a**) Total pressure ratio. (**b**) Isentropic efficiency.

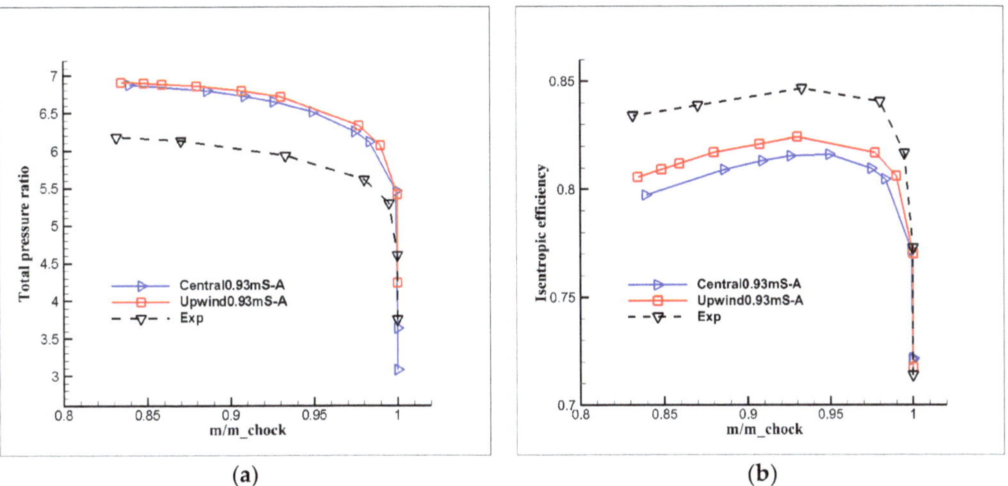

(a)   (b)

Figure 6. Compressor performance curves generated under different differential formats. (**a**) Total pressure ratio. (**b**) Isentropic efficiency.

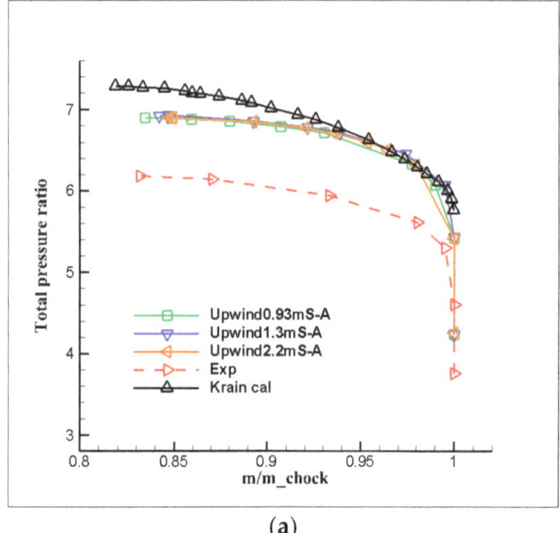

(**a**)

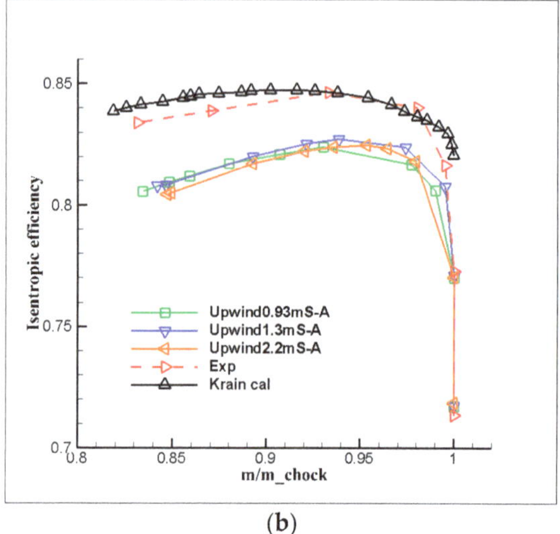

(**b**)

Figure 7. Compressor performance under different grid numbers. (**a**) Total pressure ratio. (**b**) Isentropic efficiency.

The positions of the experimental laser measurement planes in the compressor passage are shown in Figure 8. In Figure 9, the relative Mach number distributions on the different planes were compared between the experimental data and the simulation results under the design's mass flow rate conditions to further verify the accuracy and feasibility of the numerical simulation performed in this study.

The numerical results on the pressure surface side on plane "−1" were consistent with the experimental results (Figure 9a,b). The relative Mach number in the passage gradually increased with an increase in the blade height. Notably, some inconsistencies were observed between the numerical results and the experimental results at midspan and the upper right area of plane "−1". The errors can be attributed to the choice of turbulence model and the discrete format used in the numerical simulation. The secondary flows may be inconsistent between the experimental procedures and the numerical simulation.

The relative Mach number distribution trend and the specific values of calculation results on plane "4" and plane "10" were consistent with the experimental results (Figure 9c–f). However, the specific value of the relative Mach number at the blade tip was lower and the low-speed area was larger for the numerical simulation than values obtained from the experiments. The error may have resulted because the laser measurement used in the experiment did not accurately reflect the distribution of the tip leakage flow. In addition, a low-velocity area was caused by the tip leakage flow.

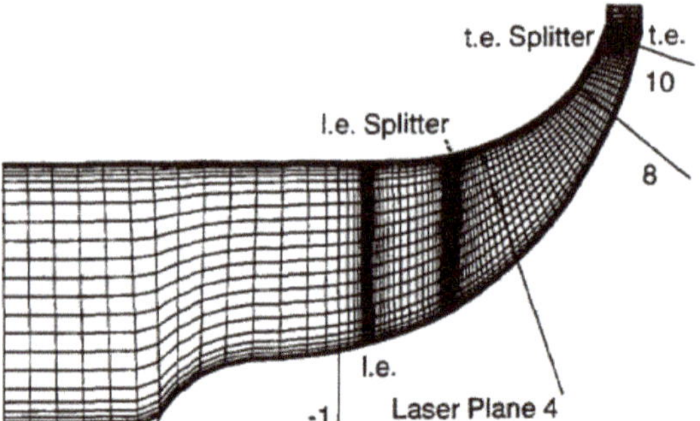

Figure 8. The experimental laser measurement planes in the compressor passage.

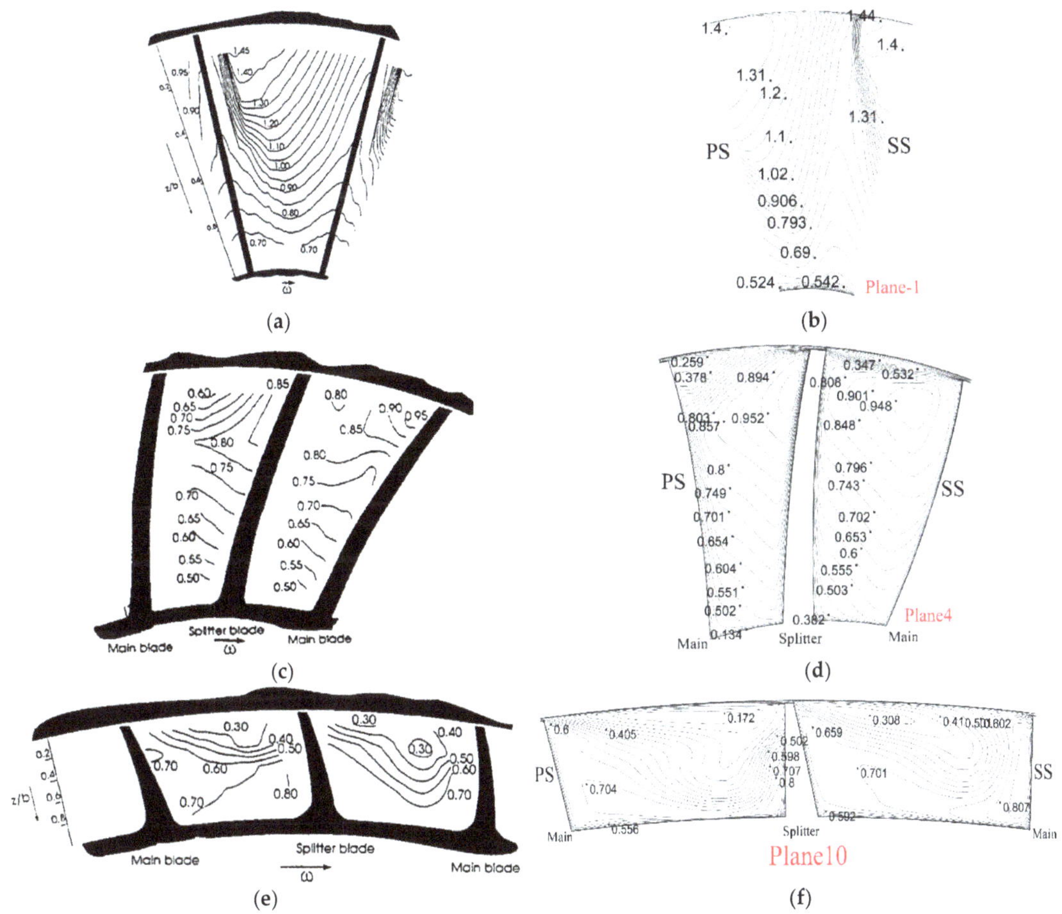

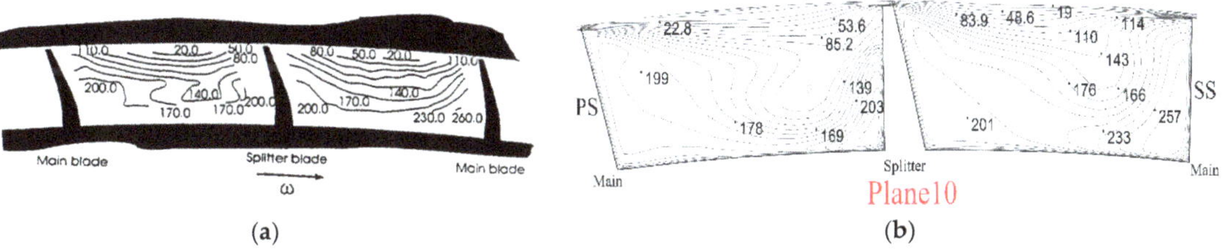

Figure 9. The relative Mach number distribution on the different planes. (**a**) Experimental data on plane "−1". (**b**) Numerical results on plane "−1". (**c**) Experimental data on plane "4". (**d**) Numerical results on plane "4". (**e**) Experimental data on plane "10". (**f**) Numerical results on plane "10".

The meridional velocities on plane "10" were shown and the results showed that the distribution of the meridional velocities under the numerical simulations was consistent with the distribution obtained through the experiment (Figure 10).

Figure 10. Distribution of meridional velocities on plane "10". (**a**) Experimental data on plane "10". (**b**) Numerical results on plane "10".

In summary, the results showed that the numerical simulations with the Second Order Upwind format using the S-A turbulence model accurately predicted the performance and exhibited the internal flow parameters of the Krain impeller. The numerical calculation method used in this study was accurate as shown by the high consistency with the experimental results. Therefore, the subsequent studies conducted in this study were based on this numerical method.

3. Simulation Results and Discussion

3.1. Compressor Performance Analysis

The performance curves of the compressor with different casing parameters are presented in Figure 11. The abscissa represents the mass flow rate, and the ordinate represents the compressor's total pressure ratio and efficiency. In the curves, "sw" represents the solid-wall casing, and the other symbols are provided above. Analysis of the compressor performance curves showed that the compressor's near-stall mass flow rate was reduced after application of the self-circulating casing treatment, and the compressor stability was increased to varying degrees. A higher circumferential coverage ratio was correlated with a lower compressor near-stall mass flow rate. The total pressure ratios of the compressor with the self-circulating casing treatment were higher compared with the total pressure ratios for the solid-walled casing under the small and medium mass flow rate conditions (Figure 11a). Self-circulating casing treatments with different circumferential coverage ratios increased the impeller's pressure-boosting capacity. Notably, the impeller's pressure-boosting capacity increased with an increase in circumferential coverage ratio under the small mass flow rate conditions. All four self-circulating casing treatments effectively improved the compressor's isentropic efficiency under the small and medium mass flow rate conditions (Figure 11b). The efficiency curves of the SRCC0.72 and SRCC0.9 designs almost overlapped under the small mass flow rate conditions. These findings indicate that the compressor efficiency increases with increase in circumferential coverage ratio under the small mass flow rate conditions.

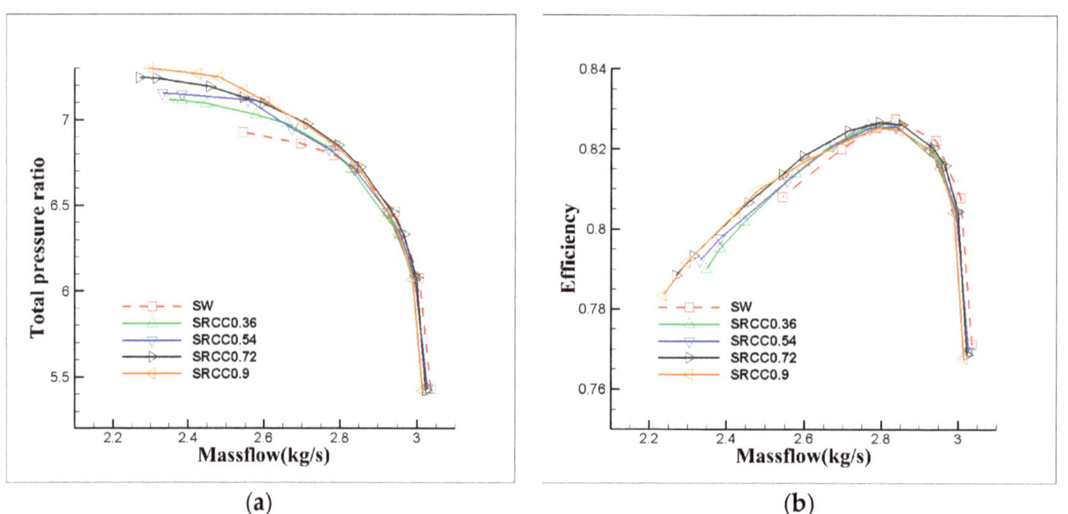

Figure 11. Performance curves of the compressor under different casing parameters. (**a**) Total pressure ratio. (**b**) Isentropic efficiency.

Stall Margin Improvement (*SMI*) and Peak Efficiency Improvement (*PEI*) were introduced in this study for a precise quantitative analysis. The *SMI* and *PEI* were expressed as follows:

$$SMI = \left(\frac{\pi^*_{SRC,stall}}{\pi^*_{sw,stall}} \times \frac{M^*_{sw,stall}}{\pi^*_{SRC,stall}} - 1 \right) \times 100\% \tag{1}$$

$$PEI = \frac{\eta^*_{SRC,peak} - \eta^*_{sw,peak}}{\eta^*_{sw,peak}} \times 100\% \tag{2}$$

where "π^*" represents the compressor total pressure ratio, "M" represents the mass flow rate and "η^*" represents the compressor isentropic efficiency. The subscripts "stall" and "peak" represent the near-stall condition and the peak efficiency condition, respectively. The specific values of the *SMI* and *PEI* of the four research schemes are presented in Table 3.

Table 3. *SMI* and *PEI* of self-circulating casings with different circumferential coverage ratios.

	SMI/%	*PEI/%*
SRCC0.36	11.26	−0.13
SRCC0.54	12.67	−0.23
SRCC0.72	18.05	−0.17
SRCC0.9	20.22	−0.27

The self-circulating casing treatments with 36%, 54%, 72% and 90% circumferential coverage ratios generated an *SMI* of 11.26%, 12.67%, 18.05% and 20.22% and a *PEI* of −0.13%, −0.23%, −0.17% and −0.27%, respectively (Table 2). Quantitative analysis showed that the *SMI* gradually increased as the circumferential coverage ratio increased and the self-circulating casing's ability to expand compressor stability also gradually increased. However, the difference in *SMI* between the 36% and 54% circumferential coverage ratios was not significant. Analysis of the compressor peak efficiency showed that the PEIs generated by different self-circulating casing treatments were all negative, and the compressor peak efficiency decreased slightly with a change in the self-circulating casing treatments.

3.2. Comparative Analysis of the Compressor Internal Flow

Analyses were conducted to explore the internal flow of the compressor at the near-stall point under the different casing configurations, and the following content is shown using time-averaged results of the unsteady calculation.

The relative Mach number contours of different research schemes on the sliced planes are shown in Figure 12. These planes, which are perpendicular to the Z axis, are located in the compressor passage, and the spacing is the same for every two adjacent planes. A large low Mach number area surrounded by the red dashed line was observed in the blade tip passage under the near-stall point. The flow velocity is low in this region, resulting in flow blockage in the blade tip passage. The results for the solid-wall casing with self-circulating casing treatments showed that all Mach number areas with a value below 0.25 disappeared, and only low Mach number areas with values more than 0.3 and less than 0.75 were observed in the mainstream blade passage. The self-circulating casing treatment significantly improved the low Mach number area and expanded the flow area in the mainstream blade passage. However, self-circulating casing treatment did not significantly suppress the low-energy areas in the splitter blade passage. Analysis of the ability of the different self-circulating casing treatments to reduce the blocked flow area revealed that the low-speed area gradually decreased from the 36% to the 72% circumferential coverage ratio, but the difference between the 72% and the 90% coverage ratio was not significant for this configuration.

The relative Mach number distribution under different schemes of 96% blade span was explored to evaluate the effect of the different self-circulating casing treatments on the flow blockage in the blade tip passage (Figure 13). The findings showed that most of the area of passage exhibited low-speed flows under the near-stall point, and the flow area in this passage was markedly reduced (Figure 13a). This finding is consistent with the previous result in Figure 12. The results showed that all the self-circulating casing treatments with the different circumferential coverage ratios improved the flow blockage in the blade tip passage (Figure 13b–d). The flow area and the distance between the blade pressure surface and the boundary of the low Mach number area marked by the dashed red line gradually increased with an increase in circumferential coverage ratio. This finding indicated that the inhibition of the self-circulating casing treatments to the low-energy flow was proportional to the circumferential coverage ratio.

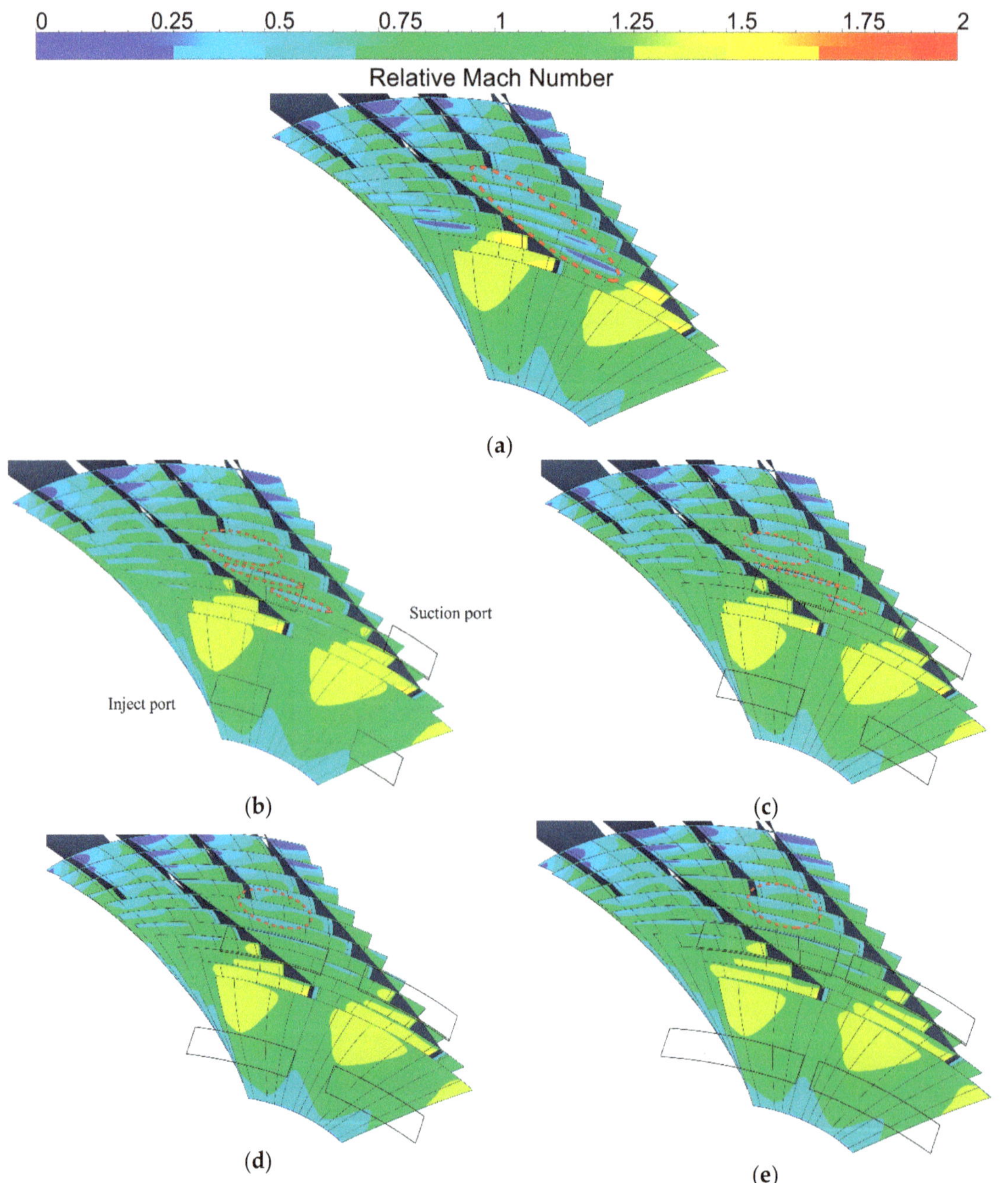

Figure 12. Relative Mach number distribution in the blade tip passage. (**a**) SW. (**b**) SRCC0.36. (**c**) SRCC0.54. (**d**) SRCC0.72. (**e**) SRCC0.9.

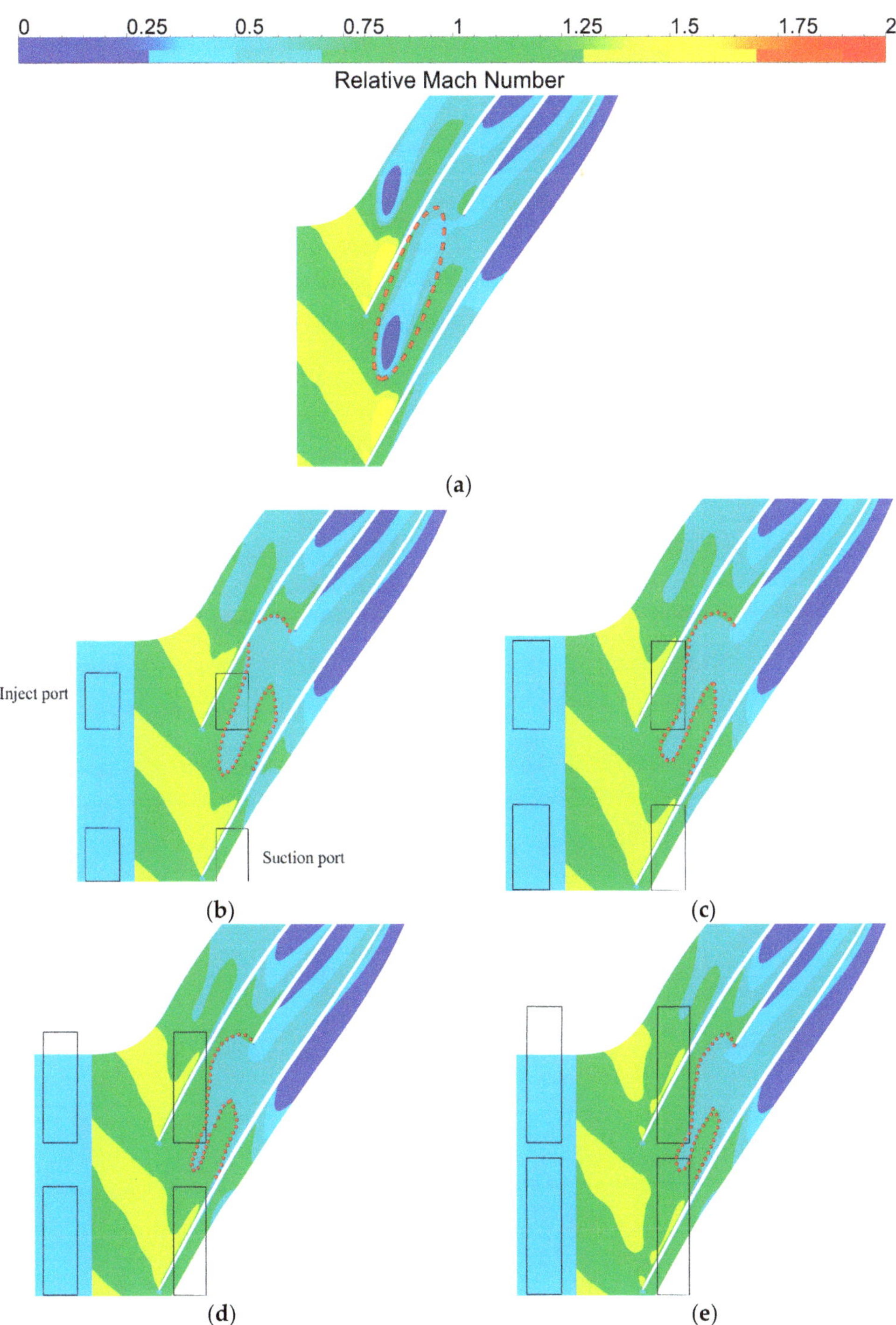

Figure 13. Contours of relative Mach number distribution of different schemes at 96% blade span. (**a**) SW. (**b**) SRCC0.36. (**c**) SRCC0.54. (**d**) SRCC0.72. (**e**) SRCC0.9.

The axial velocity distribution at 50% τ ("τ" represents the clearance height of mainstream blade leading edge at the blade tip) was explored to further indicate the improvements of the self-circulating casing treatments on the flow blockage in the blade tip passage (Figure 14). The findings indicated the presence of blockages in the passage when the axial speed w_z was negative. The backflow area was significantly reduced after application of the self-circulating casing treatments, implying that the flow area was significantly increased. The backflow area reduced with an increase in circumferential coverage ratio of the self-circulating treatments. The backflow area was smallest in the SRCC0.9 design compared with the other configurations. Analysis showed that the backflow area was not evident under the suction port of the self-circulating casing. These findings indicate that the ability of the self-circulating casing treatment to restrain the flow blockage increased with an increase in the circumferential coverage ratio.

The relative velocity vector of the airflow of different structure casings at 98% blade span was evaluated (Figure 15). A large low-speed flow area surrounded by the dashed red line in the mainstream blade passage was observed under the near-stall point. The red arrow represents the flow direction under the near-stall point. The area of low-speed flow between the blade suction surface and the adjacent blade pressure surface in the mainstream blade passage was significantly suppressed, and the angle between the flow direction and the blade suction surface was reduced after application of self-circulating magazine treatment. The self-circulating casing treatment played a significant role in improving the compressor internal flow. Analysis of the different self-circulating casing treatments showed the presence of a large low-speed flow area in the passage in SRCC0.36 and the flow direction was at an angle to the blade suction surface, whereas the flow directions of the other self-circulating casing treatments were along the blade suction surface. The low-speed flow area and the angle between the flow direction and the blade suction surface decreased with an increase in circumferential coverage ratio.

The leakage flow lines' distribution in the blade tip passage was evaluated to further illustrate the mechanism underlying the increase in the compressor internal flow by the self-circulating casing (Figure 16). The lower-speed leakage flows fill with the blade tip passage formed by the mainstream blades and the mainstream/splitter blades under the near-stall point. The leakage flow velocity rapidly decreased after the shock at the mainstream passage inlet. The leakage flows became distorted and swollen at the inlet. The leakage flows were interrupted below the suction port by the suction of the self-circulating casing. The area and intensity of the leakage flows sharply decreased downstream of the suction port, and the expansion and development of the leakage flows were effectively suppressed. The low-energy area created after the shock was also eliminated. Analysis of the different self-circulating casing treatments showed that the leakage flow lines from the mainstream blade gap became progressively less dense at the blade tip, as indicated by the red dashed line in Figure 16. The leakage flow distribution area gradually decreased under the suction port from the 36% to the 90% circumferential coverage ratio. The effect on suppressing the development of leakage flows towards adjacent blades increased with an increase in circumferential coverage ratio. The results showed that SRCC0.9 had the greatest inhibiting effect on the leakage flows, and the circumferential development of leakage flows had been completely sucked for this configuration.

The parameter of dimensionless helicity was introduced in this study to evaluate the expansion and fragmentation of the leakage vortexes at the blade tip. Dimensionless helicity is expressed as follows:

$$H_n = \frac{\vec{W} \bullet \vec{\zeta}}{\left|\vec{W}\right|\left|\vec{\zeta}\right|} \tag{3}$$

where "$\vec{W}$" represents the relative velocity vector and "$\vec{\zeta}$" represents the vortex vector. H_n denotes the tightness of the leakage flow lines around the core of the leakage vortexes, and the value of H_n is close to 1 (the range of H_n is -1 to 1). A sudden change in the value of the H_n (for example, from 1 to -1) often indicates the expansion and fragmentation of the leakage vortexes. The contours of H_n for different casings at the 98% blade span are presented in Figure 17. The results showed an abrupt change in H_n value at the mainstream passage inlet under the solid-wall casing, indicating that the leakage vortexes were broken at this location, and the resulting low-speed flows caused blockage in the passage. The self-circulating casings with the 36% and 54% circumferential coverage ratios did not eliminate the break-up of the leakage vortexes, and an abrupt change in value was observed at the passage inlet for these configurations. However, the sudden change in value disappeared under the self-circulating casing treatment with the 72% and 90% circumferential coverage ratios, and these treatments effectively inhibited the break-up of the leakage vortexes.

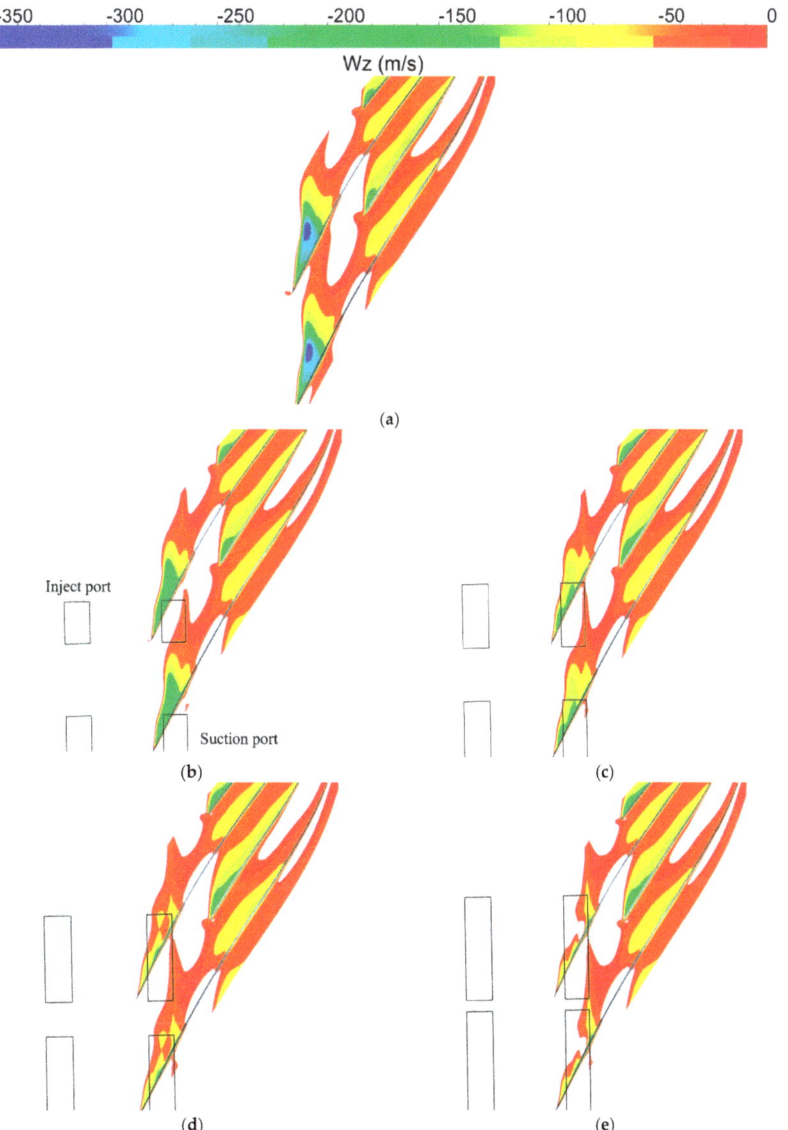

Figure 14. The axial velocity distribution at 50% τ. (**a**) SW. (**b**) SRCC0.36. (**c**) SRCC0.54. (**d**) SRCC0.72. (**e**) SRCC0.9.

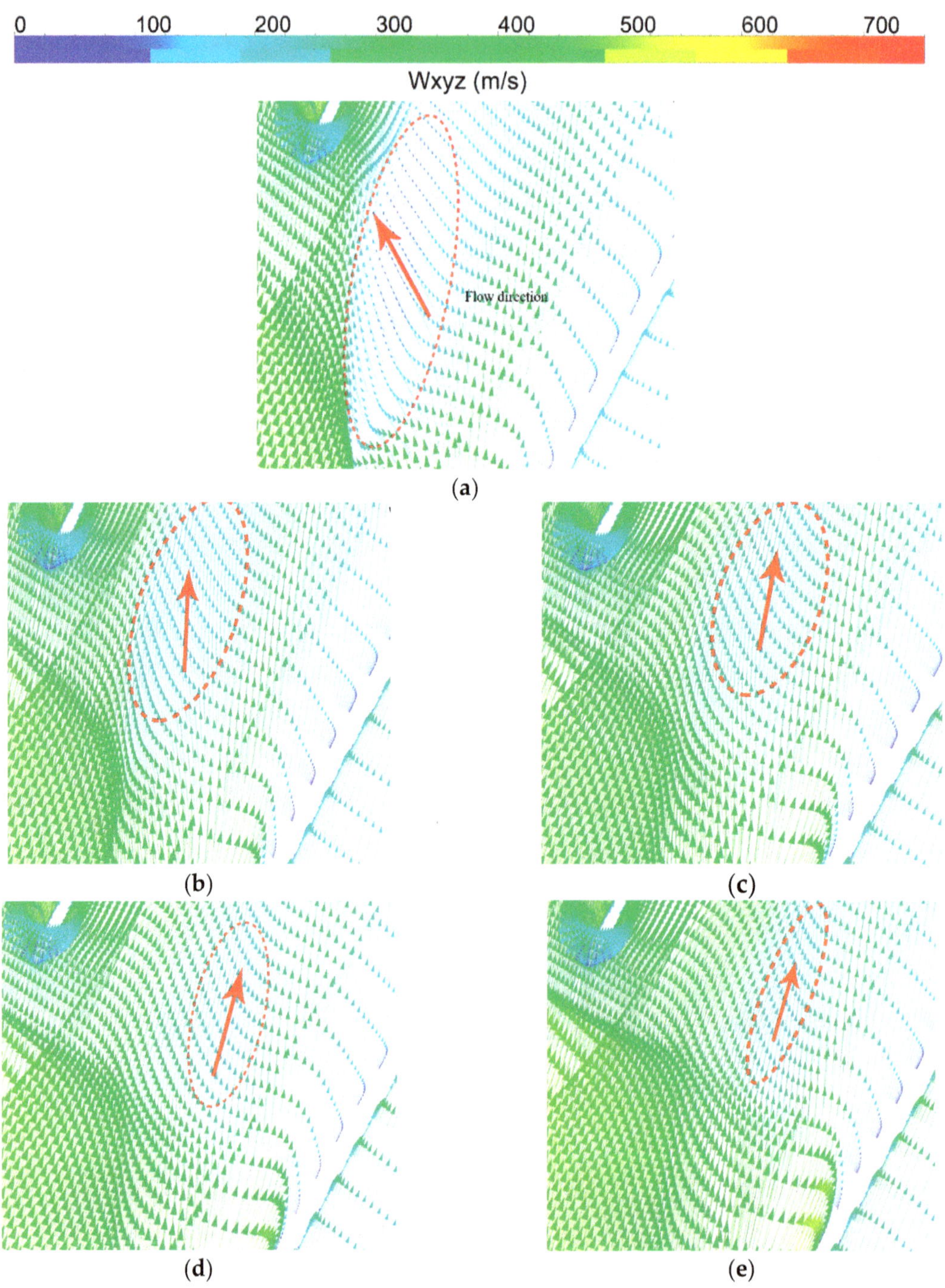

Figure 15. Relative velocity vector of different structure casings at 98% blade span. (**a**) SW. (**b**) SRCC0.36. (**c**) SRCC0.54. (**d**) SRCC0.72. (**e**) SRCC0.9.

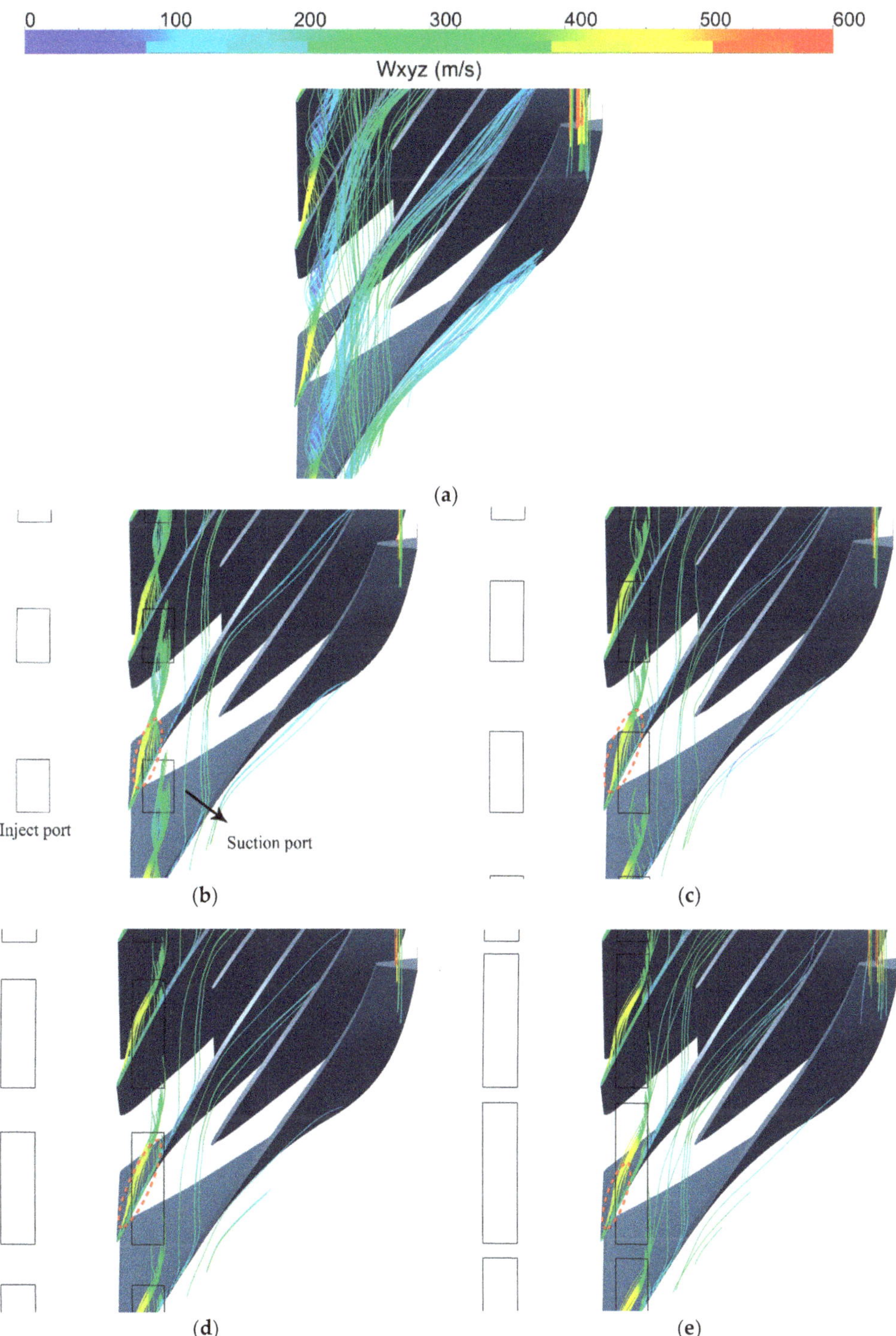

Figure 16. Distribution of leakage flows in the blade tip passage. (**a**) SW. (**b**) SRCC0.36. (**c**) SRCC0.54. (**d**) SRCC0.72. (**e**) SRCC0.9.

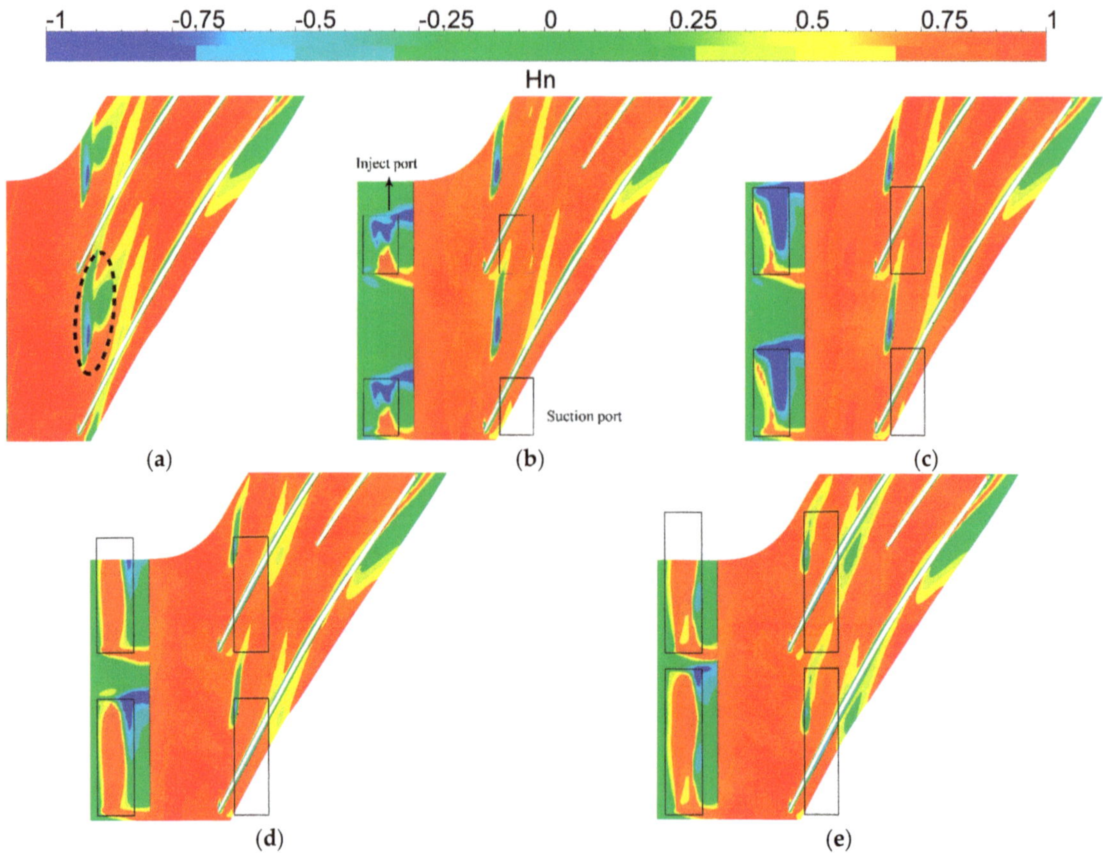

Figure 17. Non-dimensional helicity distribution of different casing structures at 98% blade span. (**a**) SW. (**b**) SRCC0.36. (**c**) SRCC0.54. (**d**) SRCC0.72. (**e**) SRCC0.9.

The absolute vorticity contours, static pressure isoline distribution, the track of the vortex core and the interface between mainstream and leakage flow were explored at the 98% blade span (Figure 18). The dashed black lines represent the track of the vortex core, and the dashed red lines represent the interface. The results indicate that the interface between the mainstream and the leakage flow almost fully blocked the front of the entire mainstream passage, and the vortex core track developed from the blade suction surface to the adjacent blade pressure surface. The interface between mainstream and leakage flow and the vortex core track at the blade tip leading edge deflect towards the blade suction surface, and the flow in the blade tip passage was improved after application of the self-circulation casing treatments. Notably, the self-circulating casing treatment with different circumferential coverage ratios had different effects on the vortex core track and interface between mainstream and leakage flow. The difference between SRCC0.36 and SRCC0.54 was not significant for the interface. However, a significant difference in the vortex core track was observed between SRCC0.36 and SRCC0.54. The vortex core track was more inclined to the suction surface in SRCC0.54 than in SRCC0.36. However, the effects of SRCC0.36 and SRCC0.54 on the vortex core track and the interface were not as effective as those observed for SRCC0.72 and SRCC0.9. The results showed that SRCC0.72 did not present a significant difference in the vortex core track compared with that of SRCC0.9. However, the interface was more inclined to the blade suction surface, and the flow area in the mainstream passage was larger in SRCC0.9 compared with SRCC0.72. In summary, the effects of the self-circulating casing treatment on the vortex core track and the mainstream/leakage flow interface were directly proportional to the circumferential coverage ratio.

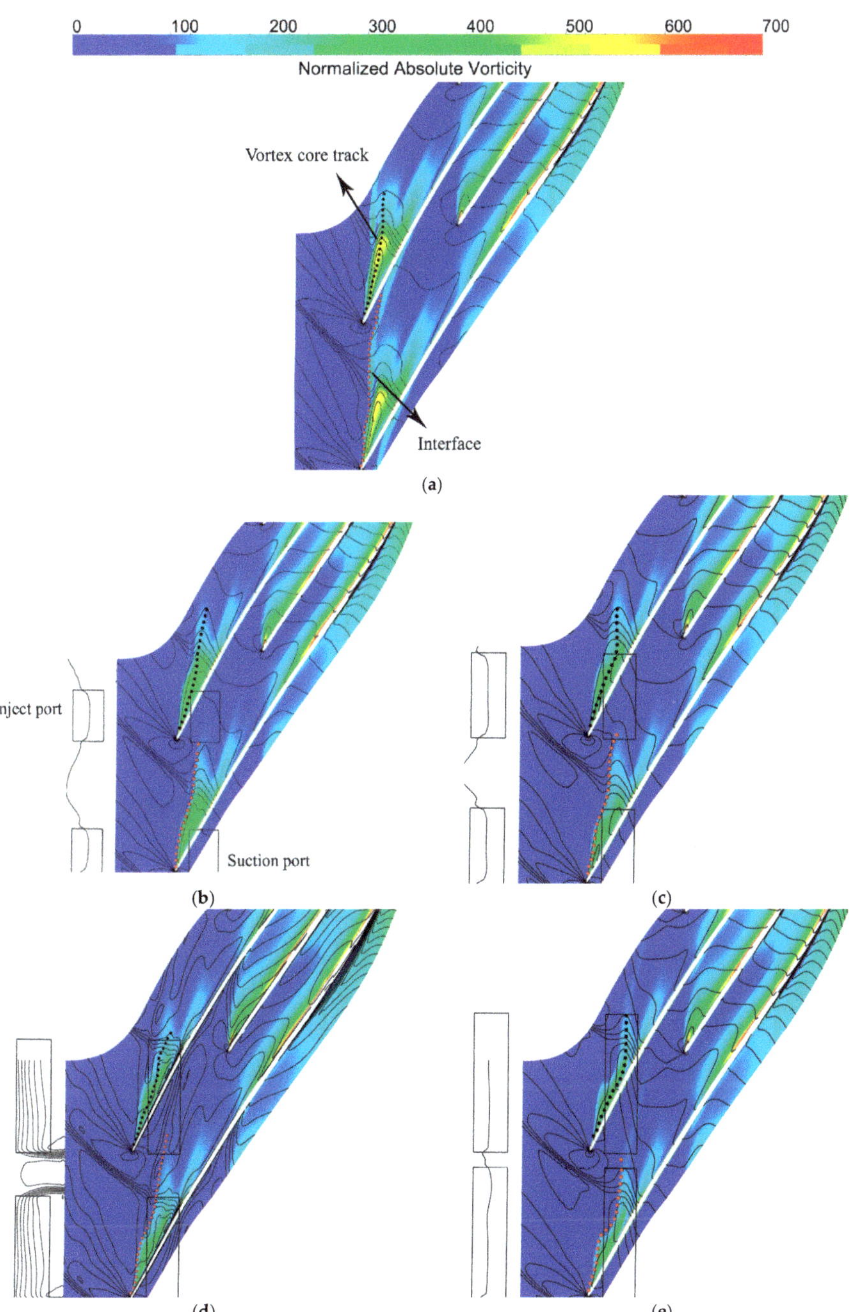

Figure 18. Contours of absolute vorticity and static pressure distributions at 98% blade span. (**a**) SW. (**b**) SRCC0.36. (**c**) SRCC0.54. (**d**) SRCC0.72. (**e**) SRCC0.9.

The relative blockage area distribution curves along the axial direction were generated to quantitatively evaluate the improvements in blockage in the blade tip passage by the four self-circulating casings (Figure 19). The horizontal axis represents the axial relative position of the blade tip passage, whereby 0–1 represents the passage from the mainstream blade leading edge to the splitter blade leading edge, and a value greater than 1 denotes the splitter blade passage. The vertical axis represents the relative blockage area. The self-circulating casing treatments effectively reduced the blockage area in the blade tip passage in most axial position ranges, especially in the 0 to 0.5 range compared with the solid-wall casing. A larger circumferential coverage ratio was correlated with a smaller blockage area and a higher ability of the self-circulating casing to improve the flow in the blade tip passage. The blockage area in the splitter blade passage was also improved by the various self-circulating casing treatments to varying degrees. These findings indicate that the self-circulating casing treatments improved the blockage in the mainstream blade passage and reduced the blockage in the splitter blade passage, which explains why the self-circulating casing treatment significantly improved the compressor's stability.

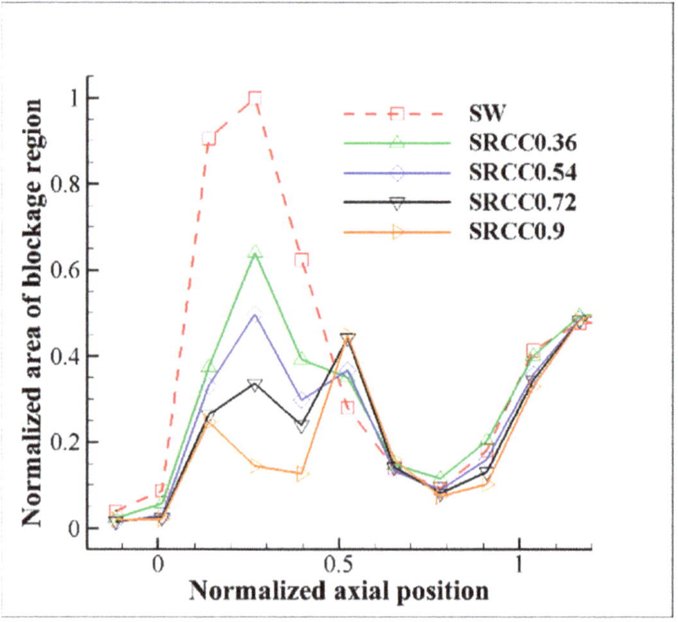

Figure 19. Distribution curve of the blocked area of the blade tip channel along the axis.

The changes in the blockage ratio of different self-circulating casings in the blade tip passage were determined over a complete period at 50% τ (Figure 20). The blockage ratio is expressed as follows:

$$B_R = \frac{A_b}{A} \times 100\% \tag{4}$$

where "A" represents the selected plane area and "A_b" represents the area with negative relative axial velocity on the selected plane. The results showed that the blockage ratios of the different self-circulating casings changed over time at 50% τ (Figure 20). Notably, the blockage area at the blade tip was smaller than the solid-wall casing under the near-stall point after application of the self-circulating casing treatment. The blockage ratio decreased with increase in circumferential coverage ratio, and the minimum occurrence time decreased. The research schemes with high circumferential coverage ratio exhibited a blockage ratio less than that of the low circumferential coverage ratio at all times, which is highly consistent with the results reported in Figure 19. The unsteady time average of the blockage distributions was determined for the different casings at 99% blade span to further evaluate the blockage ratios in the blade tip passage for the four self-circulating casing treatments (Figure 21). The red region represents the area in which W_z was negative. The contours indicate that the blockage area in the blade tip passage decreased with an increase in the circumferential coverage ratio, indicating that the qualitative analysis results were highly consistent with the quantitative results presented in Figure 20.

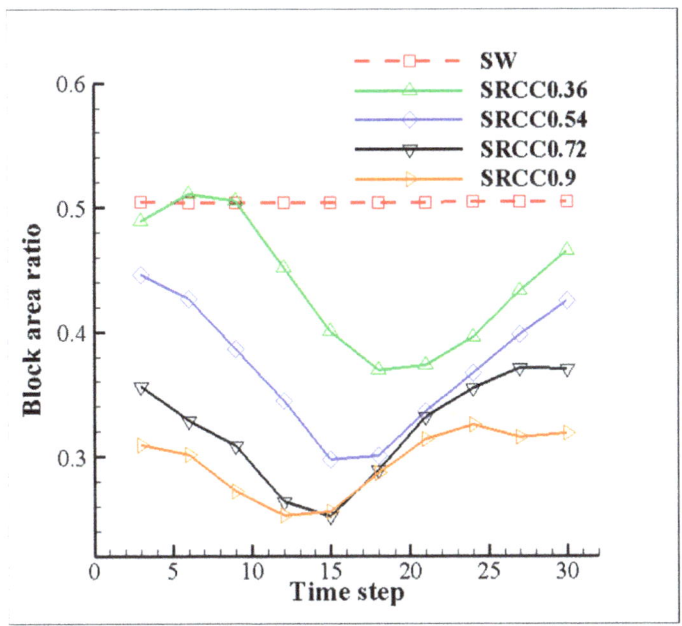

Figure 20. Changes in the blockage ratio under different self-circulating casings at 50% τ.

The meridional dimensionless radial velocity contours in and under four self-circulating casings were determined (Figure 22). The radial velocity selected in contours represented the circumferential average value of unsteady time-average simulation results, and the dimensionless radial velocity was equal to the ratio of the local radial velocity to the average axial velocity at the inlet. The numbers in the contours show the dimensionless radial velocity value at that location, and a positive value indicates an upward direction. The results showed that the absolute dimensionless radial velocities at the suction and inject port of four self-circulation casings were relatively large. The red area with the largest absolute value of radial velocity increased with an increase in the circumferential coverage ratio, indicating that the upward radial velocity at the suction port was proportional to the circumferential coverage ratio. The blue area corresponds to the downward radial velocity at the inject port, and a larger area indicates a greater and more extensive downward radial velocity. The contours show that the downward radial velocity at the inject section increased with an increase in the circumferential coverage ratio, and the area had a positive effect. The downward radial velocity area under the inject port increased with an increase in the circumferential coverage ratio, implying that a larger circumferential coverage ratio was correlated with a large velocity and the area occupied by the jet.

A histogram of the dimensionless mass flow rate at the suction and inject port of the self-circulating casings was generated under different circumferential coverage ratios (Figure 23). The abscissa represents the circumferential coverage ratio, and the ordinate represents the dimensionless suction and inject mass flow rate based on the compressor design mass flow rate. The results showed that a large circumferential coverage ratio was correlated with a large mass flow rate through the suction and inject port of the self-circulating casings. The gas from the inject port flowed into the compressor mainstream, which increased the compressor's inlet mass flow rate and delayed the occurrence of a stall.

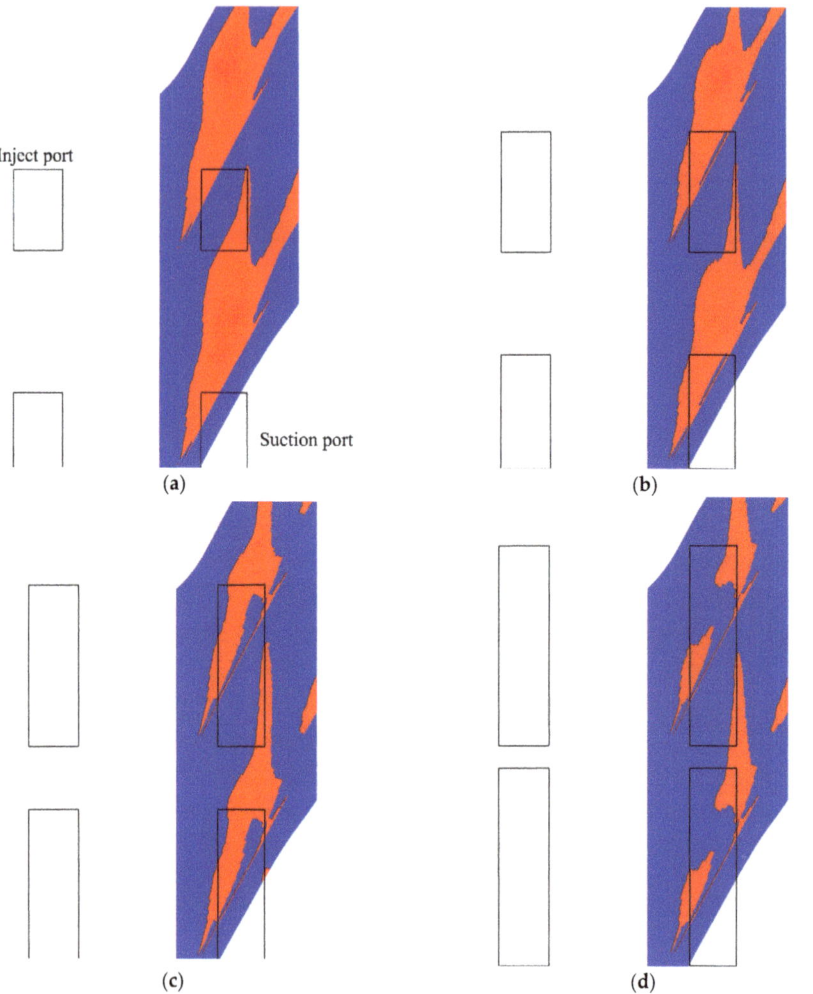

Figure 21. Contour distribution of the blockage area at 99% blade span. (**a**) SW. (**b**) SRCC0.36. (**c**) SRCC0.54. (**d**) SRCC0.72.

The absolute Mach number and streamline distribution in the self-circulating casing were determined and are presented in Figure 24. The circumferential section with the same radial height was selected for the different self-circulating casings. The arrow in the figure represents the impeller rotation direction. "BL" and "IN" represent the inject port section and suction port section, respectively. Analysis showed that increase in the circumferential coverage ratio increased the high Mach number area in the self-circulating casing, implying that a high circumferential coverage ratio significantly enhanced the flow in the self-circulating casing. This explains why the mass flow rate through the self-circulating casing increased with an increase in the circumferential coverage ratio. The airflow movement at the suction port of the self-circulating casing presented a circumferential to radial deflection, then the airflows flowed along the self-circulating casing wall to the inject port, indicating that the airflows through the self-circulating casing required a specific circumferential to radial deflection space provided by the self-circulating casing. The airflow could smoothly pass through the self-circulating casing but existed in the self-circulating casing in the form of low Mach number vortexes. A large circumferential coverage ratio significantly enhanced the airflow at the self-circulating casing suction port and increased the flow capacity of the self-circulating casing.

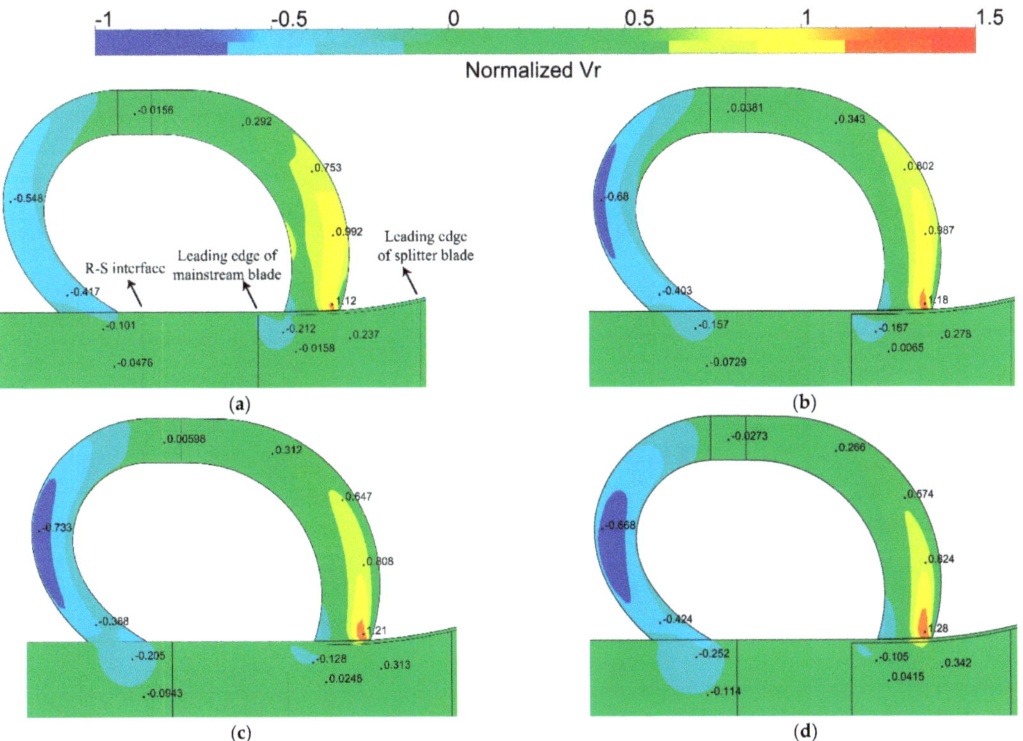

Figure 22. Contours of dimensionless radial velocity on the meridian plane. (**a**) SW. (**b**) SRCC0.36. (**c**) SRCC0.54. (**d**) SRCC0.72.

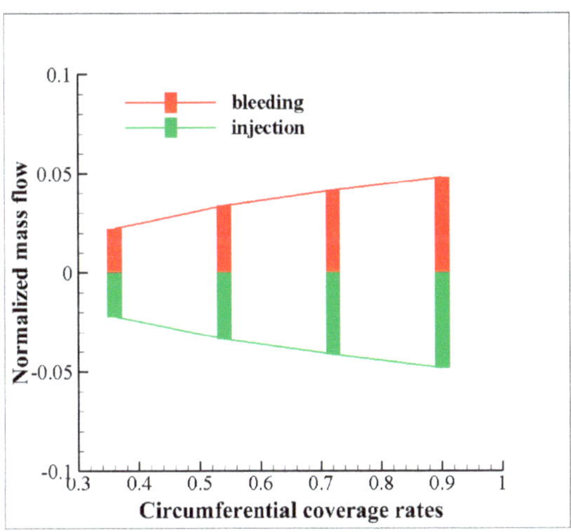

Figure 23. Variation of normalized mass flow level in suction/jet port of the different self-circulating casings.

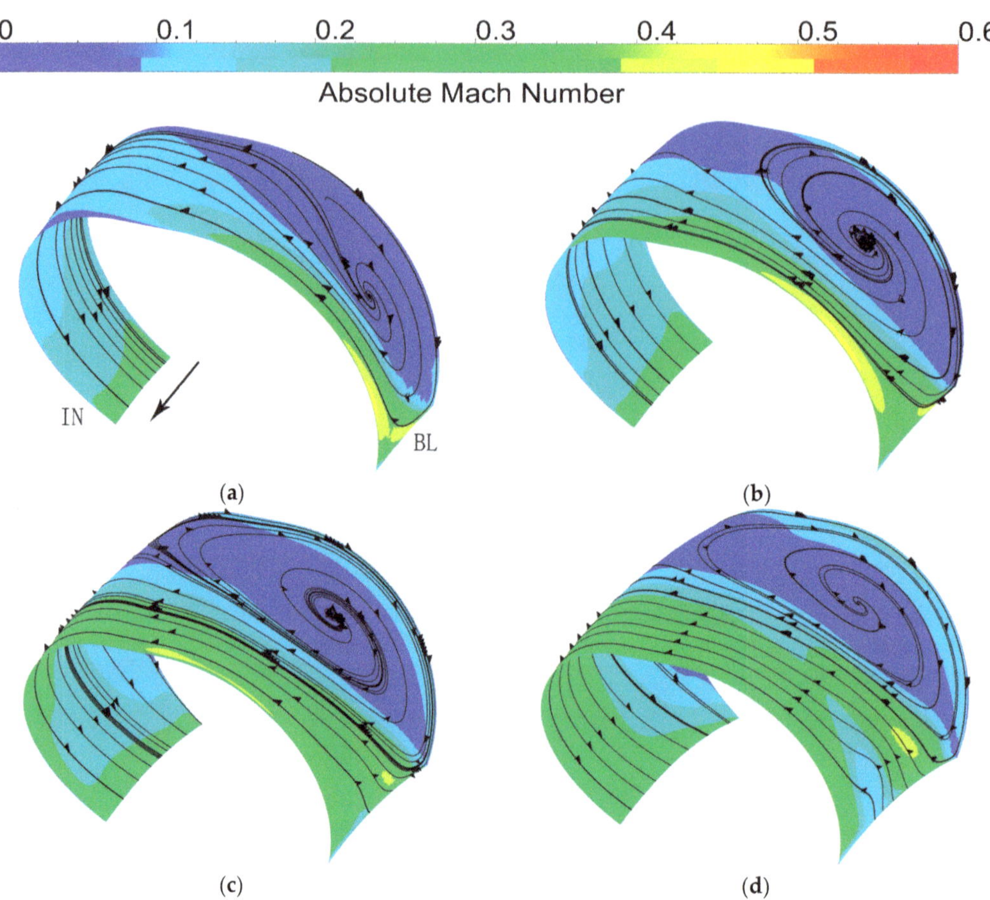

Figure 24. Absolute Mach number and streamline distributions under the different self-circulating casings. (**a**) SW. (**b**) SRCC0.36. (**c**) SRCC0.54. (**d**) SRCC0.72.

Change in dimensionless mass flow rate through the different self-circulating casings was determined over time (Figure 25). The ordinate represents the dimensionless mass flow rate based on the compressor design flow. The findings indicate that the dimensionless mass flow rate through the different self-circulating casings had significant fluctuations at different times. The mass flow rate was positively correlated with the circumferential coverage ratio. The mass flow rate increased gradually with an increase in the circumferential coverage ratio. Analysis of the mass flow rate change curve showed that the fluctuation rules of the different self-circulating casing treatments were remarkably similar, and the wave crest and wave trough basically occurred at the same time.

The entropy distribution in and under the four self-circulating casings on the meridian plane was determined as shown in Figure 26. The contours indicate that the entropy in the self-circulating casings had no significant difference between SRCC0.36 and SRCC0.54. In general, the entropy in the self-circulating casing increased with an increase in the circumferential coverage ratio, implying that the flow loss also increased. The high entropy area consistently increased in the blade tip passage with increase in circumferential coverage ratio. The maximum of entropy in SRCC0.9 was more than 250 J/K. The results in Figures 22 and 23 showed that the mass flow rate through the suction/inject port, the velocity and momentum of flow in the self-circulating casing gradually increased with an increase in the circumferential coverage ratio. Mixing of the mainstream with the jet became violent, resulting in an increase in the entropy. This finding implies that the circumferential coverage ratio was positively correlated with the flow loss under the inject port.

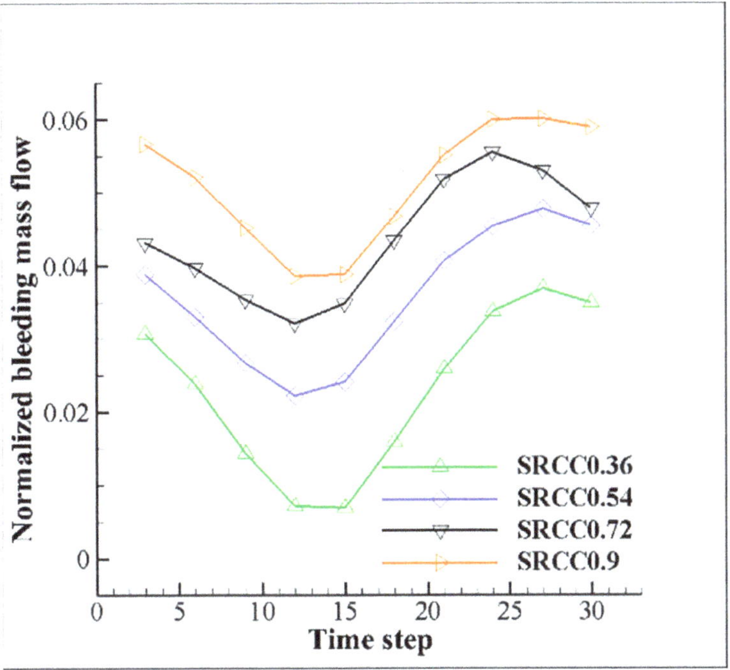

Figure 25. Change in dimensionless mass flow rate under the different self-circulating casings over time.

The previous analysis indicates that the self-circulating casing treatments improved the flow blockage but caused a higher flow loss in the blade tip passage, and the flow loss had an impact on the compressor's efficiency. The relative total pressure loss coefficient distributions of the compressor were evaluated under the different self-circulating casing treatments in a time-averaged flow field (Figure 27). The relative total pressure loss coefficient is expressed as

$$C_p = \frac{(Pt_{inlet} - P_t)}{Pt_{inlet}} \tag{5}$$

where "Pt_{inlet}" represents the average relative total pressure at the rotating domain inlet and "Pt" represents the local relative total pressure. The relative total pressure loss coefficient can be used to reflect the degree of flow loss in the rotating domain under the relative coordinate system. The relative total pressure loss of incoming flow for the solid-wall casing was small under the near-stall point, and the stratification was evident. However, a sudden increase in the relative total pressure loss was caused by the large-scale mixing of the leakage flows with the mainstream in the blade passage. The high relative total pressure loss area indicated by the dashed red line fills in the mainstream blade passage inlet and the flow loss was markedly large under the near-stall point. The results showed that the four different self-circulating casing treatments significantly reduced the high relative total pressure loss area in the passage under the near-stall point, and the relative total pressure loss coefficient was also significantly reduced (Figure 27b–e). The red high total pressure loss area was completely eliminated after application of the self-circulating casing treatments, implying that the self-circulating casing effectively sucked the leakage vortexes which would have expanded and developed in this region, which is consistent with the previous results in Figures 16 and 17. The area with high total pressure loss at the blade tip passage inlet gradually increased with an increase in the circumferential coverage ratio. Analysis of the position showed that this high entropy area was just below the suction port, which is consistent with the results presented in Figure 22. The results indicated that the total pressure loss coefficient near the mainstream blade basin at the passage behind the suction port decreased with an increase in the circumferential coverage ratio, and this effect extended to the splitter blade passage. The total pressure loss coefficient in the blade tip passage was the lowest in the SRCC0.9 design compared with the other designs. The total pressure loss coefficient was distributed in most of the passages comprising the mainstream blade pressure surface and the splitter blade, as shown in the blue area in Figure 27e. In summary, the flow loss at the blade tip was significantly reduced, and the corresponding compressor efficiency increased after application of the self-circulating casings, which is consistent with the performance results.

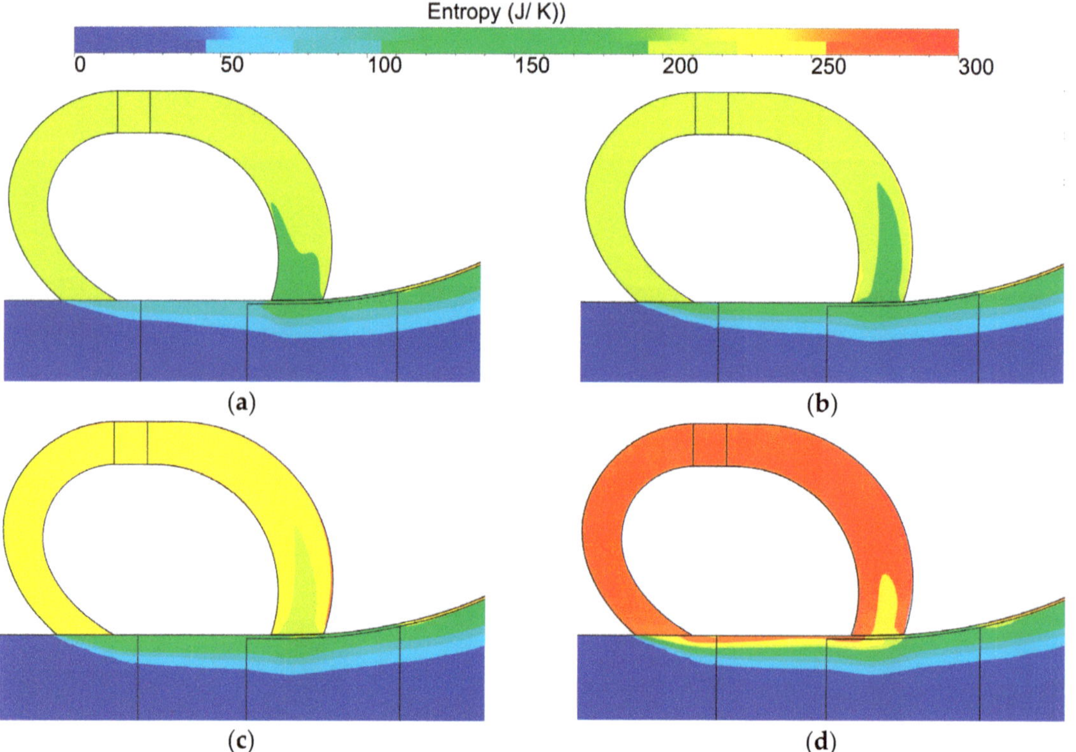

Figure 26. Contours distribution of entropy on the meridian plane under different self-circulating casings. (**a**) SW. (**b**) SRCC0.36. (**c**) SRCC0.54. (**d**) SRCC0.72.

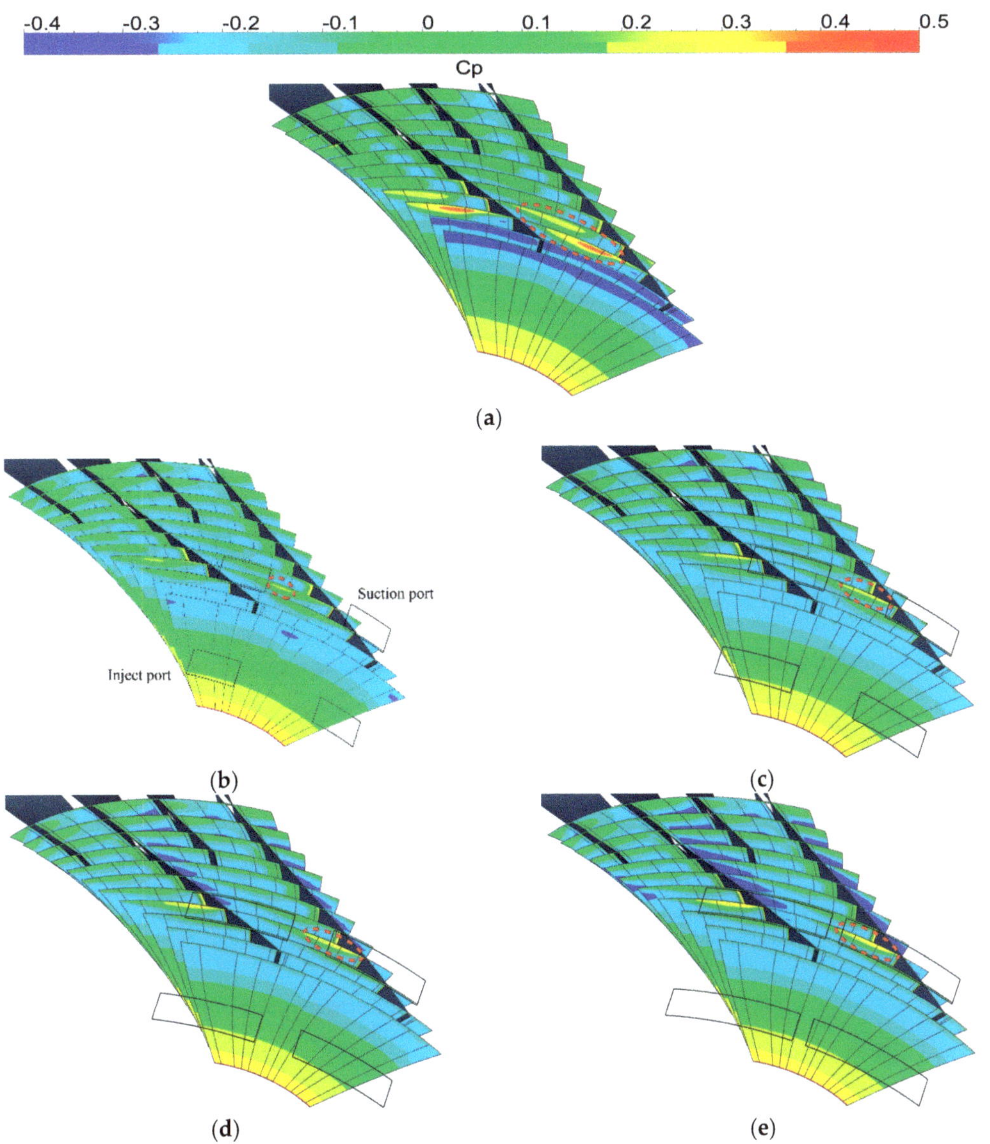

Figure 27. Contours of total pressure loss coefficient at the blade tip under different self-circulating casings. (**a**) SW. (**b**) SRCC0.36. (**c**) SRCC0.54. (**d**) SRCC0.72. (**e**) SRCC0.9.

4. Conclusions

In this study, four self-circulating casing research schemes with different circumferential coverage ratios were designed. The mechanisms underlying the effect of the self-circulating casing treatment with different circumferential coverage ratios on centrifugal compressor performance and stability were explored by evaluating the compressor performance and internal flow comparative analyses. The main conclusions of the study are presented below:

(1) Application of self-circulating casings with different circumferential coverage ratios effectively increased the compressor's stable working range and improved its boost capacity. The compressor's boost capability increased with an increase in the circum-

ferential coverage ratio. SRCC0.9 schemes, which had the largest circumferential coverage ratio, exhibited the highest increase in stall margin, with an *SMI* of 20.22%. The ability to expand compressor stability decreased with a decrease in the circumferential coverage ratio. Analysis of the compressor peak efficiency showed that the descent ranges of the four self-circulation casing treatments were less than 0.3%. The compressor isentropic efficiencies of the four schemes with different self-circulation casing treatments with a mass flow rate more than 2.8 kg/s were less than that of the solid-wall casing, whereas the isentropic efficiencies with a mass flow rate less than 2.8 kg/s were higher than that of the solid-wall casing.

(2) The leakage flows were sucked by self-circulation casing in the blade tip passage after application of the self-circulating casing treatments. The development of distortion, expansion and break-up of leakage flows was restrained, thus alleviating the flow blockage in the passage. The suction effect on the leakage flows at the blade tip increased with an increase in the circumferential coverage ratio, and the corresponding ability to eliminate flow blockage and improve the stall margin was also enhanced.

(3) The mass flow rate through the self-circulating casing significantly fluctuated with time. A larger circumferential coverage ratio was correlated with a larger circumferential to radial direction deflection space provided by the self-circulating casing for the flow in the impeller passage, and a more favorable airflow direction deflection at the suction port smoothly progressing to the self-circulating casing. The mass flow rate through the self-circulating casing, the flow velocity at suction and inject port and the flow capacity of the self-circulating casing significantly increased with an increase in the circumferential coverage ratio. Notably, the mixing of flows at the inject port was relatively intense, and the corresponding flow loss increased with an increase in the circumferential coverage ratio.

(4) The four different self-circulating casing treatments used in this study significantly reduced the high relative total pressure loss area in the passage under the near-stall point. The total pressure loss coefficient and the flow loss in the blade tip passage decreased with an increase in the circumferential coverage ratio.

Author Contributions: Conceptualization, H.Z. and W.C.; methodology, Q.L.; software, Q.L.; validation, H.W. and Q.L.; formal analysis, Q.L. and Jing, F.; investigation, H.W.; resources, Q.L.; data curation, H.W. and Q.L.; writing—original draft preparation, H.W.; writing—review and editing, H.W. and Q.L.; visualization, H.Z. and F.J. All authors have read and agreed to the published version of the manuscript.

Funding: This research was funded by [51006084] grant number [National Natural Science Foundation of China], [2017-II-0005-0018] grant number [National Science and Technology Major Projects of China] and [51536006] grant number [Key Program of National Natural Science Foundation of China].

Data Availability Statement: The authors confirm that the data supporting the findings of this study are available within the article.

Acknowledgments: The work is funded by National Natural Science Foundation of China (Grant Nos. 51006084), National Science and Technology Major Projects of China (Grant Nos. 2017-II-0005-0018) and the Key Program of National Natural Science Foundation of China (Nos. 51536006).

Conflicts of Interest: The authors declare no conflict of interest.

References

1. Li, D.; Yang, C.; Zhou, M.; Zhu, Z.; Wang, H. Numerical and experimental research on different inlet configurations of high-speed centrifugal compressor. *Sci. China Technol. Sci.* **2012**, *55*, 174–181. [CrossRef]
2. Li, P.; Gu, C.; Song, Y. Optimum design of a low-pressure centrifugal compressor for MW stage gas turbine. *J. Tsinghua Univ.* **2015**, *55*, 1110–1116.
3. Saha, S.L.; Kurokawa, J.; Matsui, J.; Imamura, H. Passive Control of Rotating Stall in a Parallel Wall Vaned Diffuser by J-Grooves. *J. Turbomach.* **2001**, *123*, 507–515. [CrossRef]

4. Lu, X.; Chu, W.; Zhu, J. Experimental and Numerical Investigation of a Subsonic Compressor with Bend Skewed Slot Casing Treatment. ASME GT 2006-90026. *Proc. Inst. Mech. Eng. Part C J. Mech. Eng. Sci.* **2006**, *220*, 1785–1796. [CrossRef]
5. Yan, S.; Chu, W.; Zhang, H.; Liu, K. Research on the effect of different axial positions of suction port of self-circulating casing. *Propuls. Technol.* **2019**, *40*, 1478–1489. [CrossRef]
6. Fisher, F.B. *Application of Map Width Enhancement Devices to Turbocharger Compressor Stages*; SAE Technical Paper; SAE International: Warrendale, PA, USA, 1989; p. 880794.
7. Hunziker, R.; Dickmann, H.P.; Emmrich, R. Numerical and experimental investigation of a centrifugal compressor with an inducer casing bleed system. *Proc. Inst. Mech. Eng. Part A J. Power Energy* **2001**, *215*, 783–791. [CrossRef]
8. Ishida, M.; Surana, T.; Ueki, H.; Sakaguchi, D. Suppression of unstable flow at small flow rates in a centrifugal blower by controlling tip leakage flow and reverse flow. *ASME J. Turbomach. Trans.* **2005**, *127*, 76–83. [CrossRef]
9. Ishida, M.; Sakaguchi, D.; Ueki, H. Optimization of Inlet Ring Groove Arrangement for Suppression of Unstable Flow in a Centrifugal Impeller. In Proceedings of the ASME Turbo Expo 2005: Power for Land, Sea, and Air, Reno, NV, USA, 6–9 June 2005; ASME Paper: New York, NY, USA, 2005; No. GT2005-68675.
10. Ishida, M.; Sakaguchi, D.; Ueki, H. Effect of Pre-Whirl on Unstable Flow Suppression in a Centrifugal Impeller with Ring Groove Arrangement. In Proceedings of the ASME Turbo Expo 2006: Power for Land, Sea, and Air, Barcelona, Spain, 8–11 May 2006; ASME Paper: New York, NY, USA, 2006; No. GT2006-90400.
11. Tamaki, H. Effect of recirculation device with counter swirl vane on performance of high pressure ratio centrifugal compressor. *J. Turbomach.* **2012**, *134*, 051036. [CrossRef]
12. Jung, S.; Pelton, R. Numerically Derived Design Guidelines of Self Recirculation Casing Treatment for Industrial Centrifugal Compressors. In Proceedings of the ASME Turbo Expo 2016: Turbomachinery Technical Conference and Exposition, Seoul, Republic of Korea, 13–17 June 2016; ASME Paper: New York, NY, USA, 2016; No. GT2016-56672.
13. Semlitsch, B.; Mihaescu, M. Flow Phenomena Leading to Surge in a Centrifugal Compressor. *Energy* **2016**, *103*, 572–587. [CrossRef]
14. Zheng, X.; Zhang, Y.; Guo, G.; Zhang, J.; Yang, M. Research on Treatment and Stability Expansion of Transonic Centrifugal Compressor Casing. *J. Eng. Thermophys.* **2010**, *31*, 2023–2026.
15. Xu, W.; Wang, T.; Ding, L. Unsteady numerical simulation of a centrifugal compressor with a holed casing. *J. Shanghai Jiaotong Univ.* **2012**, 536–539.
16. Xu, W.; Wang, T.; Gu, C. Internal flow field analysis of a centrifugal compressor with holed casing under low mass flow conditions. *Chin. J. Electr. Eng.* **2012**, 123–129.
17. Kang, J.; Huang, G.; Zhu, J. Mechanism analysis of self-circulating casing expand the stability of centrifugal compressor. *J. Aeronaut.* **2014**, 3264–3272.
18. Cao, S. Effect of casing treatment on performance of transonic centrifugal compressor. *Propuls. Technol.* **2017**, *38*, 773–777.
19. Gan, J.; Wu, Y.; Chen, Z. Study on the effect of self-circulating casing on the performance and stability of high-speed centrifugal impeller. *Fan Technol.* **2019**, *61*, 1–9.
20. Shang, P. *Research on Stabilization Enlargement Mechanism of Self-Circulating Casing of Supercritical Carbon Dioxide Centrifugal Compressor*; Tianjin University of Technology: Tianjin, China, 2022. [CrossRef]
21. Wang, W. *Research on Design Law and Flow Mechanism of Tip Jet and Self-Circulating Casing Treatment of Axial Flow Compressor*; Northwest University of Technology: Xi'an, China, 2016.
22. Krain, H.; Hoffmann, B.; Pak, H. Aerodynamics of a centrifugal compressor impeller with transonic inlet conditions. In Proceedings of the ASME 1995 International Gas Turbine and Aeroengine Congress and Exposition, Houston, TX, USA, 5–8 June 1995; ASME Paper: New York, NY, USA, 1995; No. 95-GT-79.
23. Eisenlohr, G.; Krain, H.; Richter, F.A.; Tiede, V. Investigation of the flow through a high-pressure ratio centrifugal impeller. In Proceedings of the ASME Turbo Expo 2002: Power for Land, Sea, and Air, Amsterdam, The Netherlands, 3–6 June 2002; ASME Paper: New York, NY, USA, 2002; No. GT2002-30394.
24. Kang, S. Numerical investigation of a high-speed centrifugal compressor impeller. In Proceedings of the ASME Turbo Expo 2005: Power for Land, Sea, and Air, Reno, NV, USA, 6–9 June 2005; ASME Paper: New York, NY, USA, 2005; No. GT2005-68092.

Article

Atomization Characteristics of Kerosene in Crossflow with an Incident Shock Wave

Yongsheng Zhao *, Junfei Wu and Xiangyang Mu

China Academy of Aerospace Aerodynamics, Beijing 100074, China
* Correspondence: zhaoyongshengying@163.com; Tel.: +86-13401156643

Abstract: An unsteady numerical simulation method was used in order to explore more efficient atomization methods for liquid fuel in scramjet combustors and to study the influence of different shock wave incident positions on the atomization characteristics of kerosene in crossflow. The wedge compression surface was used to generate the incident shock wave, and the incident position of the shock wave on the fuel jet was controlled by changing the angle of the wedge surface. The inlet Mach number was 2.01; the total temperature was 300 K, and the momentum ratio was 12. The research results show that as the incident position of the shock wave moves upstream, the penetration depth of the jet is essentially unchanged, but the inner edge trajectory of the jet is closer to the wall. Because the shock wave affects the Kelvin–Helmholtz instability of the jet, the unsteadiness of the jet root is strengthened, and the unsteadiness downstream of the jet is weakened. The atomization of the jet and the stability of the particle-size distribution are, thus, realized more quickly. The incident shock wave reduces the Sauter mean diameter of the jet section and makes the droplet distribution more uniform. The incident shock wave makes the atomization angle of the jet along the flow direction increase first and then decrease. The changes in the jet characteristics are determined by the changes in the reflux region, momentum transport, and pressure distribution caused by the incident shock wave.

Keywords: shock wave; crossflow; incident position; numerical simulation; atomization characteristics

Citation: Zhao, Y.; Wu, J.; Mu, X.
Atomization Characteristics of
Kerosene in Crossflow with an
Incident Shock Wave. *Aerospace* **2023**,
10, 30. https://doi.org/10.3390/
aerospace10010030

Academic Editor: Hexia Huang

Received: 10 November 2022
Revised: 24 December 2022
Accepted: 27 December 2022
Published: 30 December 2022

1. Introduction

The scramjet has become the preferred form of propulsion system for high-speed aircrafts because of its excellent performance at high flight Mach number, and it has become the focus of research in the aerospace field [1]. For scramjets, the speed of the incoming flow is fast, and the residence time of the fuel is on the order of milliseconds. Therefore, it is one of the key technologies in achieving efficient fuel mixing and stable combustion within a limited time and space. From the perspective of the entire combustion process, the atomization process of fuel occupies nearly 60% of the time [2,3]. Therefore, it is very important to explore efficient fuel atomization modes to improve the performance of scramjets. Injecting fuel by transverse jets is a promising fuel atomization method.

There have been many research studies on the aerodynamic structure, breakup mechanism, and unsteady characteristics of crossflow in supersonic flow. Hidemi [4] studied the effects of injection and main flow conditions on the turbulent structure produced by a gaseous sonic transverse injection into a supersonic airstream. Rana [5,6] studied the unsteady characteristics of the jet in crossflow. It is pointed out that the Kelvin–Helmholtz (K-H) instability in the shear layer of the windward jet is the dominant factor in the mixing of the windward jet. Génin [7] pointed out that the K-H instability caused by the velocity between the jet and air would induce large-scale turbulent vortices. In summary, we have a certain understanding of the aerodynamic characteristics of the crossflow in a super-sonic flow. To further increase the atomization efficiency, the effect of the incidence of shock waves is considered. The shock wave generated by the shock generator is used to enhance the mixing of the fuel and mainstream, and accelerate the fuel atomization process, which has been largely of interest because of its high efficiency. Nakamura et al. [8] took the transverse jet of hydrogen as the research object and focused on the influence of the incident shock wave on flame stability. The research results showed that flame stability would weaken or even extinguish as the incident point of the shock wave moved forward. Tahsini et al. [9] studied the impact of incident shock waves on combustion efficiency and total pressure loss, and pointed out that incident shock waves could improve combustion efficiency and increase the total pressure loss. At the same time, when the incident point of the shock wave was located upstream of the jet, the combustion efficiency was significantly higher. Shekarian et al. [10] used the physical model of incident shock generated by a wedge of $10°$ and the calculation model of RANS to explore the impact of incident shock position on the recirculation zone of the spray field. The results showed that different incident positions had different effects on the recirculation zone, and the incident shock changed the characteristics of the recirculation zone of the incident position. When the incident position was located downstream of the jet, it was more conducive to mixing and flame stability. Mai et al. [11] conducted a similar study by combining experimental and numerical methods, and the results showed that due to the interaction between the shock wave and jet, when the incident point was located downstream of the jet, a large-scale recirculation zone would be formed, which further enhanced the mixing and increased the residence time. Schetz [12] and Erdem et al. [13] adopted experimental research methods to further confirm that the incident point of a shock wave located downstream of the jet could effectively enhance mixing. Huang et al. [14,15] used a numerical simulation to study the influence of geometric characteristics of the shock generator (such as angle, position, size, etc.) on mixing. The results showed that the geometric characteristics of the shock generator would affect the mixing effect; the influence of the geometric characteristics on mixing was not monotonic, but there was an optimal design scheme. Gerdroodbary et al. [16] studied the coupling interference between the incident shock wave and the flow field of the transverse jet on porous walls and used shock generators of different angles to explore their influence on the mixing effect. The results showed that a large mixing efficiency was obtained downstream at $15°$.

The above studies focused on the influence of an incident shock wave on combustion performance, the structural characteristics of the flow field itself, and the mixing effect. However, the influence of different incident shock waves on the characteristics of the spray in crossflow, such as penetration depth, droplet spanwise distribution, and particle size distribution, is investigated in relatively few studies. The mechanisms of related changes also need to be explored. It is very important to further understand the effect of shock waves on fuel atomization characteristics in crossflow. Therefore, this paper uses an accurate unsteady numerical simulation method to study the influence of different incident positions of a shock wave on the characteristics of the transverse jet itself, in order to provide technical support for the mixing enhancement technology under the action of an incident shock wave.

2. Numerical Simulation Method

2.1. Calculation Model

The coupled implicit solver based on pressure was used. The SST k-ω model (Shear stress transport), which combines the standard k-ω model in the near-wall region with the k-ε model in the far-field region, was used as the turbulence model. At the same time, the low Reynolds number correction was considered. This model is more accurate and reliable than results obtained by the standard k-ε model [17]. The process of atomization, evaporation, and mixing of kerosene droplets in a supersonic flow was simulated based on a discrete phase model (DPM). The governing equation was the Navier–Stokes equation of 3D mass average [18]:

$$\frac{\partial Q}{\partial t} + \frac{\partial E}{\partial x} + \frac{\partial F}{\partial y} + \frac{\partial G}{\partial z} = \frac{\partial E_v}{\partial x} + \frac{\partial F_v}{\partial y} + \frac{\partial G_v}{\partial z} + S \tag{1}$$

In the equation, $Q = [\rho,\ \rho u,\ \rho v,\ \rho w,\ \rho E,\ \rho c_s,\ \rho k,\ \rho \omega]^T$, where E, F, and G are inviscid fluxes, E_v, F_v, G_v are viscous fluxes, and S is the source term. The equation $E = e + (u^2 + v^2 + w^2)/2$ is the total internal energy per unit mass, and c_s is the mass fraction of each component. Based on the finite volume method [19], the equations were solved; the physical quantities were discretized with second-order accuracy, and the residuals of each variable were less than 10^{-4}. The unsteady simulation time step of the flow was 5×10^{-6} s, and the unsteady method was used to track the kerosene droplets, which had a time step of 10^{-5} s. The equation for the motion of the droplet is [20]:

$$\frac{du_P}{dt} = F_D u - u_P + \frac{g(\rho - \rho_P)}{\rho_P} + F \tag{2}$$

where u_P is the droplet velocity, F_D is the drag force per unit mass, u is the continuous phase velocity, g is the gravitational acceleration, ρ is the continuous phase density, ρ_P is the droplet density, and F is the other forces. The atomization cone model was used for the spray model, and the droplet diameter distribution was a Rosin–Rammler distribution. The K-H/R-T model was used for the droplet secondary breakup model. The kerosene–air mixture was used for species transport. The density of the liquid kerosene was 780 kg/m^3, and the gas phase was set as an ideal gas.

2.2. Physical Model and Grid

The physical model of the combustor in the numerical simulation is shown in Figure 1. The size of the combustor inlet was 100 mm $\times$ 80 mm. The shock wave was generated by the wedge-shaped compression surface, and the incident position of the shock wave was changed by controlling the height of the wedge-shaped surface (H). The value of H is shown in Table 1.

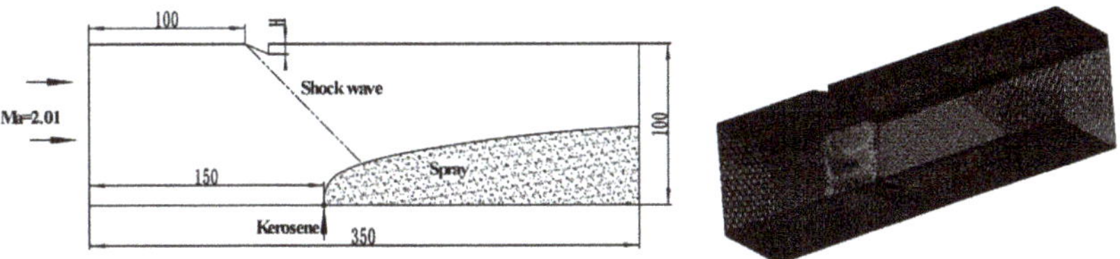

Figure 1. Physical model and grid (unit/mm).

Table 1. Values of H in different cases.

	CASE0	CASE1	CASE2	CASE3
H/mm	0	4.4	5.89	6.34

Kerosene was mixed with the mainstream in crossflow. The nozzle diameter (D) was 1 mm. The inlet Mach number was 2.01; the total temperature was 300 K, and the fuel–air momentum ratio was 12. This is a typical working condition of the combustor and can ensure that the jet has a certain penetration depth. A structured grid was used. The height of the first layer was 0.05 mm, and Y^+ was less than 3. The total number of grids was 3.8 million.

2.3. Numerical Method Verification

To verify the accuracy of the numerical method, the operating condition without an incident shock wave (CASE0) was simulated. The droplet distribution is shown in Figure 2, where the diameter and color of the circle represent the droplet size. The following conclusions can be made: (1) After the large-size droplet is ejected from the nozzle, due to the K-H unstable wave, the small-diameter droplet is gradually peeled off from the surface of the large-size droplet. These small droplets travel downstream with the airflow, forming a spray. In the process of moving, the diameter of the large droplet gradually decreases as the small droplet is gradually peeled off. When the movement time of large droplets is longer than the breaking characteristic time, the Rayleigh–Taylor (R-T) instability comes into play. The large droplet is further broken into smaller droplets, and the diameter of the droplet on the outer edge of the jet is larger than that of the droplet near the wall. (2) The simulation results of CASE0 were compared with the fitting relations of penetration depth obtained from tests conducted by Gopala [21] and Li [22], and the comparison results are shown in Figure 2. It can be seen that the numerical method adopted in this paper is in good agreement with the experimental results, and this method can be used to study the effect of shock waves on the atomization characteristics of kerosene in crossflow.

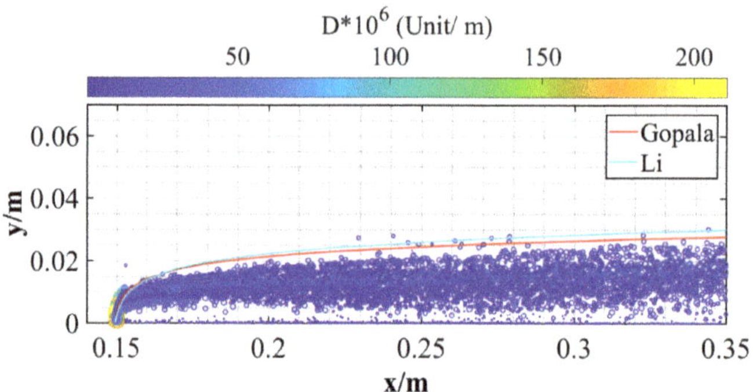

Figure 2. Comparison results between the numerical simulation and the test.

3. Results and Analysis

3.1. Influence of Shock Waves on Jet Characteristics in the Central Section

The penetration depth is a very important parameter in the study of the atomization characteristics of kerosene in crossflow. The penetration depth is the depth of the liquid jet through the main flow, which determines the atomization and mixing degree of the jet. The particle size distribution in the central section can characterize the penetration characteristics of the jet. Figure 3 shows the particle size distribution of the droplet in the central section of the jet in different cases, where the diameter and color of the circle represent the size of the droplet. The following can be seen from the results: (1) When there is no incident shock, there is a certain distance between the inner edge of the jet and the lower wall surface, and there is a long pure gas zone between the jet and the combustor wall. The species in this zone are almost all gaseous, with only a few liquid drops. The particle size upstream of the jet is larger than the size downstream, and the particle size of the inner edge of the jet is smaller than that of the outer edge. (2) When there is a shock wave, the outer-edge trajectory of the jet does not change much, but the inner-edge trajectory is close to the combustion chamber wall. Moreover, the length of the pure gas zone gets shorter. (3) With the shock wave moving upstream, the pure gas region is pushed farther towards the jet root (the outlet of the orifice) and is divided into two parts.

Figure 4 shows the influence of different incident positions on the concentration distribution of kerosene in the jet center section. Due to the K-H instability, obvious surface waves appear on the upper surface of the jet, and Lmax is the farthest distance affected by the surface wave. When X is greater than Lmax, the outer edge of the spray is relatively smooth. The following can be seen from the results: As the incident point moves upstream, (1) the surface wave moves upstream; the wavelength becomes shorter, and Lmax gradually decreases. This indicates that, due to the effect of shock waves, the unsteadiness of the jet root is strengthened, while the unsteadiness downstream of the jet is weakened. In other words, the jet accelerates atomization and achieves stability more quickly. (2) The high-concentration core area becomes shorter, and the penetration depth of the jet in front of the shock wave increases. The height of the central trajectory increases. Case0 is L, and

Case1 is L + 0.5 D. Due to the small movement of the incident position of the shock wave, the heights of the central trajectory of CASE2 and Case3 are not much different, both of which are L + 1.25 D.

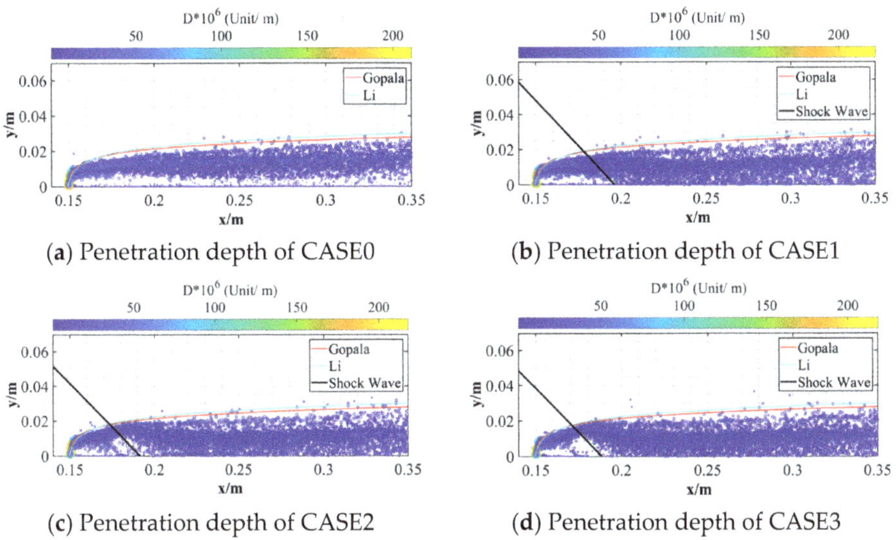

(**a**) Penetration depth of CASE0

(**b**) Penetration depth of CASE1

(**c**) Penetration depth of CASE2

(**d**) Penetration depth of CASE3

Figure 3. Influence of different incident positions on penetration depth (z = 0 m).

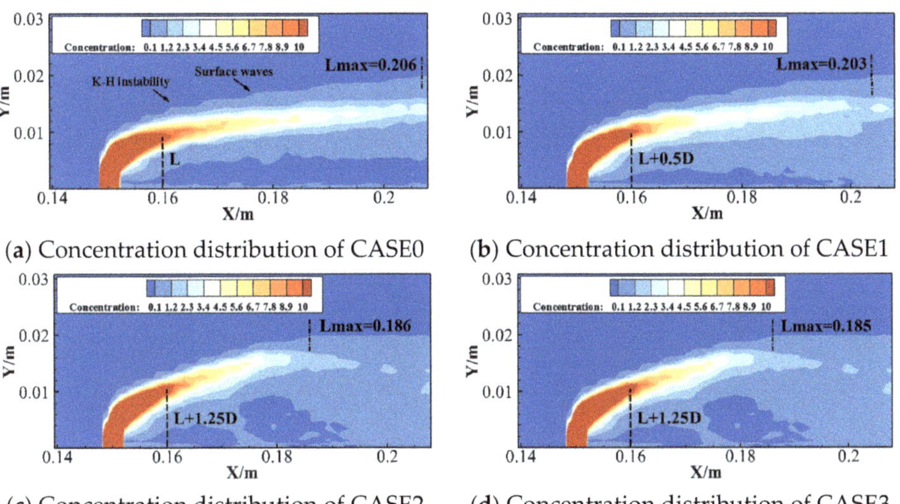

(**a**) Concentration distribution of CASE0

(**b**) Concentration distribution of CASE1

(**c**) Concentration distribution of CASE2

(**d**) Concentration distribution of CASE3

Figure 4. Concentration distribution of liquid kerosene at different incident locations (z = 0 m).

In order to analyze the causes of the above changes, the influence of the shock wave incident position on the flow-field characteristics of the jet center section were obtained, as shown in Figure 5. When the shock wave interacts with the jet, a reflux region is formed in the shock front. When the incident point of the shock wave moves upstream, the reflux region will move upstream. Due to the influence of the jet root, a new reverse reflux region will be formed, and the area of the reflux region will increase. The change in the characteristics of the reflux region leads to the change in the characteristics of the jet.

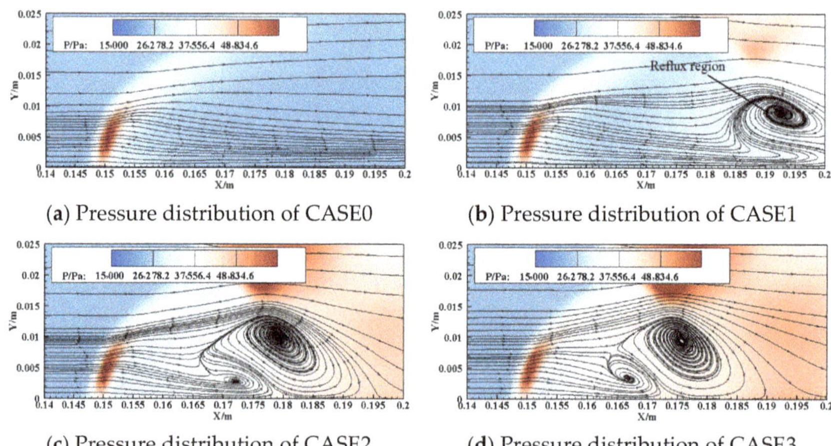

(**a**) Pressure distribution of CASE0

(**b**) Pressure distribution of CASE1

(**c**) Pressure distribution of CASE2

(**d**) Pressure distribution of CASE3

Figure 5. Influence of different incident locations on flow-field characteristics (z = 0 m).

3.2. Influence of Shock Waves on Jet Characteristics in the Transverse Section

The SMD of different sections was counted along the flow direction, and the results are shown in Figure 6. It can be seen that when there is no incident shock, the SMD does not change much after the X = 20 D section, and when there is an incident shock, the SMD does not change much after the X = 60 D section. After the incident shock wave, the SMD of the cross section becomes significantly smaller and further decreases as the shock wave moves upstream of the jet.

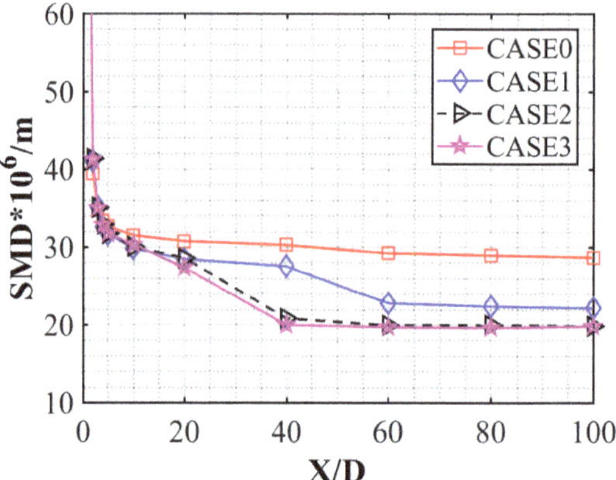

Figure 6. SMD of different sections along the flow direction.

The histograms of the particle size distribution in different sections of CASE0 and CASE4 were obtained, as shown in Figure 7. The distribution of the droplet diameter generally increases first and then decreases, and the particle size distribution is approximately Gaussian, which is similar to the results obtained by Inamura [23] and Wu [24]. Compared with the condition without an incident shock wave, when there is an incident shock wave, the bandwidth of the particle size distribution before X = 10 D is not affected, but the bandwidth of the particle size distribution downstream of the jet decreases gradually and concentrates in the area of small particle size. After X = 40 D, the particle size bandwidth is essentially unchanged.

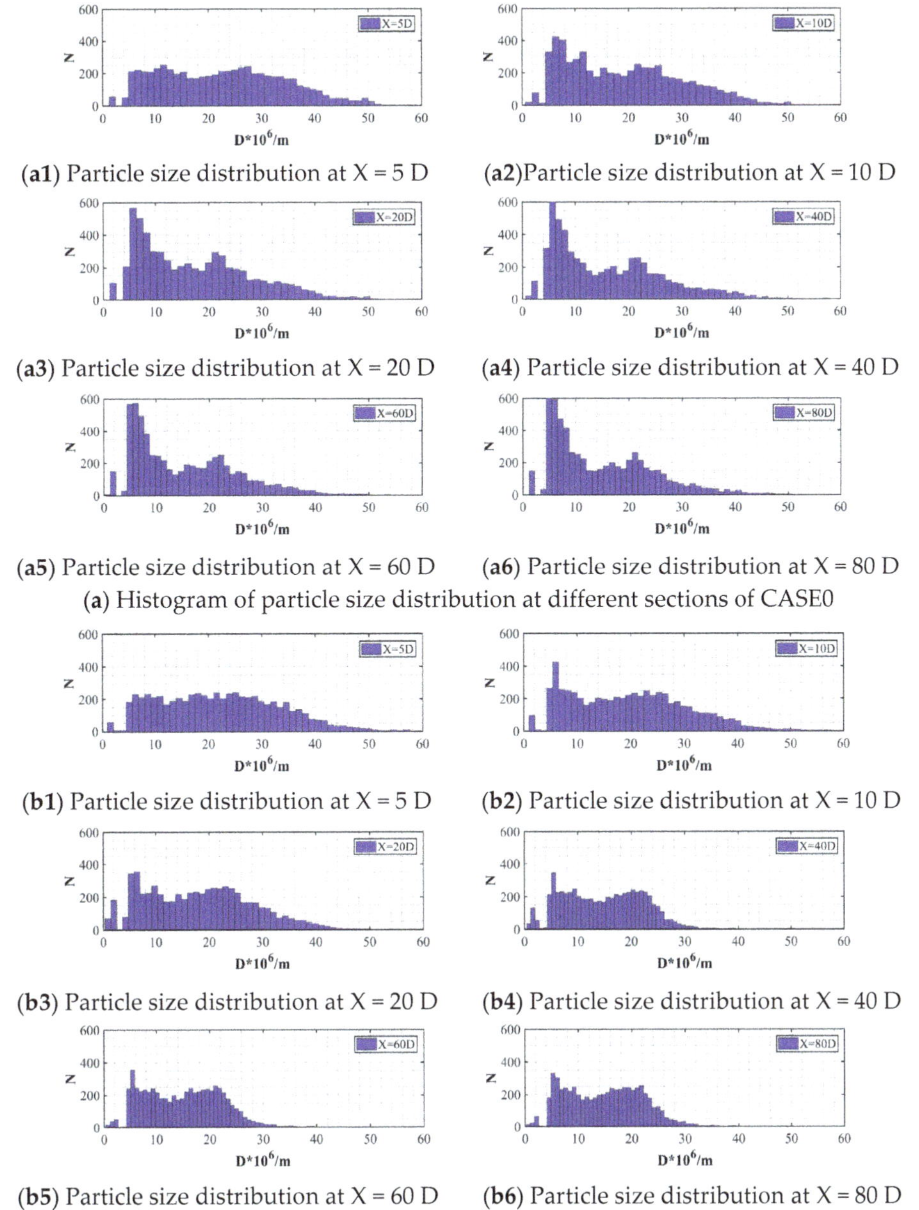

(**a1**) Particle size distribution at X = 5 D

(**a2**) Particle size distribution at X = 10 D

(**a3**) Particle size distribution at X = 20 D

(**a4**) Particle size distribution at X = 40 D

(**a5**) Particle size distribution at X = 60 D

(**a6**) Particle size distribution at X = 80 D

(**a**) Histogram of particle size distribution at different sections of CASE0

(**b1**) Particle size distribution at X = 5 D

(**b2**) Particle size distribution at X = 10 D

(**b3**) Particle size distribution at X = 20 D

(**b4**) Particle size distribution at X = 40 D

(**b5**) Particle size distribution at X = 60 D

(**b6**) Particle size distribution at X = 80 D

(**b**) Histogram of particle size distribution at different sections of CASE0

Figure 7. Histogram of the particle size distribution in different sections.

Figure 8 shows the spatial distribution of droplets spanwise at different incident positions at X = 80 D. It can be seen that large droplets are not only distributed in the periphery of the spray, but also in the middle of the spray. When a shock wave is incident, the spatial distribution of the droplet is more uniform. The droplet distribution area moves to the lower wall of the combustion chamber; as the incident point of the shock wave moves upstream of the jet, the spatial distribution of the droplets becomes more uniform.

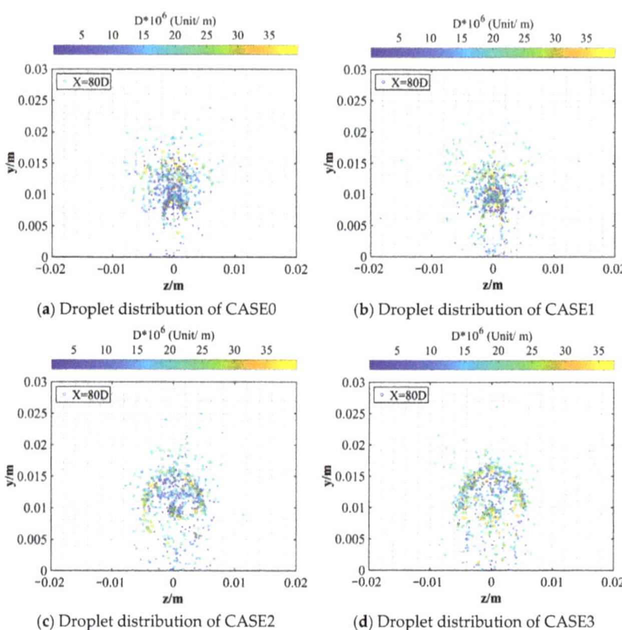

(**a**) Droplet distribution of CASE0

(**b**) Droplet distribution of CASE1

(**c**) Droplet distribution of CASE2

(**d**) Droplet distribution of CASE3

Figure 8. Influence of different incident positions on spanwise distribution of droplet (X = 80 D).

Figure 9 shows the Mach number distribution characteristics at different incident locations. When there is no incident shock wave, the flow field has an obvious subsonic region due to the momentum exchange between the droplet and the mainstream. As the shock wave moves upstream, the subsonic region moves towards the lower wall of the combustor, and the subsonic region gradually becomes smaller. This shows that the shock wave accelerates the momentum exchange between the droplet and the mainstream, making the distribution of the droplets more uniform.

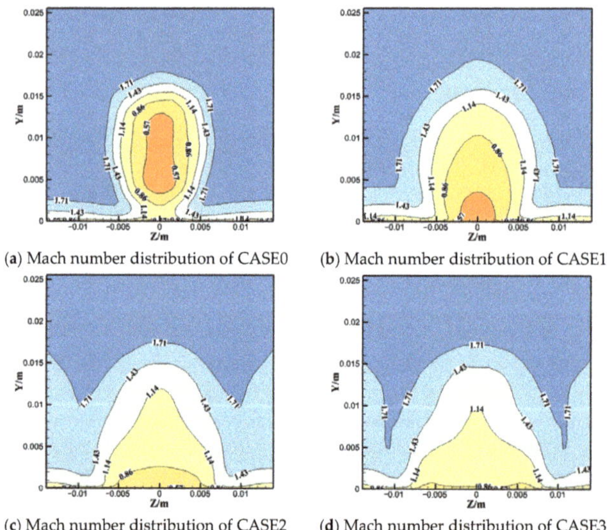

(**a**) Mach number distribution of CASE0

(**b**) Mach number distribution of CASE1

(**c**) Mach number distribution of CASE2

(**d**) Mach number distribution of CASE3

Figure 9. Influence of different incident locations on the Mach number distribution (X = 80 D).

3.3. Influence of Shock Waves on Jet Characteristics in the Spanwise Section

The particle size distribution of the y = 0.01 m section was selected to analyze the effect of shock waves on the atomization angle, and the results are shown in Figure 10. When there is no incident shock, the width of the jet increases gradually. When a shock wave is incident, the width of the jet increases first and then decreases. This change becomes more obvious as the shock wave moves upstream of the jet. In the initial stage, large droplets are mainly distributed in the middle of the jet, while small droplets are distributed at both the edge and the middle of the jet. When a shock wave is incident, large droplets will disappear more quickly, and the particle size of the jet tends to become uniform more quickly.

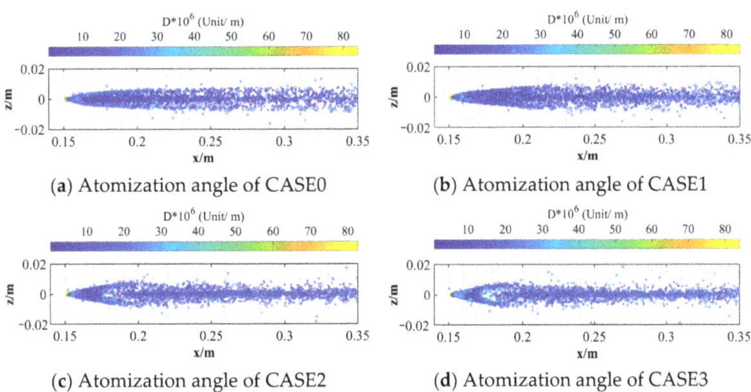

(**a**) Atomization angle of CASE0

(**b**) Atomization angle of CASE1

(**c**) Atomization angle of CASE2

(**d**) Atomization angle of CASE3

Figure 10. Influence of different incident positions on the atomization angle (y = 0.01 m).

Figure 11 shows the influence of different incident locations on the pressure distribution in the y = 0.01 m section. The following can be seen: (1) When there is no incident shock wave, due to the obstruction of the jet, a shock wave appears at the jet root and a high-pressure region is formed. As the incident shock wave moves upstream, the high-pressure region at the root of CASE1 and CASE2 is essentially unchanged, but the high-pressure region at the root of CASE3 changes. (2) When there is no incident shock wave, there is no adverse pressure gradient downstream of the jet root. When there is an incident shock wave, the adverse pressure gradient will be formed downstream of the jet root, and with the shock wave moving upstream, this leads to the change in the atomization angle.

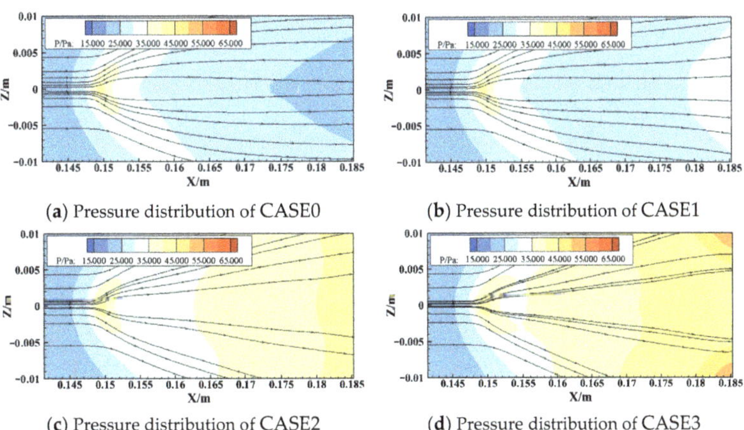

(**a**) Pressure distribution of CASE0

(**b**) Pressure distribution of CASE1

(**c**) Pressure distribution of CASE2

(**d**) Pressure distribution of CASE3

Figure 11. Influence of different incident locations on the pressure distribution (y = 0.01 m).

4. Conclusions

The effect of different incident positions of a shock wave on the atomization characteristics in crossflow was studied by a precise unsteady simulation method. It focused on the changes of atomization characteristics in the center, transverse, and spanwise sections at different incident positions. The following conclusions were obtained:

1. When there is a shock wave, the outer-edge trajectory of the jet does not change much, but the inner-edge trajectory is close to the combustion chamber wall. Additionally, the length of the pure gas zone gets shorter. With the shock wave moving upstream, the pure gas region is pushed further towards the jet root and is divided into two parts. Meanwhile, the incident shock wave changes the K-H instability of the jet. The unsteadiness of the jet root is strengthened, while the unsteadiness downstream of the jet is weakened. The main reason for the characteristic change is the change in the reflux region caused by the incident shock wave.
2. Because the incident shock wave changes the momentum exchange characteristics between the mainstream and the jet, the SMD of the transverse section of the jet decreases, and the droplet distribution becomes more uniform.
3. Due to the adverse pressure gradient caused by the incident shock wave, the width of the jet increases first and then decreases, and this change becomes more obvious as the shock wave moves upstream of the jet.

In this study, the temperature of kerosene and inflow was 300 K, and the influence of different temperatures was not considered, which weakens the influence of evaporation. This will be studied in the future.

Author Contributions: Conceptualization, Y.Z. and J.W.; methodology, Y.Z.; software, Y.Z.; validation, Y.Z., J.W. and X.M.; formal analysis, Y.Z.; investigation, Y.Z. and X.M.; resources, Y.Z.; data curation, Y.Z. and X.M.; writing—original draft preparation, Y.Z.; writing—review and editing, Y.Z.; visualization, J.W.; supervision, J.W.; project administration, J.W.; funding acquisition, Y.Z. All authors have read and agreed to the published version of the manuscript.

Funding: This research received no external funding.

Data Availability Statement: Data is available through corresponding author.

Conflicts of Interest: The authors declare no conflict of interest.

References

1. Charles, R. McClinton. X-43-scramjet power breaks the hypersonic barrier Dryden lectureship in research for 2006. In Proceedings of the 44th AIAA Aerospace Sciences Meeting, Reno, NV, USA, 9–12 January 2006.
2. Ronald, S. A Century of Ramjet Propulsion Technology Evolution. *J. Propuls. Power* **2004**, *20*, 27–58.
3. Seiner, J.; Dash, S.; Kenzakowshi, D. Historical Survey on Enhanced Mixing in Scramjet Engines. *J. Propuls. Power* **2001**, *17*, 1273–1286. [CrossRef]
4. Takahashi, H.; Masuya, G.; Hirota, M. Effects of Injection and Main Flow Conditions on Supersonic Turbulent Mixing Structure. *AIAA J.* **2010**, *48*, 1748–1756. [CrossRef]
5. Rana, Z.A.; Thornber, B.; Drikakis, D. Transverse jet injection into a supersonic turbulent cross-flow. *Phys. Fluids* **2011**, *23*, 585. [CrossRef]
6. Rana, Z.A.; Thornber, B.; Drikakis, D. On the importance of generating accurate turbulent boundary condition for unsteady simulations. *J. Turbul.* **2011**, *12*, N35. [CrossRef]
7. Génin, F.; Menon, S. Dynamics of sonic jet injection into supersonic crossflow. *J. Turbul.* **2010**, *11*, 1–30. [CrossRef]
8. Nakamura, H.; Sato, N.; Kobayashi, H.; Niioka, T.; Masuya, G. Combustion of transverse hydrogen injection with shock wave in a supersonic airstream. In Proceedings of the 5th Asia-Pacific Conference on Combustion, Adelaide, SA, Australia, 17–20 July 2005.
9. Tahsini, A.M.; Mousavi, S.T. Investigating the supersonic combustion efficiency for the jet-in-cross-flow. *Int. J. Hydrogen Energy* **2015**, *40*, 3091–3097. [CrossRef]
10. Shekarian, A.A.; Tabejamaat, S.; Shoraka, Y. Effects of incident shock wave on mixing and flame holding of hydrogen in supersonic air flow. *Int. J. Hydrogen Energy* **2014**, *39*, 10284–10292. [CrossRef]
11. Mai, T.; Sakimitsu, Y.; Nakamura, H.; Ogami, Y.; Kudo, T.; Kobayashi, H. Effect of theincident shock wave interacting with transversal jet flow on the mixing and combustion. *Proc. Combust. Inst.* **2011**, *33*, 2335–2342. [CrossRef]
12. Schetz, J.A.; Maddalena, L.; Burger, S.K. Molecular weight and shock-wave effects on transverse injection in supersonic flow. *J. Propuls. Power* **2010**, *26*, 1102–1113. [CrossRef]

13. Erdem, E.; Saravanan, S.; Lin, J.; Kontis, K. Experimental investigation of transverse injection flowfield at Mach 5 and the influence of impinging shock wave. In Proceedings of the 18th AIAA/3AF International Space Planes and Hypersonic Systems and Technologies Conference, Tours, France, 24–28 September 2012.
14. Huang, W.; Tan, J.G.; Liu, J.; Yan, L. Mixing augmentation induced by the interaction between the oblique shock wave and a sonic hydrogen jet in supersonic flows. *Acta Astronaut.* **2015**, *117*, 142–152. [CrossRef]
15. Huang, W.; Wang, Z.G.; Wu, J.P.; Li, S.B. Numerical prediction on the interaction between the incident shock wave and the transverse slot injection in supersonic flows. *Aerosp. Sci. Technol.* **2013**, *28*, 91–99. [CrossRef]
16. Gerdroodbary, M.B.; Ganji, D.D.; Amini, Y. Numerical study of shock wave interaction on transverse jets through multiport injector arrays in supersonic crossflow. *Acta Astronaut.* **2015**, *115*, 422–433. [CrossRef]
17. Menter, F.R. Two-Equation Eddy-Viscosity Turbulence Models for Engineering Applications. *AIAA J.* **1994**, *32*, 1598–1605. [CrossRef]
18. Xing, J.-W.; Yang, Y. Three-dimensional simulation of H2O vitiation effects on combustor performance for scramjet. *J. Propuls. Technol.* **2011**, *32*, 5–10.
19. Thomas, R.A.; Bussing Murmant, E.M. Finite-Volume Method for the Calculation of Compressible Chemically Reacting Flows. *AIAA J.* **1988**, *26*, 1070–1078.
20. Zhai, X.F.; Bai, H.C.; Li, C. Numerical simulation of atomization of liquid jet in supersonic flow. *Phys. Gases* **2021**, *6*, 23–29.
21. Gopala, Y.; Zhang, P.; Bibik, O.; Lubarsky, E.; Zinn, B. Liquid fuel jet in crossflow-trajectory correlations based on the column breakup point. In Proceedings of the 48th AIAA Aerospace Sciences Meeting Including the New Horizons Forum and Aerospace Exposition, Orlando, FL, USA, 4–7 January 2010.
22. Li, L.; Lin, Y.; Xue, X.; Gao, W.; Sung, C.J. Injection of liquid kerosene into a high-pressure subsonic air crossflow from normal temperature. In Proceedings of the ASME Turbo Expo 2012: Turbine Technical Conference and Exposition, Copenhagen, Denmark, 11–15 June 2012. [CrossRef]
23. Inamura, T.; Nagai, N. Spray Characteristics of Liquid Jet Traversing Subsonic Airstreams. *J. Propuls. Power* **1997**, *13*, 250–256. [CrossRef]
24. Wu, P.K.; Kirkendall, K.A.; Fuller, R.P.; Nejad, A.S. Spray Structures of Liquid Jets Atomized in Subsonic Crossflows. *J. Propuls. Power* **1998**, *14*, 173–182. [CrossRef]

 *aerospace*

Article

Plasma Actuation for the Turbulent Mixing of Fuel Droplets and Oxidant Air in an Aerospace Combustor

Zhengqi Tai [1,2], Qian Chen [1,2,*], Xiaofei Niu [1,2], Zhenhua Lin [1,2] and Hesen Yang [3]

1   School of Aeronautics and Astronautics, Shenzhen Campus of Sun Yat-sen University, Shenzhen 518107, China

2   School of Aeronautics and Astronautics, Sun Yat-sen University, Guangzhou 510275, China

3   Science and Technology on Plasma Dynamics Laboratory, Air Force Engineering University, Xi'an 710038, China

*   Correspondence: chenq289@mail.sysu.edu.cn

Abstract: In order to explore plasma-assisted turbulent mixing in aerospace engines, the dielectric barrier discharge plasma actuation for the turbulent mixing of fuel droplets and oxidant air in a ramjet combustor was studied using computational fluid dynamics. A two-way coupling of turbulent air and discrete droplets was realized by Eulerian–Lagrangian simulation, and the dielectric barrier discharge plasma action on flow was modeled by body force. The results show that the plasma actuation can rearrange the recirculation zone behind the evaporative V-groove flameholder, and the main mechanism of actuation is to increase the local momentum of the fluid; the actuation dimension, actuation intensity, and actuation position of the dielectric barrier discharge plasma have strong effects on the turbulent mixing of fuel droplets and oxidant air; and a relatively optimal turbulent mixing can be achieved by adjusting the actuation parameters.

Keywords: plasma actuation; turbulent mixing; fuel droplets; aerospace combustor; dielectric barrier discharge

Citation: Tai, Z.; Chen, Q.; Niu, X.; Lin, Z.; Yang, H. Plasma Actuation for the Turbulent Mixing of Fuel Droplets and Oxidant Air in an Aerospace Combustor. *Aerospace* **2023**, *10*, 77. https://doi.org/10.3390/aerospace10010077

Academic Editors: Hexia Huang, Sergey Leonov, Dan Zhao and Chenzhen Ji

Received: 28 November 2022
Revised: 6 January 2023
Accepted: 9 January 2023
Published: 12 January 2023

1. Introduction

Ramjet engines include subsonic combustion ramjet engines and scramjet engines. For the subsonic combustion ramjet engines, although the flow velocity is lower than that of the scramjet engines, the airflow velocity into the combustion chamber can generally reach a Mach number between 0.2 and 0.3. In order to achieve reliable ignition and stable combustion with high efficiency at higher velocity airflow conditions, various flameholders have been proposed and developed, including a V-groove flameholder, dune standing vortex flameholder, flat-plate flameholder, pneumatic flameholder, and evaporative flameholder, among which the evaporative flameholder has better ignition capacity, wider working range, higher combustion efficiency, and the ability to ignite separately [1]. Due to the above advantages, evaporative flameholders have gained wide application in both turbine and ramjet engines [2–6].

Plasma actuation technology generates plasma by ionizing the gas through a high voltage and frequency power to induce additional flow of fluid and accelerate chemical reactions by applying kinetic, thermal, and chemical actions to the neutral gas through the plasma. According to Li et al. [7], this technology is expected to provide breakthrough technology support for advanced aircraft/engine development based on active flow and combustion control. Due to the superior nature of plasma actuation, with a short response time, wide actuation band, and no moving parts, it is being widely studied by scholars, and the main directions of the research are dielectric barrier discharge (DBD), surface arc discharge, etc. Among them, DBD plasma actuation technology has received extensive attention in simulations [8–14], experiments [15–22], and mechanism studies [23–29]. The main working principle of this technology is flow control by inducing additional velocity of the fluid in the actuation region through the momentum effect of plasma actuation.

Recently, Ombrello et al. [30] studied the effect of plasma discharge on the ignition process, ignition delay time, flame propagation velocity, and combustion chamber flow field. Jin et al. [31] developed a two-dimensional mathematical model of the effect of DBD plasma on the atmospheric combustion of methane–air mixtures, which can well predict the adiabatic flame temperature in the combustion chamber. In addition, it was found that plasma can accelerate the diffusion and mixing of reactants, thus reducing the turbulent mixing time. Cui et al. [32] investigated the optimal delay time for the flammability limit at different plasma frequencies and fuel types. It was found that for methane and propane, the optimal delay time for plasma enhancement precedes the flux pulsation when the flame starts to extinguish. The ignition limit of a methane–air flame can be extended from 0.63 to 0.45 at a specific frequency. Huang et al. [33] showed that plasma actuation can promote kerosene atomization, achieve uniform distribution of kerosene, and improve the ignition limit by forming a rotating arc discharge plasma column between the electrodes and the evaporative flameholder fins in a ramjet combustor. Based on their work, it is necessary to explore other plasma actuation technology potential in this kind of combustor. In the present paper, the cold flow field of a kerosene injection combustor with an evaporative V-groove flameholder is used as the basic flow field, and the modulation effect of DBD plasma actuation on the cold flow field of a kerosene injection combustor with an evaporative V-groove flameholder is investigated by introducing plasma actuation of different dimensions, plasma actuation with different intensities, and plasma actuation at different positions.

The remainder of this paper is organized as follows: We first outline the numerical simulation method in Section 2, including the two-phase-flow mathematical model and the DBD mathematical model. To determine whether the DBD mathematical model can provide accurate flow field characteristics, a numerical simulation on flat-plate flow field is performed in this section and the results are verified with the literature. Section 3 gives results on numerical simulation of ramjet combustor and discussions on turbulent mixing of fuel droplets and oxidant air with different plasma actuation parameters. Finally, the conclusions are summarized in Section 4.

2. Numerical Simulation Method

2.1. Mathematical Model of Two-Phase Flow

The focus of this paper is on the plasma actuation for the turbulent mixing of fuel droplets and oxidant air. The numerical simulation contains gas and liquid droplets. The gaseous phase compressible Reynolds-averaged equations are as follows:

$$\frac{\partial \rho}{\partial t} + \frac{\partial}{\partial x_i}(\rho u_i) = 0 \tag{1}$$

$$\frac{\partial}{\partial t}(\rho u_i) + \frac{\partial}{\partial x_j}(\rho u_i u_j) = -\frac{\partial p}{\partial x_i} + \frac{\partial}{\partial x_j}(-\overline{\rho u_i' u_j'}) + F_i + \frac{\partial}{\partial x_j}[\mu(\frac{\partial u_i}{\partial x_j} + \frac{\partial u_j}{\partial x_i} - \frac{2}{3}\delta_{ij}\frac{\partial u_l}{\partial x_l})] \tag{2}$$

$$\frac{\partial}{\partial t}(\rho E) + \frac{\partial}{\partial x_i}[u_i(\rho E + p)] = \frac{\partial}{\partial x_j}\left[(k + \frac{c_p \mu_t}{\mathrm{Pr}_t})\frac{\partial T}{\partial x_j} + u_i(\tau_{ij})_{eff}\right] + S_h \tag{3}$$

$$\frac{\partial(\rho Y_s)}{\partial t} + \frac{\partial(\rho Y_s u_j)}{\partial x_j} = \frac{\partial}{\partial x_j}\left(\rho D_s \frac{\partial Y_s}{\partial x_j}\right) + \dot{S}_{Y_s} + \dot{\omega}_s \tag{4}$$

where ρ, t, p, μ, E, k, T, and c_p denote density, time, pressure, dynamic viscosity, total internal energy, heat transfer coefficient, temperature, and specific heat capacity, respectively. Here, u_i is the velocity component in three directions (i = 1,2,3); δ_{ij} is the Kronecker function; $(\tau_{ij})_{eff}$ is the strain rate tensor; F_i is the body force source term in the i direction; Y_s and

D_s are the mass fraction and diffusion coefficients, respectively, of the s component; the component number $s = 1, 2, 3, \ldots, N_s-1$; and N_s is the total number of components.

The Reynolds stress term in the above gaseous phase compressible Reynolds-averaged equations needs to be closed by a turbulence model, and the standard k–ε turbulence model is used in this paper.

The liquid phase equation is given as follows:

$$\frac{dX_d}{dt} = U_d \tag{5}$$

$$\frac{dU_d}{dt} = \frac{F_d}{m_d} = \frac{f_1}{\tau_d}(U_{seen} - U_d) \tag{6}$$

$$\frac{dT_d}{dt} = \frac{Nu}{3Pr}\frac{C_{p,g}}{C_{p,l}}\frac{f_2}{\tau_d}(T_{seen} - T_d) + \frac{L_V}{C_{p,l}}\frac{\dot{m}_d}{m_d} \tag{7}$$

$$\frac{dm_d}{dt} = \dot{m}_d = -\frac{Sh}{3Sc}\frac{m_d}{\tau_d}\ln(1 + B_M) \tag{8}$$

where X_d, U_d, U_{seen}, T_{seen}, and $C_{p,g}$ denote the position vector of the droplet in the flow field, the velocity vector, and the flow velocity, temperature, and specific heat capacity of the gas around the droplet, respectively. Here, Nu, Pr, Sc, $C_{p,l}$, L_V, f_2, and B_M are the Nusselt number, Sherwood number, Prandtl number, Schmidt number, liquid-phase-specific heat capacity, latent heat of evaporation of liquid droplets, heat transfer correction factor due to evaporation, and Spalding mass transport coefficient, respectively. In this paper, the KH/RT model is used to calculate the breakdown of kerosene droplets.

2.2. DBD Plasma Actuation Model

A schematic diagram of the structure of DBD plasma actuation is shown in Figure 1. Exposed and covered electrodes were installed above and at the bottom of the insulating media barrier layer, respectively. When the voltage and frequency (amplitude of 5–40 kV and frequency of 1–20 kHz [34]) applied at both ends of the electrode were high enough, the air on the upper surface of the covered electrode was weakly ionized with a large number of charged particles, which were driven by the electric field force and collided with the neutral gas, inducing the gas to flow downstream. The mainstream view on the mechanism of DBD plasma actuation induced flow is the "momentum injection effect." According to this mechanism, researchers carried out a corresponding simplification study; the simplification method was mainly conducted by decomposing the momentum exchange rate in unit volume of the DBD induced injection flow field into x direction component and y direction component, and adding to the momentum equations through the form of source terms, to simulate the DBD plasma actuation effect.

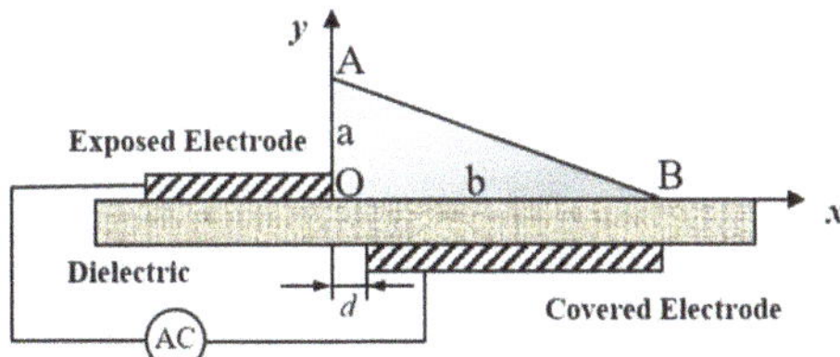

Figure 1. Schematic diagram of DBD plasma actuation structure.

The phenomenological modeling proposed by Shyy et al. [9] is one of the commonly used modeling methods, with the advantages of fast calculation and accurate results, which were used in this study to simulate the effect of plasma on the modulation of the cold flow field in a kerosene-injected combustion chamber with an evaporative V-groove flameholder. As shown in Figure 1, with the Shyy model, the electric field force is located in a triangular area on the upper surface of the covered electrode and is uniformly linearized. The electric field intensity is greatest near the edge of the exposed electrodes as $E_0 = U_0/d$, U_0 is the voltage applied between the electrodes, and $d = 0.25$ mm is the distance between the two electrodes. As the distance away from the O point increases, the electric field intensity gradually decreases. The equation of the electric field intensity is as follows:

$$\left|\overrightarrow{E}\right| = E_0 - n_1 x - n_2 y \tag{9}$$

where n_1 and n_2 are constants. These are defined as follows:

$$n_1 = (E_0 - E_b)/b \tag{10}$$

$$n_1 = (E_0 - E_b)/a \tag{11}$$

where $E_b = 30$ kV/cm is the breakdown electric field intensity, the height a of the triangular area is 1.5 mm, and the length b is 3 mm. The electric field intensity in the x and y direction components can be expressed as:

$$E_x = \frac{E n_2}{\sqrt{n_1{}^2 + n_2{}^2}} \tag{12}$$

$$E_y = \frac{E n_1}{\sqrt{n_1{}^2 + n_2{}^2}} \tag{13}$$

The body force generated by the DBD plasma actuation in the x and y directions are given as:

$$F_x = E_x \rho_c e_c \alpha \delta \vartheta \Delta t \tag{14}$$

$$F_y = E_y \rho_c e_c \alpha \delta \vartheta \Delta t \tag{15}$$

where ρ_c (= 10^{17} m^{-3}) is the charge density, e_c (= 1.602×10^{-19}) is the meta-charge, α is the charge collision efficiency factor, δ is the Dirac function (located in the triangular region where the body force exists, otherwise it is 0), ϑ is the frequency of the applied voltage, and Δt is the time of the plasma action in one actuation cycle of a radio frequency AC voltage.

2.3. Numerical Simulation Implementation Method

For the numerical simulation of the combustor, the continuity equation, momentum equation, energy equation, and component transport equation of the gaseous phase were solved simultaneously and implicitly. Next, the turbulence model transport equations were solved implicitly; then, the liquid-phase equations were solved, and the source terms of the gas-phase equations were updated; these steps were repeated until convergence of the flow field was reached. The main transport equation and the turbulence model transport equation were spatially discretized using the second-order upwind scheme. The gradient was calculated using the least-squares method based on grid cells. The Roe flux differential split scheme was adopted for the convection flux.

In the numerical simulation of the plasma actuation of the spray field in the combustor carried out in Section 3, in order to enhance the stability of the numerical simulation of the two-phase flow field, the numerical simulation of the pure airflow in the combustor was carried out first. After obtaining the converged flow field, the liquid kerosene fuel was injected, and the numerical simulation studies on the spray field (without and with plasma actuation) in the combustor were carried out.

2.4. Validation of Numerical Simulation Method

In order to obtain satisfactory plasma actuation, the phenomenological model proposed by Shyy et al. [9] was validated numerically. The computation domain and geometry model are consistent with that study, as shown in Figure 2.

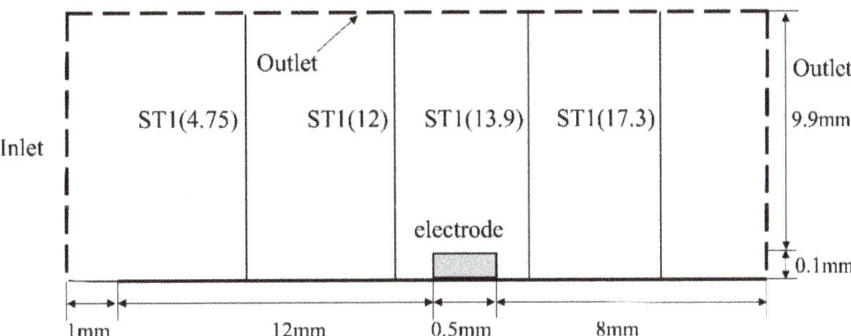

Figure 2. Schematic diagram of calculation domain.

The height of the computational domain was 10 mm, and the distance between the inlet and outlet was 21.5 mm. The flat plate, at a distance of 1 mm from the inlet, constituted the bottom of the computational domain. The exposed electrode, with a length of 0.5 mm and a height of 0.1 mm, was mounted at a distance of 12 mm from the top of the plate. The mesh of the computational domain is shown in Figure 3, and the grid near the electrodes was refined to observe the effect of DBD plasma actuation on the flow field. The number of cells for the plate was 180 thousand. The velocity inlet was used as the inlet boundary condition, and the pressure outlet was used as the outlet boundary condition. The upper wall surface was set as the slip boundary condition, and the lower wall surface had two boundary conditions: the 1 mm length area from the front edge of the plate to the left boundary of the grid was set as the slip boundary condition, and the plate was set as the no-slip boundary condition.

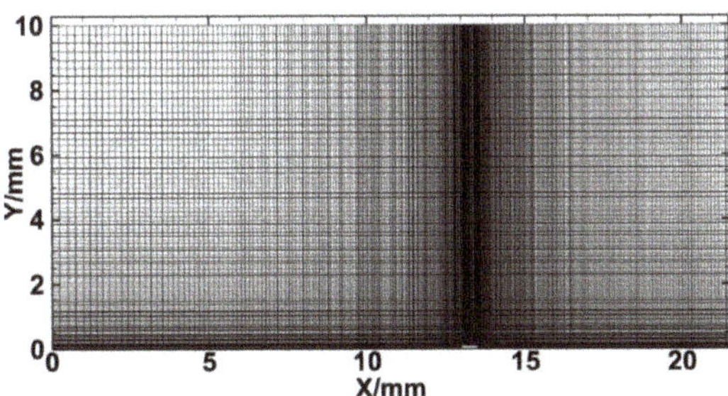

Figure 3. Computational grid.

The velocity distributions at four different locations for inlet parameters of 3 kHz, 4 kV, and 5 m/s are provided in Figure 4, and the results are in agreement with the paper by Shyy et al. [9]. It can be seen that the velocity reaches a maximum downstream from the electrode, and the extent to which the velocity exceeds the free flow velocity indicates the intensity of the plasma actuation; and, in general, the velocity structure downstream from the electrode is similar to that of a wall jet. In addition, it can be seen from Figure 5 that the wall shear stress in the presence of the plasma actuation is larger than that without the plasma actuation. The flow velocity gradually decreases to the free flow velocity away from the wall, which explains the reason that the wall shear stress is negative. The wall shear stress profiles at different frequencies and voltages are given in Figure 5a,b, respectively, and it can be seen that the peak value of the corresponding shear stress becomes larger as the voltage and frequency increase. Wall shear stress profiles for different inlet flow velocities (5 m/s and 2 m/s) at the same actuation voltage and frequency are presented in Figure 5c,d, where the shear stress is dimensionless in terms of the corresponding maximum shear stress without plasma actuation. Here, $V_\infty = 2$ m/s achieves a larger peak, indicating that the corresponding effect of the induced generated wall jet increases with decreasing free flow velocity.

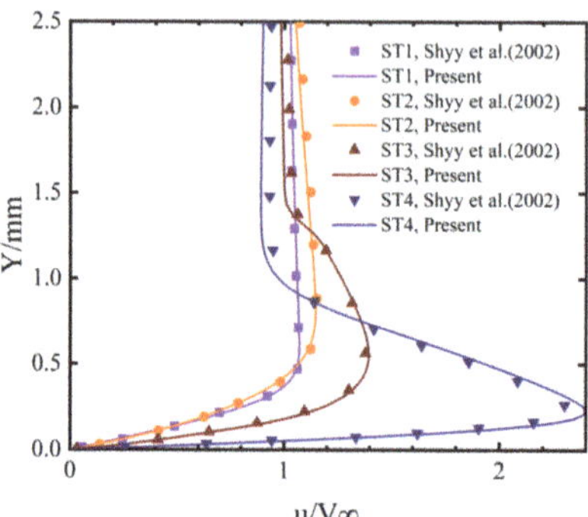

Figure 4. Velocity profiles at four different locations.

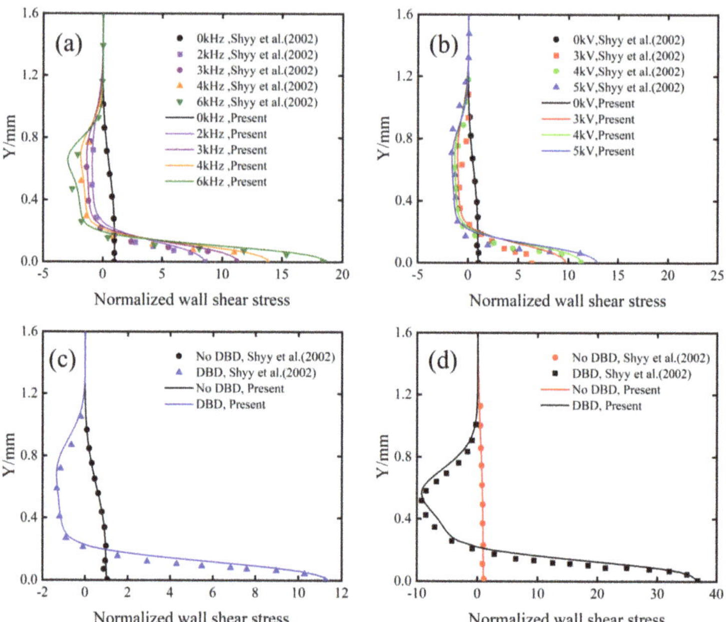

Figure 5. Effects of different parameters on wall shear stress at ST4. (**a**) Frequency, (**b**) voltage, (**c**) inlet velocity with 5 m/s, (**d**) inlet velocity with 2 m/s.

The root mean square error (RMSE) was used to assess the distance between the current simulation results and Shyy et al. [9]. RMSE is given as follows:

$$RMSE = \sqrt{\frac{\sum_{i=1}^{n}\left(X_{present,i} - X_{Shyy,i}\right)^2}{n}} \tag{16}$$

where X denotes normalized velocity u of Figure 4 and normalized wall shear stress of Figure 5, respectively. Table 1 provides the velocity RMSE of the baseline model at different flow locations; the RMSE at different locations is very small, which implies that the current numerical simulation method is quite reliable. At the same time, the RMSE for shear stresses with different parameters in Table 2 is also very small relative to the normalized wall shear stress results. Overall, the results are in better agreement with those in the paper of Shyy et al. [9]. For the combustor flow field simulation in Section 3, computation results are not compared with experiment results, because there are no turbulent mixing experiment data available for these aerospace combustors.

Table 1. The RMSE of velocity profiles at four different locations.

Location	ST1	ST2	ST3	ST4
RMSE	0.0101	0.0097	0.0254	0.0567

Table 2. RMSE of normalized wall shear stress with different parameters.

Frequency	RMSE	Voltage	RMSE	V_∞ = 5 m/s	RMSE	V_∞ = 2 m/s	RMSE
0 kHz	0.0491	0 kV	0.0491	No DBD	0.0104	No DBD	0.0654
2 kHz	1.0853						
3 kHz	1.5048	3 kV	1.6115				
4 kHz	1.6202	4 kV	1.3871	DBD	1.2720	DBD	4.4923
6 kHz	2.5094	5 kV	0.6133				

3. Results and Discussion

3.1. Basic Characteristics of the Spray Field in the Combustor

In the subsonic combustion ramjet combustor, although the incoming Mach number ahead of the holder was less than 0.3, the velocity at the trailing edge of the holder was generally greater than Mach number 0.3 because of the relatively large blockage of the holder, so the flow field was treated as a flow of compressible viscous fluid [1].

The flameholder model is illustrated in Figure 6b, and the local and overall grids of the flow field are shown in Figure 6a,c. The length and width of the combustion chamber were 1487 mm and 140 mm, respectively, and the evaporation tube was located in the center of the combustion chamber. The rest of the flow field grid parameters were similar to those used in Xu [1]. Since the fuel mixing and combustion occurred mainly in the evaporation tube location (purple area) and its wake area (red area), these two parts were refined. To accommodate the complex shape of the flameholder, the computational domain was divided using an unstructured grid. The refining of the regions was beneficial for capturing more flow details. The number of cells in the whole combustion chamber was 0.3 million. The maximum dimensions of the purple, red, and green regions were 0.5 mm, 1 mm and 5 mm, respectively, and the boundary layer of the flameholder wall was also refined. To make the simulation results more general, the working conditions of Huang et al. [33] were used. The inlet Mach number was 0.2, the inlet static pressure was 70 kPa, the inlet static temperature was 357 K, the kerosene injection position was 11 mm upstream of the evaporation tube circle, and the kerosene injection used the upper and lower holes simultaneously to ensure that the kerosene was evenly distributed up and down with the airflow movement into the evaporation tube, which is more consistent with the actual situation. The numerical simulation of the plasma on the spray mist field of the evaporation tube was constant. The air flow field velocity contour, velocity vector, and kerosene droplets distribution in the situation of the equivalent ratio of 0.4 for injection are shown in Figure 7, from which it can be seen that the kerosene droplets are uniformly distributed on the upper and lower sides of the V-groove, and the evaporation tube plays the role of slowing down the kerosene droplets and increasing their evaporation time.

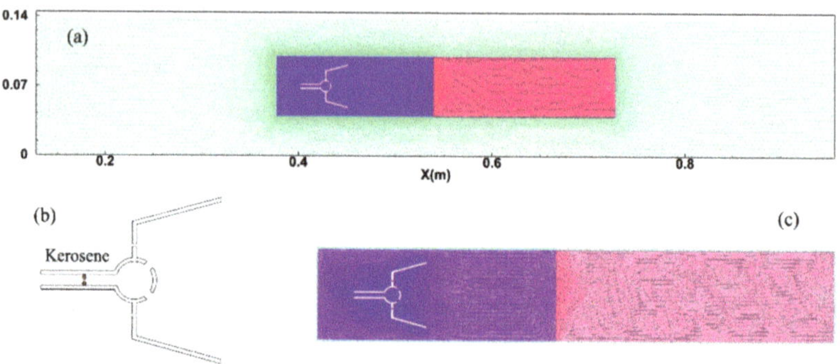

Figure 6. Flameholder structure and grid: (**a**) whole grid of combustor, (**b**) flameholder structure, (**c**) local grid of flameholder.

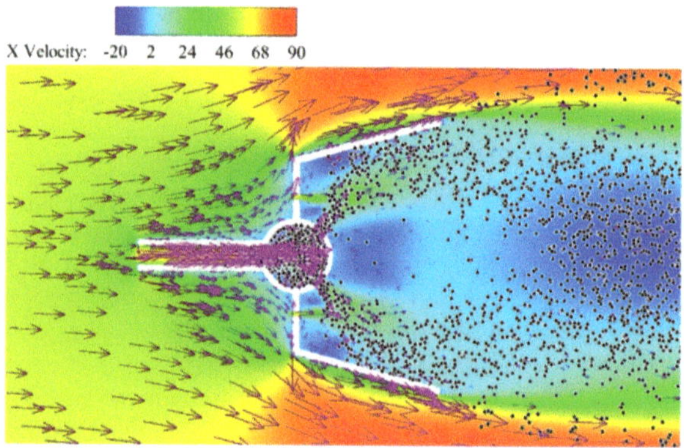

Figure 7. The basic characteristics of the spray field.

3.2. Plasma Actuation of the Spray Field in the Combustor

This section focuses on the effects of different plasma actuation dimensions, intensities, and positions on the air flow field and kerosene droplet distribution.

3.2.1. Actuation Dimensions

The different actuation dimensions are depicted schematically in Figure 8, where DBD plasma actuation with covered electrode lengths of 20 mm, 10 mm, and 5 mm are evenly arranged on the upper and lower wall inner surface of the V-groove.

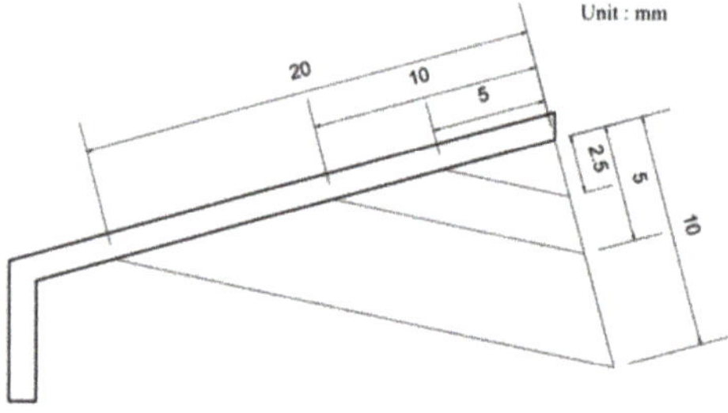

Figure 8. Different actuation dimensions (only the upper wall of the V-groove is shown).

We have analyzed the effects of different actuation dimensions on the spray field in the combustor under three different actuation intensities, and the actuation parameters are shown in Table 3.

Table 3. Different actuation dimensions.

Actuation Intensity	Actuation Dimensions on Inner Surface		
	Length 1 (mm)	Length 2 (mm)	Length 3 (mm)
85 kV, 14 kHz	20	10	5
45 kV, 14 kHz	20	10	5
5 kV, 14 kHz	20	10	5

The effects of different actuation dimensions on the actuation of the spray field are demonstrated in Figures 9–11. In Figure 9, it can be seen that the region and velocity of the induced jet increase with the increase in the actuation dimensions at the same actuation intensity. In the case of the inner surface actuation length of 20 mm, the plasma actuation forms a recirculation zone inside the flameholder. Due to the excessive velocity, the kerosene droplets all gather to the midline position of the flameholder, with the result that most of the droplets cannot flow out from the outlet of the evaporation tube and can only flow out from the inlet of the evaporation tube in the reverse direction, which has a harmful impact on stabilizing the flame and improving the combustion efficiency. When the inner surface actuation length is 10 mm, the region affected by the DBD plasma actuation and the induced velocity is significantly reduced. At the same time, kerosene droplets flowing out of the evaporation tube nozzles are entrained by the recirculation zone to the middle position before intersecting at the position downstream of the evaporation tube, and finally, the droplets are propelled out of the flameholder by the recirculation zone. With the inner surface actuation length of 5 mm, the DBD plasma actuation-induced jet can only affect the shape of the recirculation zone after the evaporation tube in a small region. The kerosene droplets do not touch the V-groove fins because of the recirculation zone near the wall, and they flow directly out of the flameholder. Finally, similar results to those in Figure 9 can be observed in Figures 10 and 11. With other plasma actuation conditions being the same, the larger the actuation dimensions, the larger the induced velocity and recirculation zone generated, and ultimately the effect on the flow field is more significant. It is interesting that in Figure 11b a fine mixing is obtained.

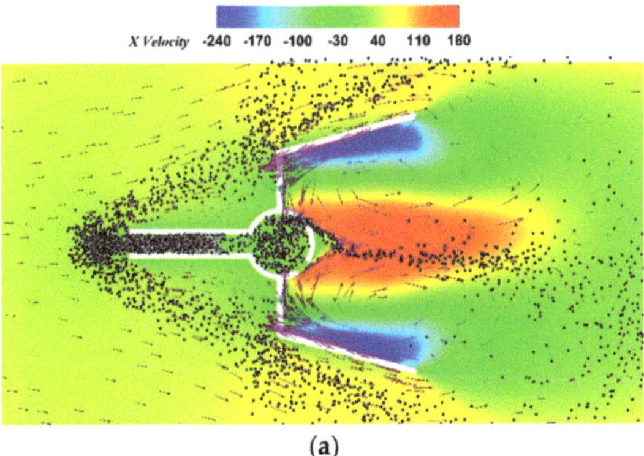

(a)

Figure 9. *Cont.*

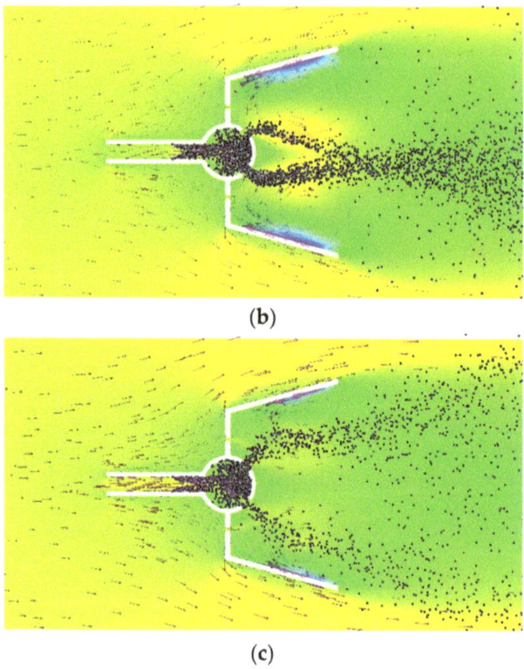

(b)

(c)

Figure 9. Spray field at different actuation dimensions: (**a**) 20 mm, (**b**) 10 mm, (**c**) 5 mm with 85 kV and 14 kHz.

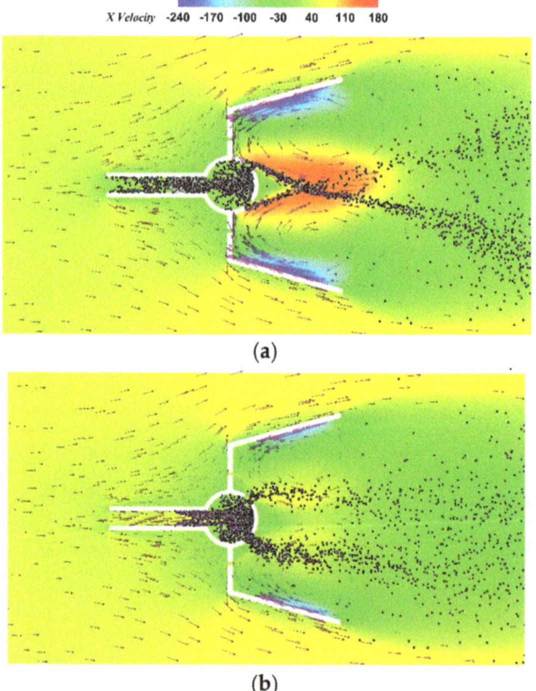

(a)

(b)

Figure 10. *Cont.*

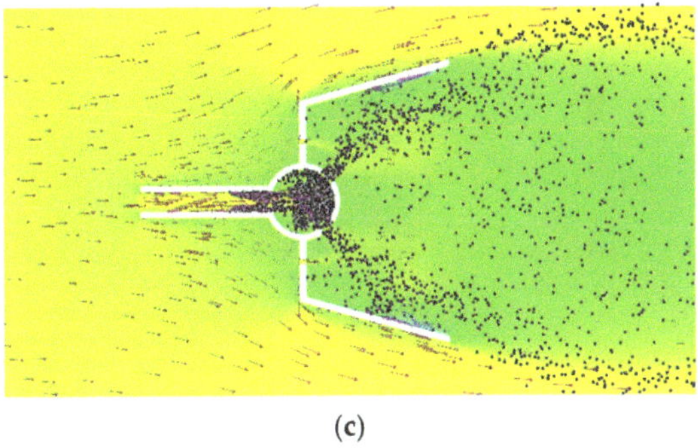

(c)

Figure 10. Spray field at different actuation dimensions: (**a**) 20 mm, (**b**) 10 mm, (**c**) 5 mm with 45 kV and 14 kHz.

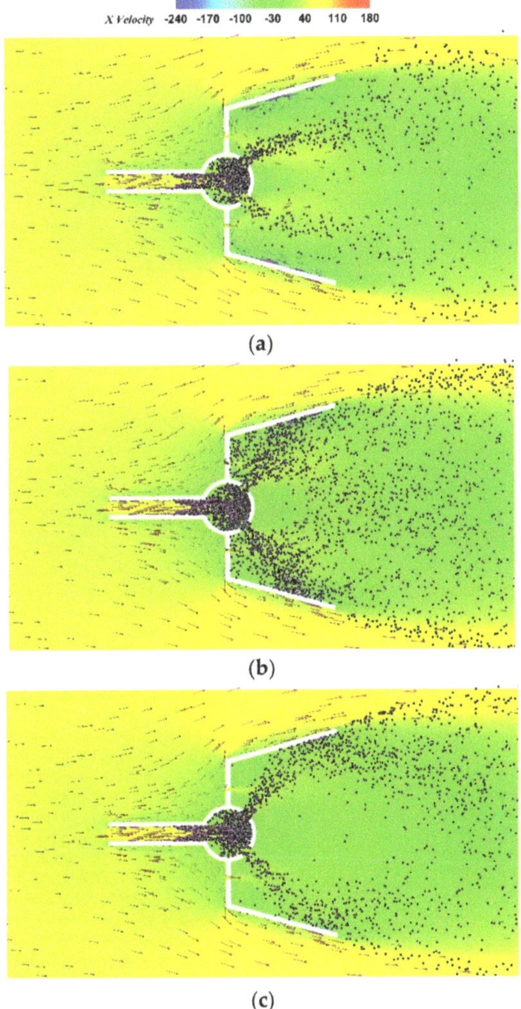

(a)

(b)

(c)

Figure 11. Spray field at different actuation dimensions: (**a**) 20 mm, (**b**) 10 mm, (**c**) 5 mm with 5 kV and 14 kHz.

3.2.2. Actuation Intensities

In this section, the effects of different actuation intensities on the gas-phase flow field and kerosene droplet distribution for actuation lengths of 20, 10, and 5 mm at the inner surface of the flameholder (the actuation position is shown in Figure 10) are investigated. The actuation parameters for the different intensities are summarized in Tables 4 and 5.

Table 4. Different actuation intensities with 20 mm actuation length on inner surface.

Work Conditions	Voltage (kV)	Frequency (kHz)
a	85	14
b	85	9
c	85	5
d	45	14
e	45	9
f	45	5
g	5	14
h	5	9
i	5	5

Table 5. Different actuation intensities with 10 mm and 5 mm actuation lengths on inner surface.

Work Conditions	Voltage (kV)	Frequency (kHz)
a	5	14
b	45	14
c	85	14

The spray field of different actuation intensities is shown in Figure 12, whereas the case in Figure 12a has been discussed in Section 3.2.1 and will not be repeated here. As shown in Figure 12a,b, with large actuation intensity, the induced velocity generated by the plasma actuation far exceeds the velocity at the outlet of the evaporation tube, forcing the kerosene droplets to flow out from the evaporation tube inlet instead of the exit, which is not conducive to the evaporation and combustion of kerosene droplets. For operating conditions (c) and (d), it can be seen that the velocity generated by DBD plasma actuation is still slightly on the high side, and the kerosene droplets are gathered in the center of the V-groove by the induced airflow, which accelerates the velocity of the kerosene droplets and reduces the contact time between the kerosene and air. At the same time, the concentration of a large number of droplets in the center leads to a local fuel enrichment, which is unfavorable to ignition. Observing the remaining five working conditions (Figure 12e–i), it can be found that as the actuation intensity gradually decreases, the DBD-induced velocity also gradually decreases, and the airflow returns from the end of the V-groove fin plate to the exit position of the evaporation tube, which is conducive to increasing the heat exchange between high-temperature air and kerosene, increasing the air–fuel ratio and reducing the ignition delay time.

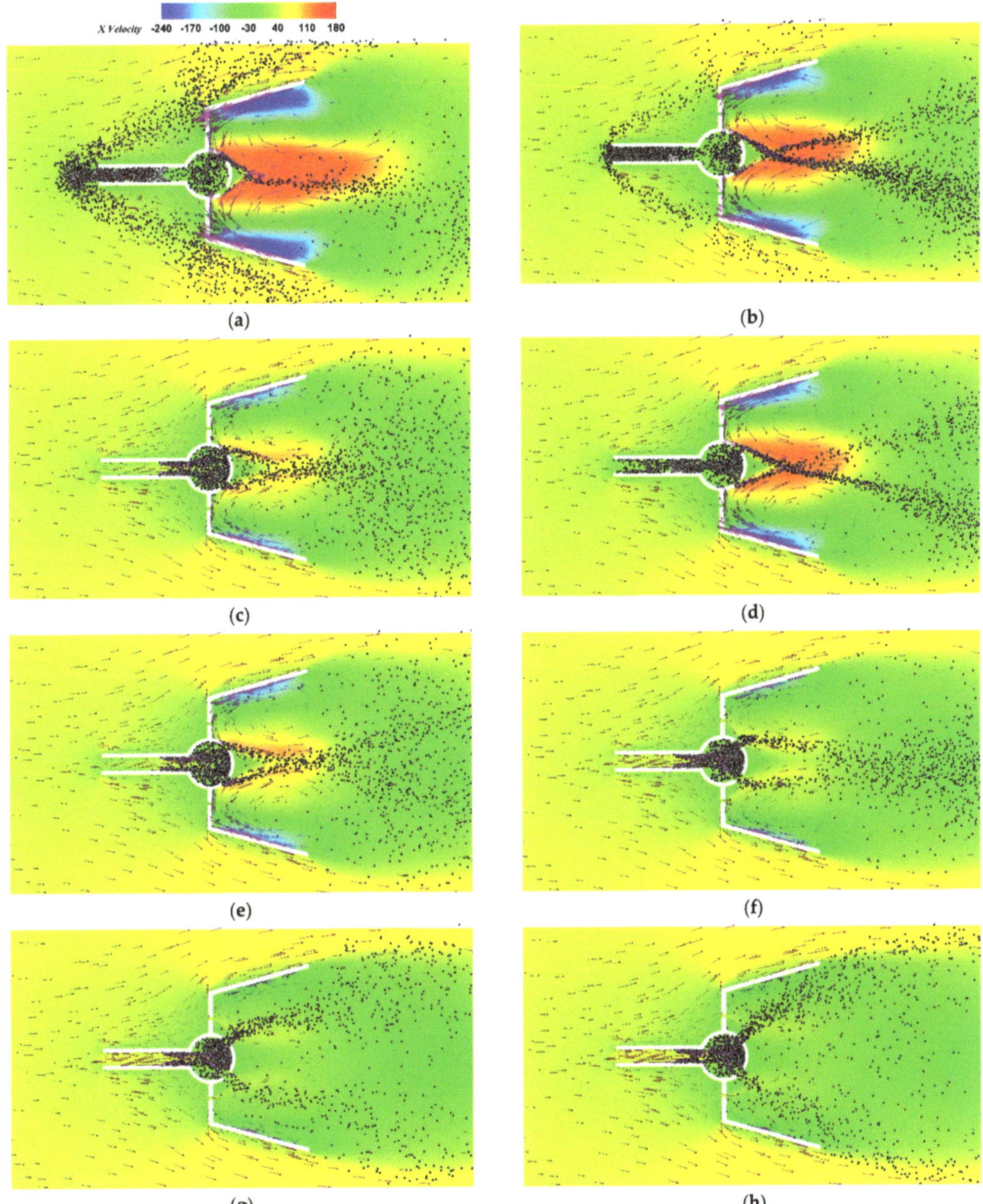

Figure 12. *Cont.*

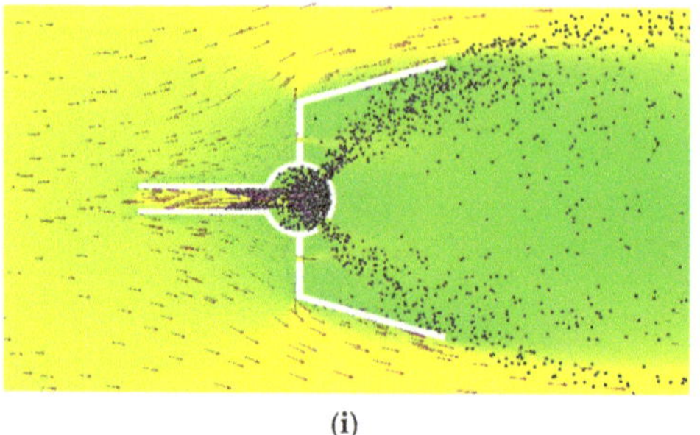

(i)

Figure 12. Spray field under different actuation intensities with 20 mm actuation length on inner surface: (**a**) 85kV, 14kHz, (**b**) 85kV, 9kHz, (**c**) 85kV, 5kHz, (**d**) 45kV, 14kHz, (**e**) 45kV, 9kHz, (**f**) 45kV, 5kHz, (**g**) 5kV, 14kHz, (**h**) 5kV, 9kHz, (**i**) 5kV, 5kHz.

Figures 13 and 14 provide the spray fields for inner surface actuation lengths of 10 mm and 5 mm, respectively (actuation positions are shown in Figure 8), and the actuation parameters for the different intensities are given in Table 5. As illustrated in Figure 13, the DBD plasma actuation induces an increase in the wall jet velocity as the plasma actuation intensity increases. Figure 13a demonstrates that the distribution region of kerosene droplets is greater compared to other operating conditions, and some of the kerosene droplets flow back inside the flameholder, which can increase the contact time between the kerosene droplets and the air, thus further facilitating kerosene ignition and combustion. For the contour of the spray field in the flameholder in Figure 13b,c, it can be seen that the kerosene droplets cannot be uniformly distributed in the rear region of the V-groove due to the increase in the induced velocity caused by the excessive actuation intensity and the formation of kerosene droplet aggregation. The heating effect of high temperature backflow on kerosene droplets is reduced, which is not helpful for the evaporation of kerosene. At the same time, the backflow generated by the actuation induction accelerates the speed of the kerosene droplet outflow, and the contact time with the air becomes shorter, which is not beneficial for ignition and combustion.

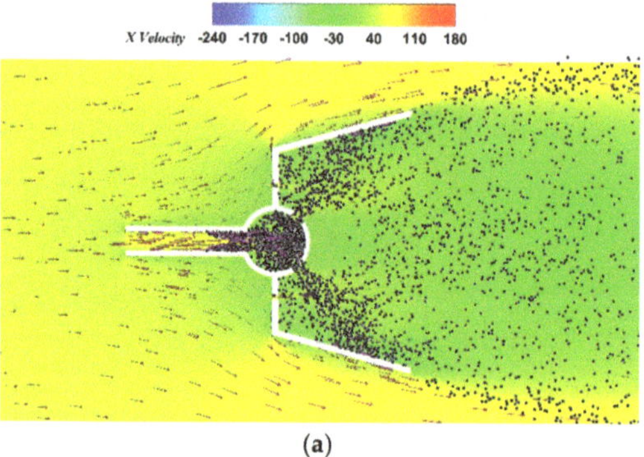

(a)

Figure 13. *Cont.*

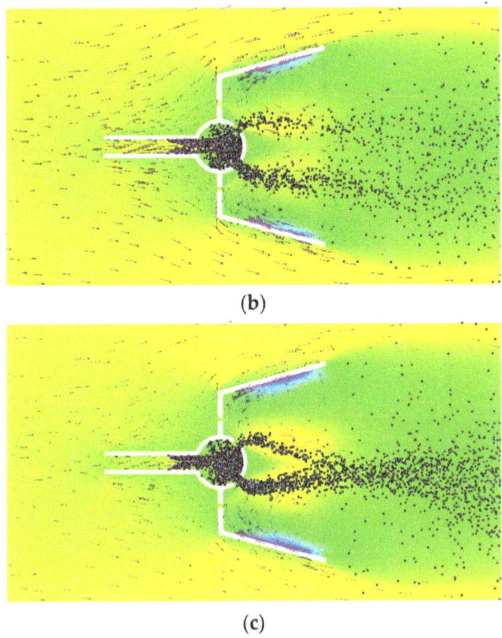

(b)

(c)

Figure 13. Spray field under different actuation intensities with 10 mm actuation length on inner surface: (**a**) 5kV, 14kHz, (**b**) 45kV, 14kHz, (**c**) 85kV, 14kHz.

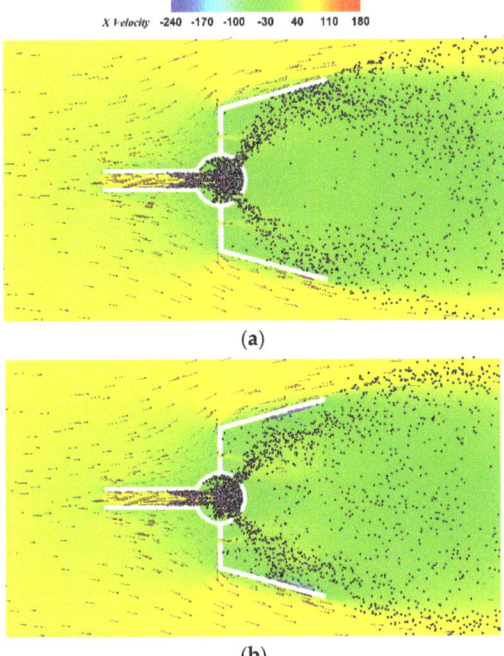

(a)

(b)

Figure 14. *Cont.*

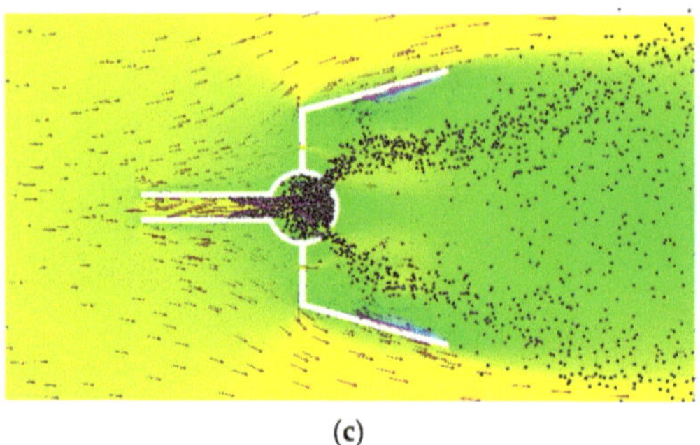

(**c**)

Figure 14. Spray field under different actuation intensities with 5 mm actuation length on inner surface: (**a**) 5kV, 14kHz, (**b**) 45kV, 14kHz, (**c**) 85kV, 14kHz.

As can be seen in Figure 14, the inner surface actuation length of 5 mm has the same trend as the inner surface actuation lengths of 20 mm and 10 mm, and the induced velocity generated by the DBD plasma actuation increases with the increase in the actuation voltage. However, the excellent effect of plasma actuation in Figure 14a is not obvious due to the small actuation dimension. In addition, the kerosene droplets in Figure 14b appear on the inner side of the V-groove fin plate under the action of the induced jet, which is beneficial for the contact between the kerosene and the air and the improvement of the efficiency of kerosene evaporation and combustion. For working condition (c), the induced velocity is too large, resulting in the kerosene droplets failing to reach the V-groove fin plate wall surface under the action of the induced jet. In addition, the kerosene droplets are propelled away from the flameholder by the plasma induced jet, which is not conducive to ignition and combustion.

3.2.3. Actuation Positions

This section examines the effects of different DBD plasma actuation positions on the spray field, and it compares and analyzes the gas-phase flow field and kerosene droplet distribution at different actuation positions.

For the plasma actuation with an inner surface actuation length of 10 mm, two different positions are arranged on the inner surface of the V-groove fin plate (actuation parameters are shown in Table 6), as presented in Figure 15.

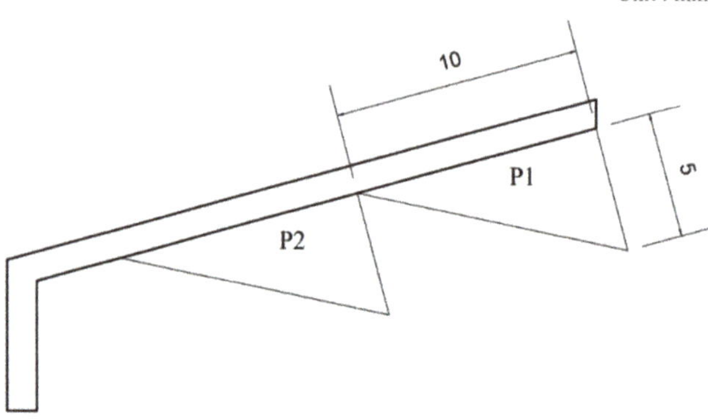

Figure 15. Schematic diagram of different actuation positions with 10 mm actuation length on inner surface (upper part).

Table 6. Parameters of different actuation positions with 10 mm actuation length on inner surface.

Positions	Voltage (kV)	Frequency (kHz)
P1	5	14
P2	5	14

The contours of the kerosene droplet and gaseous flow field at different actuation locations for an inner surface actuation length of 10 mm are provided in Figure 16. It can be seen that the actuation at position P1 in Figure 16a makes the kerosene droplets extensively return to cover the whole flameholder, which is conducive to the sufficient contact between the kerosene droplets and the high temperature gas, which is beneficial for the evaporation and combustion of the droplets. As for the actuation at position P2 in Figure 16b, the useful effect of the actuation is less on the spray field, because the induced recirculation region has less interaction with the kerosene droplets.

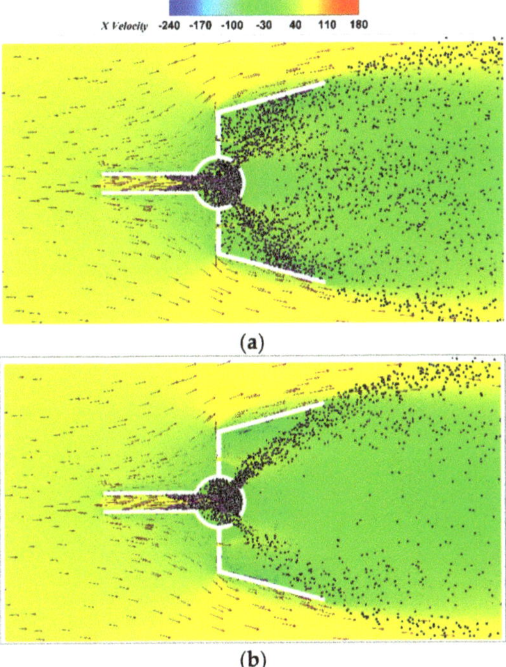

Figure 16. Spray field under different actuation positions with 10 mm actuation length on inner surface: (**a**) P1 position, (**b**) P2 position.

For the plasma actuation with the inner surface actuation length of 5 mm, four different positions are arranged on the inner wall of the V-groove fin plate (actuation parameters are provided in Table 7), as shown in Figure 17.

Table 7. Parameters of different actuation positions with 5 mm actuation length on inner surface.

Positions	Voltage (kV)	Frequency (kHz)
P1	5	14
P2	5	14
P3	5	14
P4	5	14

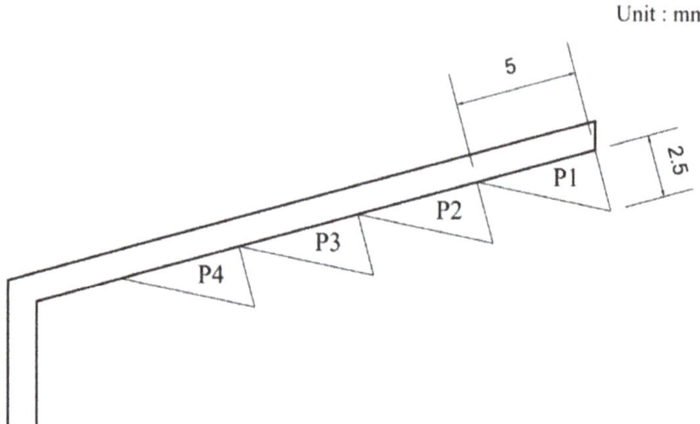

Figure 17. Schematic diagram of different actuation positions with 5 mm actuation length on inner surface (upper part).

The contours of the spray field at different actuation locations for an inner surface actuation length of 5 mm are provided in Figure 18. It is worth noting that the effect of different positions of the DBD plasma actuation (with 5 mm inner surface actuation length) on the gaseous flow field and kerosene droplets is not significant, which is mainly due to the actuation dimensions being too small, resulting in the velocity of the induced jet and the area of influence not being large.

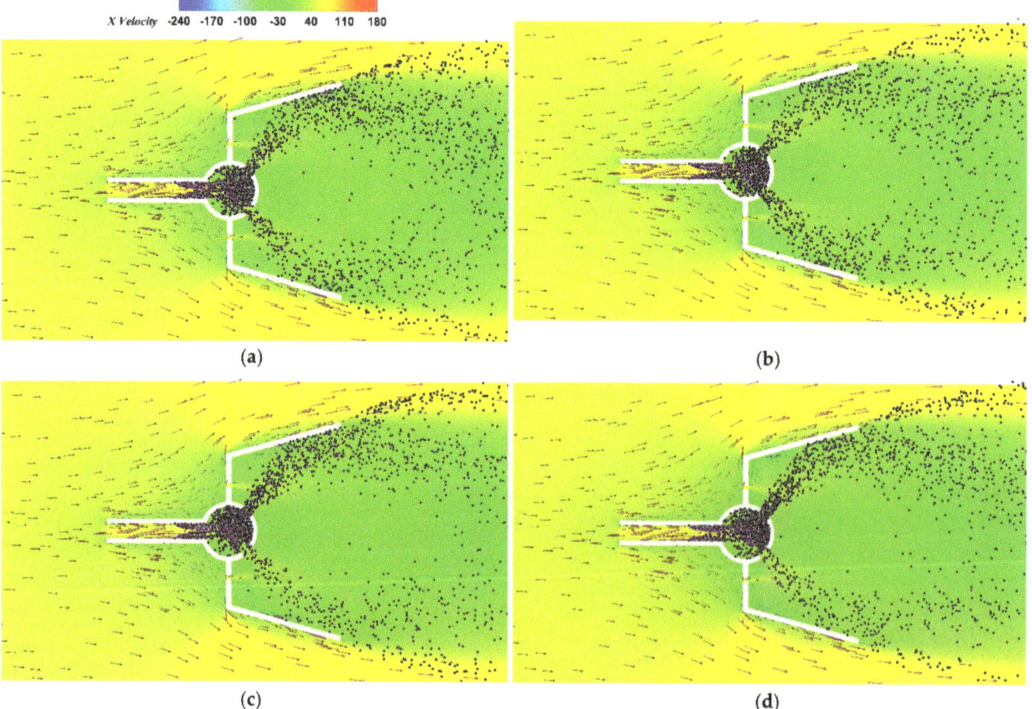

Figure 18. Spray field under different actuation positions with 5 mm actuation length on inner surface: (**a**) P1 position, (**b**) P2 position, (**c**) P3 position, (**d**) P4 position.

In the above study on DBD plasma actuation dimension, intensity, and position effects on the spray field in the subsonic ramjet combustor, it can be found that to obtain a better actuation effect, the velocity of the flow induced by the DBD plasma actuation must be close to the velocity of the flow out of the evaporation tube, and the position of the induced jet must coincide with the position of the kerosene spray reaching the inner surface of the V-groove fin plate. The relative optimal parameters are an inner surface actuation length of 10 mm at position P1, a voltage of 5 kV, and a frequency of 14 kHz. The overall contour of the flow field is shown in Figure 16a and the local flow field is shown in Figure 19.

It is observed in Figure 19a that kerosene droplets are fully distributed in the V-groove near the vertical walls when plasma actuation is applied. The kerosene droplets stay in the V-groove for a long time and are in contact with the high-temperature gas for a long time, which is convenient for the evaporation and combustion of kerosene. In Figure 19b, the wall jet velocity generated by the plasma actuation is about -40 m/s. Since the actuation position is at P1, and the actuation dimension is not too large, the influence of the induced jet on the flow is smaller near the vertical wall of the V-groove, and the same conclusion can be drawn from the fact that the velocity in the y-direction in this region is close to 0 m/s in Figure 19c. In Figure 19d, the recirculation region—under the joint action of the actuation-induced wall jet, the vertical wall inlet jet, and the evaporation tube outlet jet—is once again demonstrated by velocity vector as is shown in Figure 19b,c. The low-temperature region in Figure 19f indicates the denser region of kerosene droplets, and the kerosene droplets with actuation also flow to the upper right after exiting the evaporation tube under the effect of the recirculation region, but the angle is smaller compared to the one with non-actuation, which is beneficial for the better distribution of kerosene droplets. In all, a relatively optimal turbulent mixing of fuel droplets and oxidant air is achieved using the above-mentioned set of DBD plasma actuation parameters.

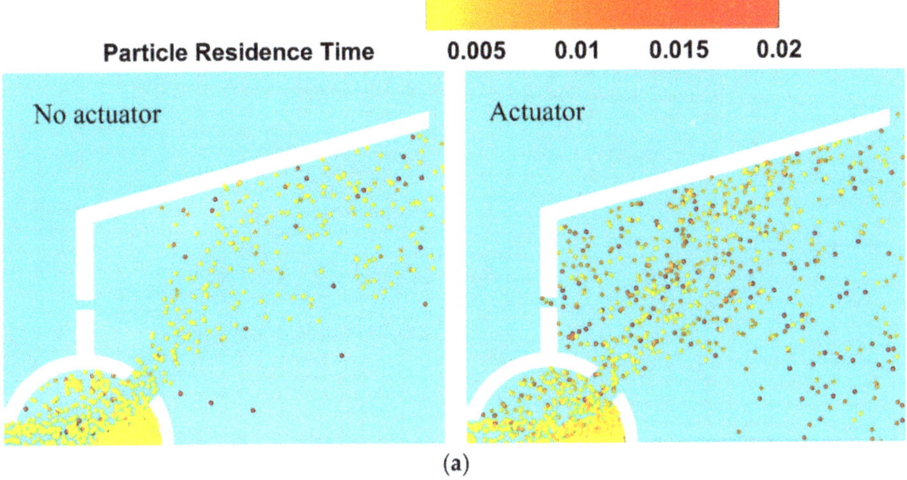

Figure 19. *Cont.*

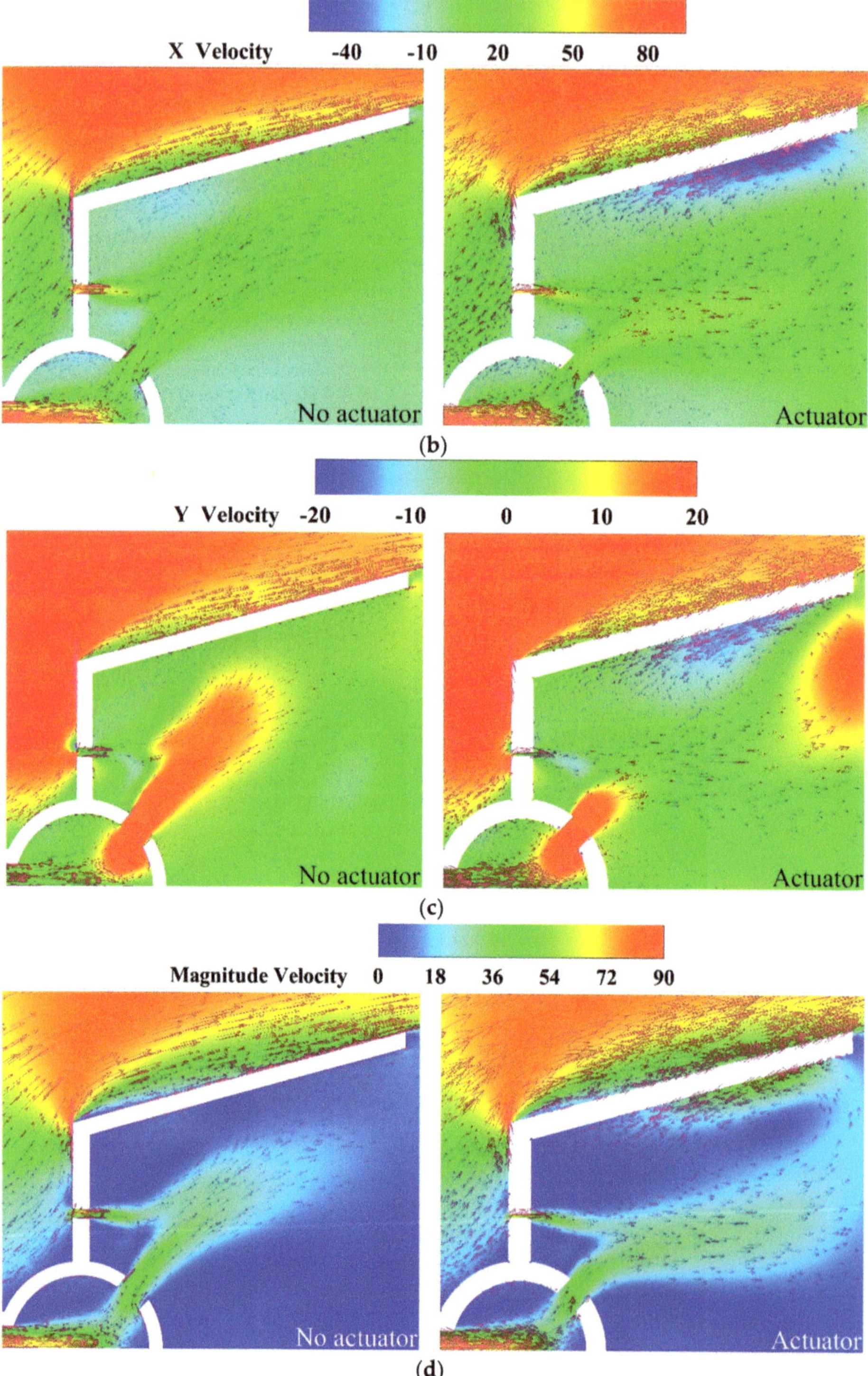

Figure 19. *Cont.*

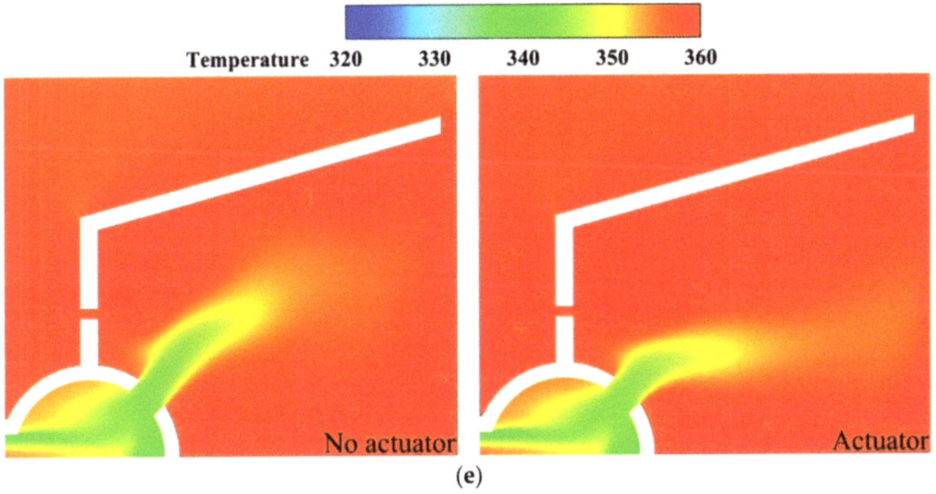

Figure 19. Comparison under non-actuation and actuation. (**a**) Kerosene droplet distribution; (**b**) X direction velocity; (**c**) Y direction velocity; (**d**) magnitude velocity; (**e**) temperature contour.

4. Conclusions

Herein, a DBD plasma actuation study was conducted for the turbulent mixing of fuel droplets and oxidant air in a subsonic combustion ramjet combustor, and the conclusions are as follows:

(1) Comparing the cases with and without plasma actuation, it can be observed that DBD plasma actuation has a significant effect on the spray field because DBD plasma actuation causes a significant change in the velocity near the inner wall surface of the evaporative V-groove flameholder fin plate, which in turn causes an important change in the recirculation regions inside the flameholder and causes a prominent change in the turbulent mixing of fuel droplets and oxidant air.

(2) DBD plasma actuation dimension, intensity, and position all affect the air flow and droplet movement in the evaporative V-groove flameholder. When the actuation dimension decreases, the effect is reduced; when the actuation intensity increases, the effect is an increase; and when the actuation position changes, the air flow and droplet movement vary accordingly.

(3) By adjusting the DBD plasma actuation parameters, relatively optimal turbulent mixing of fuel droplets and oxidant air can be achieved.

Author Contributions: Investigation, formal analysis, and writing—original draft, Z.T.; conceptualization, methodology, formal analysis, writing—review, writing—revision, and supervision, Q.C.; editing—original draft, X.N.; editing—original draft, Z.L.; investigation, H.Y. All authors have read and agreed to the published version of the manuscript.

Funding: This work was supported by the National Natural Science Foundation of China (Grant No. 91741102), the Shenzhen Fundamental Research Program (Grant No. JCYJ20190807160413162), and the Fundamental Research Funds for the Central Universities (Grant No. 19lgzd15).

Institutional Review Board Statement: Not applicable.

Informed Consent Statement: Not applicable.

Data Availability Statement: Not applicable.

Conflicts of Interest: The authors declare no conflict of interest.

References

1. Xu, J.M. Experimental Study and Numerical Simulation of Flameholder in Ramjet Combustors. Master's Thesis, Xidian University, Xi'an, China, 2007.
2. Hong, L.; Yang, G.H. Effects of Incoming Mach Number and Air Excess Coefficient on the Combustion Characteristics of Flameholder. *Adv. Aeronaut. Sci. Eng.* **2010**, *1*, 173–177.
3. Ding, Z.B.; Jin, J. Investigation on Aerodynamic and Combustion Characteristics of an Evaporating Flame-holder. *J. Rocket. Propuls.* **2013**, *39*, 27–31.
4. Ding, Z.B.; Jin, J. Numerical Investigation on Combustion Characteristics and Fuel Supply Matching for an Evaporative Flameholder. *J. Rocket. Propuls.* **2012**, *38*, 43–48.
5. Jin, L.; Tan, Y.H. Numerical Prediction of Cold Flow Field behind a Vapor Flameholder. *J. Rocket. Propuls.* **2007**, *33*, 23–27.
6. Liu, R.; Liu, Y.Y.; Gao, Z. Structural Parameters Effects on Lean Blowout Performance and Prediction Method for Piloted Vaporization Flameholder. *J. Propuls. Technol.* **2017**, *38*, 2753–2760.
7. Li, Y.; Wu, Y. Research Progress and Outlook of Flow Control and Combustion Control Using Plasma Cctuation. *Sci. Sin. Technol.* **2020**, *50*, 1252–1273. [CrossRef]
8. Zhang, P.; Liu, A.; Wang, J. Flow Structures in Flat Plate Boundary Layer Induced by Pulsed Plasma Actuation. *Sci. China Technol. Sci.* **2010**, *53*, 2772–2782. [CrossRef]
9. Shyy, W.; Jayaraman, B.; Andersson, A. Modeling of Glow Discharge-induced Fluid Dynamics. *J. Appl. Phys.* **2002**, *92*, 6434–6443. [CrossRef]
10. Zhang, D.; Wang, Y.; Wang, D. The Nonlinear Behaviors in Atmospheric Dielectric Barrier Multi Pulse Discharges. *Plasma Sci. Technol.* **2016**, *18*, 826–831. [CrossRef]
11. Yu, J.; Yu, J.; Chen, F.; Wang, C. Numerical Study of Tip Leakage Flow Control in Turbine Cascades Using the DBD Plasma Model Improved by the Parameter Identification Method. *Aerosp. Sci. Technol.* **2019**, *84*, 856–864. [CrossRef]
12. Yu, J.; Liu, H.; Wang, R.; Chen, F. Numerical Study of the flow Structures in Flat Plate and the Wall-Mounted Hump Induced by the Unsteady DBD Plasma. *Plasma Sci. Technol.* **2017**, *19*, 015502. [CrossRef]
13. Shan, H.; Lee, Y.T. Numerical Modeling of Dielectric Barrier Discharge Plasma Actuation. *J. Fluids Eng.* **2016**, *138*, 051104. [CrossRef]
14. Iqbal, M.M.; Turner, M.M. Influence of Gap Spacing between Dielectric Barriers in Atmospheric Pressure Discharges. *Contrib. Plasma Phys.* **2015**, *55*, 444–458. [CrossRef]
15. Wang, C.; Zhang, G.; Wang, X. Comparisons of Discharge Characteristics of a Dielectric Barrier Discharge with Different Electrode Structures. *Vacuum* **2012**, *86*, 960–964. [CrossRef]
16. Sokolova, M.V.; Voevodin, V.V.; Malakhov, J.I.; Aleksandrov, N.L.; Anokhin, E.M.; Soloviev, V.R. Barrier Properties Influence on the Surface Dielectric Barrier Discharge Driven by Single Voltage Pulses of Different Duration. *J. Phys. D Appl. Phys.* **2019**, *52*, 324001. [CrossRef]
17. Moussaoui, A.; Kachi, M.; Zouaghi, A.; Zouzou, N. Neutralization of Charged Dielectric Materials Using a Dielectric Barrier Discharge. *J. Electrost.* **2017**, *87*, 102–109. [CrossRef]
18. Malanichev, V.E.; Malashin, M.V.; Moshkunov, S.I.; Khomich, V.Y. Dielectric Barrier Discharge Plasma Reactor. *High Energy Chem.* **2016**, *50*, 304–307. [CrossRef]
19. Liu, Z.; Yang, L.; Wang, Z.; Sang, L.; Zhu, Q.; Li, S. Atmospheric Pressure Radio Frequency Dielectric Barrier Discharges in Nitrogen/Nrgon. *Plasma Sci. Technol.* **2013**, *15*, 871. [CrossRef]
20. Kim, J.; Kim, S.-J.; Lee, Y.N.; Kim, I.T.; Cho, G. Discharge Characteristics and Plasma Erosion of Various Dielectric Materials in the Dielectric Barrier Discharges. *Appl. Sci.* **2018**, *8*, 1294. [CrossRef]
21. Khomich, V.Y.; Malanichev, V.E.; Malashin, M.V.; Moshkunov, S.I. Dielectric Barrier Discharge Uniformity Enhancement by Air Flow. *IEEE Trans. Plasma Sci.* **2016**, *44*, 1349–1352. [CrossRef]
22. Aerts, R.; Somers, W.; Bogaerts, A.J.C. Carbon Dioxide Splitting in a Dielectric Barrier Discharge Plasma: A Combined Experimental and Computational Study. *ChemSusChem* **2015**, *8*, 702–716. [CrossRef] [PubMed]
23. Zhu, P.; Dong, L.; Yang, J.; Li, B.; Zhang, C. Vibration of Discharge Filaments in a Dielectric Barrier Discharge. *IEEE Trans. Plasma Sci.* **2014**, *42*, 1990–1994. [CrossRef]
24. Zhang, S.; Chen, Z.; Zhang, B.; Chen, Y. Numerical Investigation on the Effects of Dielectric Barrier on a Nanosecond Pulsed Surface Dielectric Barrier Discharge. *Molecules* **2019**, *24*, 3933. [CrossRef] [PubMed]
25. Peeters, F.J.J.; Rumphorst, R.F.; Van de Sanden, M.C.M. Dielectric Barrier Discharges Revisited: The Case for Mobile Surface Charge. *Plasma Sources Sci. Technol.* **2016**, *25*, 03LT03. [CrossRef]
26. Pang, L.; He, K.; Di, D.X. Effect of Pulse Polarity on Nanosecond Surface Dielectric Barrier Discharge. *IEEE Trans. Plasma Sci.* **2014**, *42*, 2340–2341. [CrossRef]
27. Osawa, N.; Yoshioka, Y.; Hanaoka, R.; Mochizuki, Y.; Kobayashi, Y.; Yamada, Y. Generation of Uniform Discharge by Dielectric Barrier Discharge Device in Atmospheric-Pressure Air. *Electr. Eng. Jpn.* **2012**, *180*, 1–9. [CrossRef]
28. Huang, Z.; Yang, L.; Hao, Y.; Li, L. Dynamic Characteristics of Dielectric Barrier Columnar Discharge During Its Decay. *IEEE Trans. Plasma Sci.* **2016**, *44*, 2568–2575. [CrossRef]
29. Bouchikhi, A. Dielectric Barrier Discharge Effect on Capacitively Coupled RF Argon Glow Discharge. *Indian J. Pure Appl. Phys.* **2022**, *60*, 163–170.

30. Ombrello, T.; Won, S.H.; Ju, Y.; Williams, S. Flame propagation enhancement by plasma excitation of oxygen. Part II: Effects of O2 (a1Δg). *Combust. Flame* **2010**, *157*, 1916–1928. [CrossRef]
31. Jin, D.; He, L.; Liu, X.; Chen, Y. Numerical simulation of plasma-assisted combustion of methane-air mixtures in combustion chamber. *Plasma Sci. Technol.* **2018**, *20*, 125502.
32. Cui, W.; Ren, Y.; Li, S. Stabilization of premixed swirl flames under flow pulsations using microsecond pulsed plasmas. *J. Propuls. Power* **2019**, *35*, 190–200. [CrossRef]
33. Huang, S.; Wu, Y.; Zhang, K.; Sun, J.; Jin, D.; Li, Y. Experimental Investigation on Spray and Ignition Characteristics of Plasma Actuated Bluff Body Flameholder. *Fuel* **2022**, *309*, 122215. [CrossRef]
34. Abdollahzadeh, M.; Páscoa, J.; Oliveira, P. Numerical modeling of boundary layer control using dielectric barrier discharge. In Proceedings of the IV Conferência Nacional em Mecânica dos Fluidos, Termodinâmica e Energia, MEFTE, Lisbon, Portugal, 28–29 May 2012.

Article

Axial Turbine Performance Enhancement by Specific Fluid Injection

Razvan Edmond Nicoara [1,2,*], **Daniel Eugeniu Crunteanu** [1] **and Valeriu Alexandru Vilag** [2]

1   Faculty of Aerospace Engineering, University Politehnica of Bucharest, 060042 Bucharest, Romania
2   Romanian Research and Development Institute for Gas Turbines COMOTI, 061126 Bucharest, Romania
*   Correspondence: razvan.edmond@gmail.com

Abstract: Extensively used in modern gas turbine engines in various applications, ranging from aerospace, marine and terrestrial propulsion to power generation and gas pumping, the axial flow turbines have been continuously updated and are now capable of high performances and reliability. One drawback that has not yet been resolved is the poor performance of the axial turbines at lower-than-nominal regimes. To solve these shortcomings, a new method to improve the performances at partial regimes by specific fluid injection is proposed in this paper. The influence of the injection system is determined by conducting a numerical analyze, studying the influence of different parameters (i.e., number, dimensions and position of the of injection orifices) on the overall performances of the turbine. The study is completed on a single stage 1300 KW turbine with the injection system being applied to different power settings across the working line. The results show that the power generated by the turbine can be enhanced by as much as 30% for different configurations of the injection system (i.e., high number of small size orifices) and different partial regimes.

Keywords: axial turbine; injection system; performance enhancement; partial regimes; CFD

1. Introduction

Due to large-scale use of turbomachinery in different industries such as aviation, marine and terrestrial propulsion, power generation, gas pumping etc. and due to constant tightening of regulations regarding greenhouse, these systems have been constantly improved. Modern axial flow turbines are capable of withstanding high temperatures and mechanical loads achieving high performances without sacrificing reliability. In a modern gas turbine engine, it is common that the service life of the axial turbines should match those of the respective engines. As the turbines are designed for nominal regimes, thus for a set of input parameters and requirements, the performances at different regimes (different input parameters) may be less than desirable. Connecting the turbine with the compressor in a gas turbine engine assembly or with a power consumer (in the case of a power turbine) which can have different power requirements depending on the rotational speed, results in low efficiencies at partial regimes. Thus, a need to adjust the flow through the turbine, to better adapt at the respective requirements and inlet parameters, can be identified. In order to achieve this, it is necessary to introduce a regulation factor through which the flow regime and implicitly the performance of the turbine can be controlled depending on the engine regime or the requirements of the mechanical power consumer. One such control system was proposed as early as 1966 in form of variable vanes [1]. By controlling the vane angle it is possible to modify the minimal section and the exit flow angles, thus modifying the flow through the vanes to obtain higher performances for the respective inlet parameters. A solution such as this has yet to be implemented on a modern high temperature engine due to issues regarding the reliability of a movable mechanism in an extreme temperature environment.

The method discussed in this paper targets the same control of the flow through the vanes but using an injection system instead of mobile elements. By specific fluid injection,

Citation: Nicoara, R.E.; Crunteanu, D.E.; Vilag, V.A. Axial Turbine Performance Enhancement by Specific Fluid Injection. *Aerospace* **2023**, *10*, 47. https://doi.org/10.3390/aerospace10010047

Academic Editor: Dan Zhao

Received: 23 November 2022
Revised: 16 December 2022
Accepted: 27 December 2022
Published: 3 January 2023

a deviation of the main flow through the vane near the minimal section is anticipated, thus reducing the flow section.

The use of injection in turbine profiles is not a new concept, this process being used for long time for cooling purposes. Due to the extreme temperature environment (modern gas turbines work at a turbine inlet temperature around 1800 °C [2], which is significantly higher than the melting point of the blades' material), an efficient cooling is needed. The methods currently in use are classified as internal and external cooling. The cooling fluid that passes through cooling channel inside the profiles (a process known as internal cooling) is injected in the boundary layer on the profile exterior in order to form a fluid film that protects the blades material from the high temperature gases of the main stream (process known as external cooling or film cooling). The difference in temperature of the blade material respective to the hot gases is a few hundred degrees. The flow structure complexity of coolant ejected from the film holes is determined by the blowing ratio, distribution of the film holes, direction of the film holes and the pressure gradient in the main stream direction [2]. The film cooling method was the subject of many studies (e.g., [3–8]) which analyzed the cooling efficiency, the flow around the injection orifices, and the interaction of cooling flow with the hot gases.

Another related application of fluid injection in the boundary layer is the system applied for aircraft wings in order to increase lift and decrease the aerodynamic drag. By injecting a fluid with a specific momentum, it is possible to delay or even entirely remove the separation phenomena. This method of fluid injection near the leading edge followed by suction in the boundary layer is called the co-flow jet active control method. The results [9] showed that the wing efficiency can be increased by more than 30% using this method.

A number of studies have been conducted on the active flow control for axial turbines. These studies have been focused on a reduction of turbine losses at low Reynolds numbers [10–15]. At high attitudes a large drop in Reynolds number results when compared to take-off conditions which causes the flow through the low-pressure turbine to become laminar, thus separates more quickly form the suction sides of the blades. This phenomenon leads to a considerable reduction in turbine efficiency and aircraft engine performances. A number of different flow control methods have been proposed to prevent boundary layer separation, such as dielectric barrier discharges, synthetic jets and vortex generators. In the paper Optimisation in Active Flow Separation Control in Low-Pressure Turbines [11] the authors showed that using flow injection at the right axial distance the boundary layer detachment process can be completely mitigated with minimal pressure losses. The paper also finds that the efficiency of the reattachment process is strongly dependent on the injection location and the blow rate. Use of pulsating injection in the turbine flow was studied in the paper Efficiency of Pulsating Base Bleeding to Control Trailing Edge Flow Configurations [16], the authors showed, using numerical simulations, that a pulsed injection configuration can result in similar gains but with a more efficient use of the energy compared with steady blowing.

The use of fluid injection can also be used as a method of reducing endwall losses for high-lift aft loaded profiles. In the paper Secondary Flow Loss Reduction Through Blowing for a High-Lift Front-Loaded Low Pressure Turbine Cascade a row of injection jets are used on the suction side of the blade near the endwall. The injection jets were designed with a 30° angle respective to blade surface in order to reduce mixing losses. This flow control method was studied at high Reynolds number resulting in a maximum reduction of 42% in the area averaged total pressure loss coefficient. The same blade has been tested with midspan vortex generator jets resulting in a 21% reduction in the area averaged total pressure loss for a Reynolds number of 20,000 [17]. In a similar study [18], the authors showed that similar reductions are possible but using a lower mass flow rate (up to 22%) by using unsteady injection jets.

The use of fluid injection to increase turbine performances at off design regimes is a less studied domain with few articles describing this technology. Rohr and Yang [19] described in their paper a method of turbine efficiency enhancement at partial load through fluid injection into boundary layer on the pressure side of the vanes near the trailing edge. The numerical study was conducted with STAR-CCM+ software on a Pak-B LPT blade, two configurations of the injection jet being tested. The first case numerically simulated was with the injection jet normally to blade with the second configuration using a 45° tilt injection jet relative to blade pressure side. Tilting the injection jet yields considerable advantages in loss reduction. The previous best case of 14.3% total pressure loss for the right-angled jet can be almost reduced to zero for aninjection jet mass flow rate of 3.5%.

In this paper the study of a new injection system, designed as a method of enhancing the turbine performances at partial regimes, is presented. This study represents a continuation on the work started in paper [20], where the influence of a number of parameters was determined on a simplified 2D model and aims at determining the influence of geometrical parameters as well as the influence of the injection system at different regimes across the turbine working line. The 3D numerical model construction and the definition of the injection system are presented, and analyses are conducted to determine the influence of the number and dimension of injection orifices on the overall performances. The injection system is then applied for other partial regimes of lower respective higher power settings.

2. Materials and Methods

This chapter presents the research methodology used for this study, construction of the injection model, numerical grid parameters and influence as well as the boundary conditions and numerical model.

In order to study the influence of various geometrical and gas dynamic parameters on the performances of the injection system and the overall performances of the turbine a study methodology has been created. For this study an axial turbine for which the geometry and performances are known was defined and the flow field was determined by numerical simulation for both nominal and a partial regime. A simplified 2D model was created and the influence of a number of parameters (i.e., angle of injection, axial position on the suction side of the blade, mass flow and orifice diameter) was determined. This paper focuses on the analysis on a 3D model and the influence of related parameters such as orifice dimensions, number of orifices and axial distances as well as performances at different partial regimes. The study methodology for the entire work on the injection method is presented in Figure 1, the work presented in this paper being described by the steps 7–11.

The reference turbine used in this study is a single stage free power turbine with a nominal mass flow of 8 kg/s and a rated power of approximately 1350 KW. The profile of the vane and the rotor are presented in Figure 2 and the turbine parameters in Table 1.

Table 1. Reference turbine main parameters.

Parameter	Measurement Unit	Value
Vane shroud diameter	mm	381
Rotor shroud diameter	mm	470
Vane maximum height	mm	63.5
Rotor maximum height	mm	94.4
Number of vanes	-	44
Number of rotor blades	-	53
Rotational speed	rpm	22,000
Nominal power	KW	1351
Pressure ratio	-	2.1
Isentropic efficiency	%	87.8

In order to construct the turbine 3D model which incorporates the injection system it is necessary to determine the dimension and position of the critical section. For the reference turbine the critical section was determined by geometrical measurements at 7 different radiuses along the vane height and by joining these 7 points the minimal section projection curve on the vanes suction sidewas generated. The curve was then translated upstream with a specific value (depending on the studied case). For the injection orifices, a circular injection surface with a specific diameter was generated perpendicular to the suction side For this study the injection angle was 90° as it was found that the best results were obtained in the case of perpendicular injection [20]. The orifices were displaced equidistant on the translated curve, the number of orifices being different from case to case. The logical steps for injection system generation are presented in Figure 3.

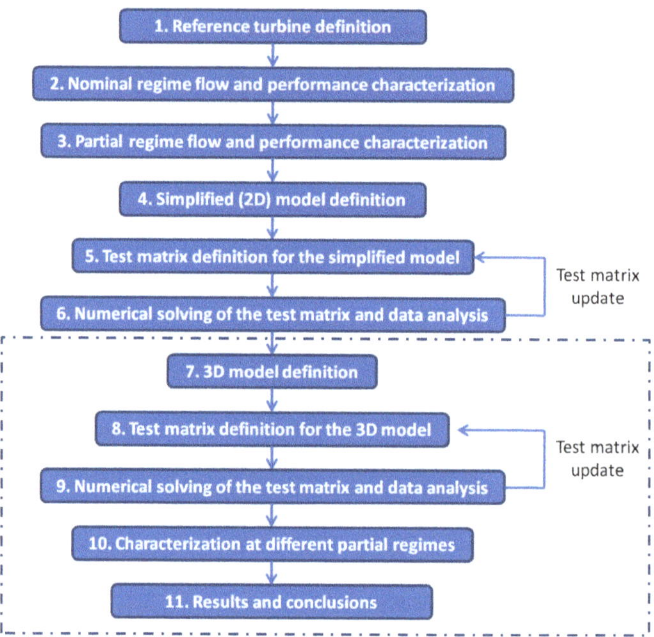

Figure 1. Study methodology for injection method characterization.

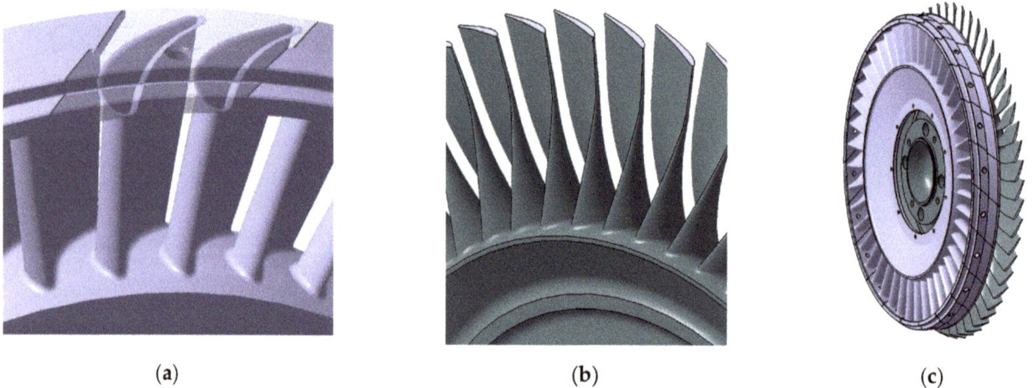

(a)   (b)   (c)

Figure 2. (**a**) Reference turbine vane profiles; (**b**) Reference turbine rotor blade profiles; (**c**) reference turbine assembly.

Numerical simulations were conducted to determine the influence of the injection system on the turbine flow and performances using ANSYS CFX software. The flow through the turbine was considered uniform, thus a single channel was used for vane and rotor respectively, using periodic conditions. The numerical grid, generated using ANSYS Mesh, is unstructured. A grid dependency analysis was conducted using 4 configurations with the number of elements increasing with a factor of approximately 2.5 between the first and the last configuration. The results of the analysis are presented in Table 2. The influence of the numerical grid is negligible as the power generated, the main parameter for this study, and the isentropic efficiencies for all the configurations are similar.

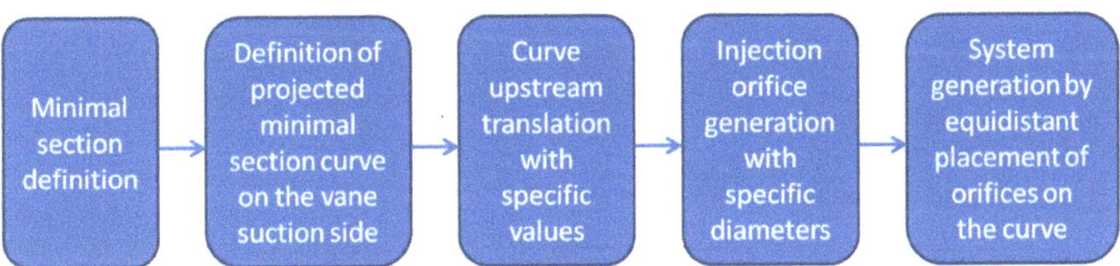

Figure 3. Steps for injection model generation.

Table 2. Numerical grid analysis results.

Configuration	Number of Elements	Generated Power [KW]	Isentropic Efficiency
1	982,282	538.46	0.8423
2	1,086,756	540.38	0.8420
3	1,855,315	538.30	0.8400
4	2,676,438	536.36	0.8410

The second configuration was chosen in order to lower the computational demands and because, for this configuration, the values of the y^+ parameter are in the recommended interval [21]. The numerical volumes and grid for both the vane and the rotor are presented in Figure 4. The general mesh size is 1 mm with 0.5 mm near the walls, 0.2 mm in the vicinity of the injection orifices and 0.05 mm for injection orifices. The maximum values of the y^+ parameter are below 150, with an average value of 80.

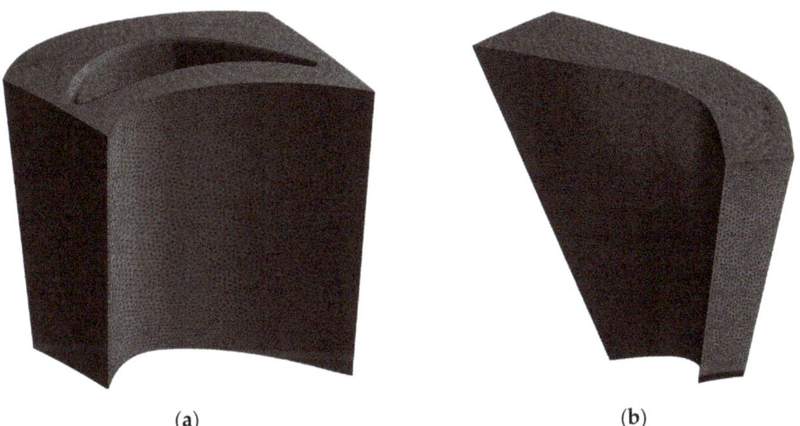

(a)   (b)

Figure 4. (**a**) Numerical grid for the vane; (**b**) Numerical grid for the rotor blade.

The numerical case is typical for axial turbine simulations; with the mass flow being set at the rotor exit, using outlet boundary conditions, and the pressure and temperature set at the vane inlet, using inlet boundary conditions. For the turbine walls, adiabatic no slip conditions were assumed. At the vane-rotor interface stage conditions were selected, as well as periodic condition for rotational periodicity for both vane and rotor blade. Furthermore, for the injection sections, inlet boundary conditions were selected, fixing the temperature and the injection mass flow. The turbulence model used for these simulations is the k–ε. The model is used many industrial processes to predict the flow in turbulent conditions due to its robustness, lower computational cost and reasonable accuracy. The superior performances of the k–ε model in turbulent flows are of particular interest for the presented study due to the turbulent nature of fluid interactions near the injection sections [21–24].

For the convergence criteria 3 parameters were monitored: residuals, mass flow imbalance and the rotor blade torque. The residual target was fixed at 1×10^{-6} reaching at least 1×10^{-4} for each case (with mass flow residuals achieving smaller values). In terms of mass flow imbalance, the values were smaller than 0.1% for both stator and rotor volumes at each of the simulated cases. The blade torque represents the final parameter that was monitored, the torque curve flattened with the criteria being a variation smaller than 0.2% at 300 iterations.

For the boundary conditions, these were set identically for all the studied cases.Verification was completed at the nominal point, the parameters at this point being similar with those for which the turbine was designed.

The influence of the injection system on the turbine performances was determined at a partial regime of approximately 89% of the engine speed. The input parameters for this regime were determined using the turbine map and working line; the parameters for both nominal and partial regime are presented in Table 3.

Table 3. Nominal and partial regime parameters.

Parameter	Measurement Unit	Nominal Regime	Partial Regime
Total Inlet Pressure	[BarA]	2.55	2.34
Total Inlet Temperature	[K]	977	868
Mass flow	[kg/s]	8	7
Rotational speed	[rpm]	22,000	20,000
Expansion ratio	-	2.12	1.37
Isentropic efficiency	[%]	87.8	87.2
Generated power	[KW]	1351	466.6

3. Results and Discussion

This study will focus on the influence of the injection section diameter and number of orifices on the turbine performances. For the orifice diameter four values were studied (0.4, 0.6, 0.8 and 1 mm) and different axial distances. As it was determined in previous studies that the lower values of the diameter lead to better results, lower than 1 mm dimensions were considered. The diameter was limited to 0.4 mm as lower values will not be feasible from a construction point of view. In case of the number of injection sections these values were determined respectively to orifice dimensions raging from 19 to 100. The injection mass flow was also limited to 2.5% of the working fluid as the fluid is considered to be drawn from the engine compressor.Using outlet boundary condition and fixing the mass flow at the rotor outlet leads to a decrease of the inlet mass flow with the respective injected mass flow.

3.1. Orifice Diameter Influence

In order to compare the influence of the injection diameter on the axial turbine performance, it is essential to maintain the inlet conditions constant for each injection section. Thus, for each diameter studied, a number of orifices were determined in order to achieve approximately the same speed and Mach number at the injection section. The Mach number values in the injection orifices for each diameter and axial distance are presented in Table 4.

The axial distance (Za) represents a dimensionless parameter that describes the axial position of the injection orifice on the suction side of the blade. It is defined as the axial distance between the injection orifice center and the projection of the minimal section of the vane on the suction side divided by the length of the minimal section.

Table 4. Mach number in the injection section.

Injection Diameter $\varnothing_{inj}$[mm]	Number of Orifices	Mach Number			
		$Za = 0.1$	$Za = 0.2$	$Za = 0.317$	$Za = 0.5/0.6$
0.4	100	0.673	0.686	0.691	0.671
0.6	50	0.681	0.691	0.7	0.674
0.8	30	0.67	0.676	0.683	0.673
1	19	0.687	0.695	0.697	0.691

From the result analysis it was determined that the power generated by the turbine has increased by as much as 33% after the injection process, with the best result obtained for a smaller diameter and a higher number of orifices. The injection system influence on the axial turbine performances decrees for higher dimension of the orifices, as can be determined from Figure 5.

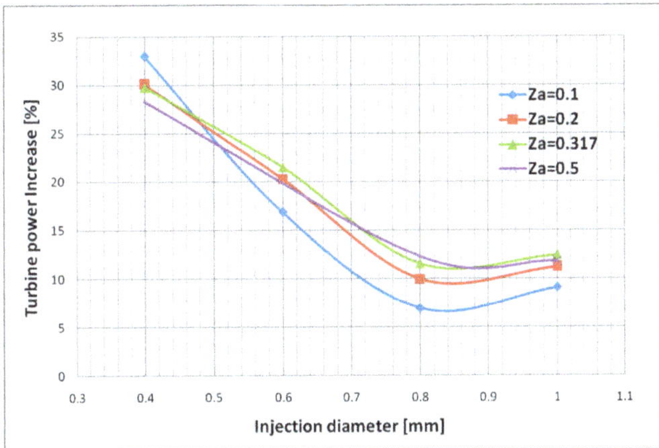

Figure 5. Turbine power increase after the injection process.

The injected fluid acts as a barrier which diverts the working fluid from the suction side of the vane near the injection section. This deviation in the near vicinity of the critical section leads to a smaller flow section for the working fluid which translates into a higher speed and a higher power generated by the rotor. After the interactions between the fluids, the injected fluid assumes the direction of the main flow and attaches to the suction side of the blade. As a result of the injection process, a low-pressure zone downstream of the injection section is created due to the main fluid deviation. This effect can be observed in Figure 6, which presents the total pressure distribution at the mean radius of the vane for the cases with and without injection. Without the injection process the flow section of the vanes is approximately equal to the geometrical section, a small low-pressure zone being created by the boundary layer on the suction side. After the injection process the low-pressure zone is greatly enhanced which determines a noticeable smaller flow section. As a result, the effective geometry of the flow channel is modified. This effect is obtained across the vane height where the system is active, as can be determined from Figure 7. As the boundary conditions were set with a fixed mass flow at the rotor outlet, the power enhancement is not a result of a mass flow increase but a result of different flow geometry.

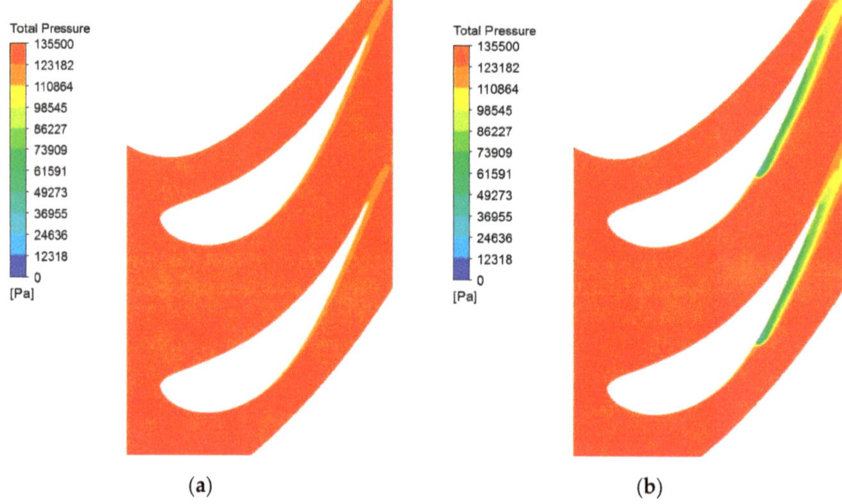

Figure 6. (**a**) Total pressure distribution at mean radius without injection; (**b**) Total pressure distribution at mean radius after the injection process.

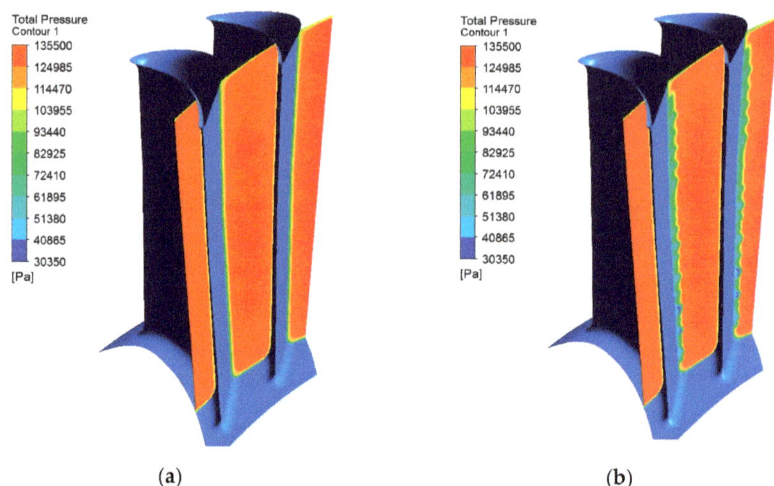

Figure 7. (**a**) Total pressure distribution without injection; (**b**) Total pressure distribution after the injection process.

3.2. Number of Orifices Influence

With the increase in injection diameter and the decrease in number of orifices (as previously explained) the distance between adjacent injection sections grows. This leads to lower effectiveness of the injection system as working fluid passes between the orifices.This effect is visible in Figure 8, for the 0.8 and 1 mm diameters, where the number of injection sections is not adequate for the vane height.

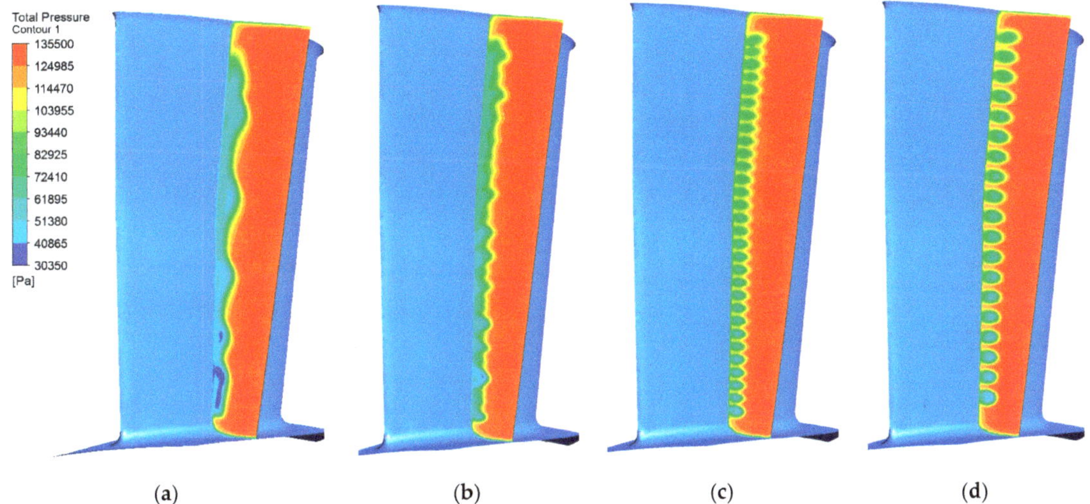

Figure 8. Total pressure distribution downstream of the injection section for: (**a**) $\varnothing_{inj} = 0.4$ mm; (**b**) $\varnothing_{inj} = 0.6$ mm; (**c**) $\varnothing_{inj} = 0.8$ mm; (**d**) $\varnothing_{inj} = 1$ mm.

In order to determine the influence of the number of orifices, the diameter and the inlet parameters of the injection fluid were maintained constant. The number of injection sections was increased from 50 to 70 for a diameter of 0.6 mm and from 30 to 45 for a diameter of 0.8 mm. The results showed, in both cases, a steady increase in the effectiveness of the injection system with the increase in number of injection sections. To better quantify the influence of the number of injection sections, relative to vane dimensions, a new dimensionless parameter, named coverage degree, has been introduced. The parameter, described by Equation (1), represents the percent of the vane height covered by injection sections.

$$\tau = \frac{n\varnothing_{inj}}{R_S - R_H} \tag{1}$$

where: τ—coverage degree, n—number of injection orifices, $\varnothing_{inj}$—orifice diameter, R_S—shroud radius, R_H—hub radius.

Applying the coverage degree to cases studied in this paper enables the comparison of cases with different dimensions and number of injection orifices. By plotting these results, Figure 9, a linear tendency can be observed. The increase effectiveness of the injection system at higher values of the coverage degree can be explained by a better isolation of the gap between two adjacent injection sections. This effect is visible in Figure 10 where a homogenization of the low-pressure zone across the vane height can be observed at the increase of the coverage degree. With the increase of this parameter from 53.6% to 75%, the influence of the injection system has grown from +20% to +30% in generated power.

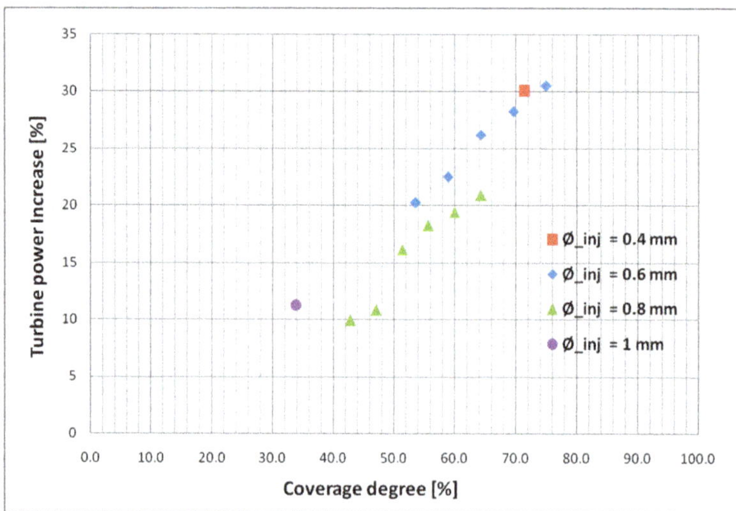

Figure 9. Turbine power increase for coverage degree variation.

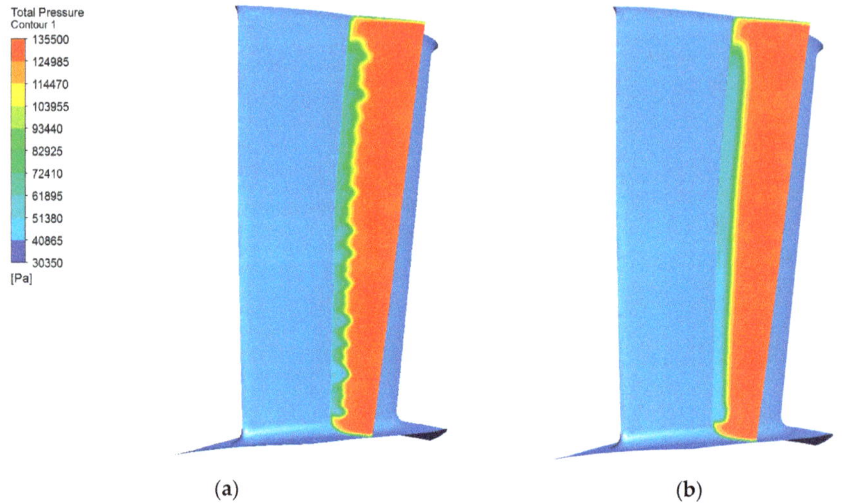

Figure 10. (**a**) Total pressure distribution for $\tau = 53.6\%$; (**b**) Total pressure distribution for $\tau = 75\%$.

3.3. Injection System Influence at Different Partial Regimes

The influence of the injection system on the turbine performances has been determined at different partial regimes by calculating the flow and the performances before and after injection for other regimes, regimes of lower respective higher power. The results previously presented have been determined for a partial regime of 89%, named "partial 1". Using the turbine map and working line, the inlet parameters for lower power regimes of approximately 84.5% and 80% and higher regimes of 91% and 92.5% have been determined. With these inlet parameters the flow and performances of the axial turbine were calculated using the same numerical model as the one previously discussed. The parameters for the partial regimes are presented in Table 5.

Table 5. Partial regimes parameters.

Name	Regime [%]	$\dot{M}$ [kg/s]	P^*_{in}[bar]	T^*_{in}[K]	Speed [rpm]	Power [KW]
Partial 4	80	5.3	1.93	753	18,000	119.2
Partial 2	84.5	6.1	2.11	810	19,100	253.1
Partial 1	89	7	2.34	868	20,000	466.6
Partial 3	91	7.4	2.42	898	20,650	587.3
Partial 5	93	7.6	2.50	925	21,000	679.3

The injection system used for this analysis consists of 55 injection orifices with a diameter of 0.6 mm resulting in a coverage degree of approximately 59%. The axial distance is 0.2, the injection temperature is similar to the turbine inlet temperature and the injection mass flow represents 2.5% of the working flow, the injection system configuration is presented in Figure 11. The injection fluid is considered to be extracted from the engine compressor and injected in the turbine vane, thus the rotor outlet mass flow (set in the numerical case with the outlet boundary conditions) is constant for the cases with and without injection.

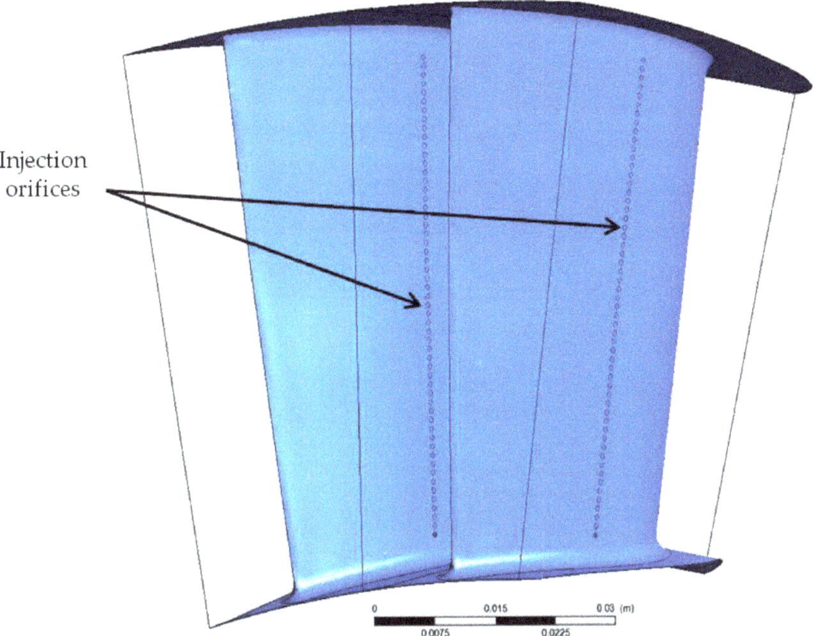

Figure 11. Selected injection configuration.

For all the partial regimes studied the generated power has increased by approximately 23% with slightly higher values at higher regimes. These increases lead to equivalent levels of power 1.5% to 2% higher.The results of the partial regimes analysis are presented in Table 6 and by plotting the results, Figure 12, it can be determined that the injection process leads to an increase of curve slope.

From this analysis it can be determined that the performance of the injection system is not limited to certain regimes but can be used for multiple power settings. The influence of the system is slightly smaller at lower power settings and increases with the engine regime.

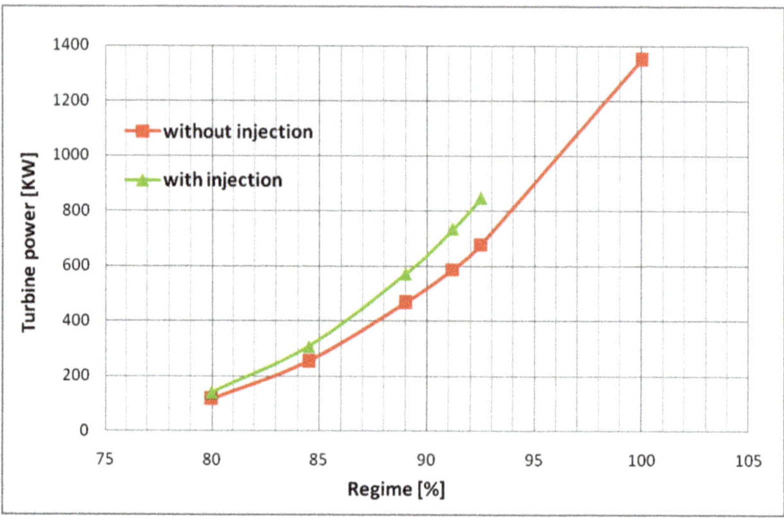

Figure 12. Turbine working line with and without injection.

Table 6. Partial regimes analysis results.

Name	Regime [%]	Power without Injection [KW]	Power after Injection [KW]	Power Increase [%]	Equivalent Regime [%]
Partial 4	80	119.2	146.0	22.5	81.5
Partial 2	84.5	253.1	306.8	21.2	86
Partial 1	89	466.6	571.0	22.4	91
Partial 3	91	587.3	733.6	24.9	93
Partial 5	92.5	679.3	847.4	24.8	94

The results of the present study are not easily comparable with other studies on performance enhancement at partial regimes as the injection system presented in this study focuses on increasing the power output of the turbine by modifying the flow geometry and not by reducing specific flow losses in the turbine by injection. The studies identified in the literature focus on loss reductions by increasing the flow momentum in certain areas of the blades using fluid injection. Thus, the injection jets are places at a low angle to reduce the resulting mixing loses. The main effect of these jets is the reattachment of the boundary layer on the suction side. The system presented in this paper focuses on changing the flow geometry of the vanes working channel to better adapt the turbine to inlet conditions for the respective partial regime. The injection angle is high, when compared with existing studies, in order to obtain an increased flow deviation in specific locations.

The placement of injection jets is also different from the studies discussed in this paper. For loss reduction, the studies showed that the injection jets should be positioned near the separation inception as for the performance enhancement, the injection orifices are placed close to the critical section. The placement of injection orifices upstream of the critical section ensures that the low-pressure bubble, formed as a result of injection, determines a decrease in minimal flow section leading to an acceleration of the main fluid.

The overall effect of the injection system can be determined in terms of fuel consumption. Thus, for the 89% partial regime presented ("partial 1") and an 22% power increase, by injection the flow consumption for the respective regime has decreased by approximately 11.5% and can reach 15% when calculated for higher partial regimes.

The results of this study should be of great interest especially for industries where long duration partial regimes operation are common. For marine propulsion the cruise gas turbine engines could operate for long periods of time (i.e., tens of hours) at these regimes, thus the 11.5% fuel savings represents an important reduction in cost of operation and an increase in ship operational range. For aviation gas turbines, the injection system can be used at partial regimes even that these regimes are reduced in term of time of operation. Furthermore, the use of the injection system could offset the impact of high ambient temperature on the engine performances.

Future research should concentrate on using multiple injection curves to simplify the manufacturing complexity of the system. By using curves to position the orifices it may be possible to obtain the same results with higher distances between the injection sections. The influence of the injection system on the engine cycle should also be studied, as by modifying the minimal section of the turbine other elements, such as the compressor, other turbine stages and burning chamber, could be affected. Another interesting topic is the interaction between the injection fluid and the cooling flow (if the turbine requires film cooling).

4. Conclusions

The purpose of this paper is to continue the research into the novel performance enhancement method by determining the influence of geometrical parameters as well as the influence of the injection system at different partial regimes. The specific fluid injection in the axial turbine flow was found to be a possible method to enhance the performances at partial regimes. The injection process leads to a deviation of the working fluid flow through the turbine vanes, which determines a decrease in the minimal flow section, thus an acceleration of the flow in the vanes and an increase power output.

From the orifice dimensions analyses it was determined that the influence of the injection system decreases at higher values of the orifices diameter for the same mass flow, as a result of higher spacing between injection sections which limits the flow deviation across the vane height. For the cases studied in this paper, the turbine power increase varied from approximately 30% for small diameter orifices (0.4 mm) to around 10% for larger orifices (0.8 and 1 mm).

From the number of injection orifices analyses it was determined that high numbers of injection sections lead to higher powers. The power increase was found to be as much as 33%, depending on the injection system configuration, with smaller dimensions and high number of injection section having the greatest influence. The number and dimensions of the injection sections must be analyzed together when designing the injection system, thus a new parameter, named the coverage degree, was introduced to incorporate both. The parameter was found to have a linear tendency when plotting against the power increase. A high value of this parameter leads to closer injection sections which limits the working fluid flow between the orifices, thus the system acts as a barrier which deviates the flow near the minimal section of the vanes leading to a lower flow section.

From a technological point of view the injection system is similar to film cooling systems. The orifice dimensions do not pose a considerable challenge but the high number of injection sections and the higher values of the coverage degree may be difficult to achieve, due to high number and close proximity of the injection orifices. This system is most suited to low-pressure turbines and free power turbines as these elements are usually not cooled and the necessary injection pressures are lower. Furthermore, the injection system can better adapt the generated power to the consumer needs with minimum influence on the gas generator. The injection system can be used for high-pressure turbines but further studies on interactions between injection flow and cooling flow, engine control and stability and available injection pressures are needed.

The injection system is not limited to a specific partial regime and can be applied to multiple power settings with similar performances in terms of percentage power increase. The power increase was determined to be around 22% achieving approximately 25% at higher regimes.

The results of the studies conducted in this paper suggest that the turbine mass flow is better used at partial regimes in an injection configuration (97.5% turbine inlet mass flow and 2.5% injection mass flow, values used in this study) than a no injection configuration (100% turbine inlet mass flow). Thus, additional power is extracted from the available energy at the turbine inlet by using a more efficient turbine geometry for the respective partial regime, geometry obtained as a result of fluid injection. The additional power extraction leads to an increase in overall engine efficiency which can improve the inherent low efficiencies of the gas turbines engines at partial regimes.

The work presented in this paper contributes to further characterize the novel performance enhancement method by specific fluid injection. In theory, the results can be used to determine the influence variation for different parameters in order to further develop the method and achieve an optimum configuration. In practice the optimum configuration might not be technically feasible due to manufacturing limitations, thus the final configuration must take into account these limitations.

Author Contributions: Conceptualization, R.E.N.; methodology, R.E.N.; formal analysis, R.E.N.; investigation, R.E.N.; resources, R.E.N.; data curation, R.E.N.; writing—original draft preparation, R.E.N.; writing—review and editing, D.E.C. and V.A.V.; supervision, D.E.C. All authors have read and agreed to the published version of the manuscript.

Funding: This research was funded by the Romanian Research Innovation and Digitalization Ministry through NUCLEU program under National Plan for Research Innovation and Development 2022–2027 grant number PN.23.12.01.01.

Data Availability Statement: Not applicable.

Conflicts of Interest: The authors declare no conflict of interest.

Nomenclature

$\varnothing_{inj}$	injection orifice diameter
Za	axial distance of the injection orifice
τ	coverage degree
n	number of injection orifices
R_s	shroud radius
R_H	hub radius
$\dot{M}$	working fluid mass flow
P_{in}^*	total pressure at the turbine inlet
T_{in}^*	total temperature at the turbine inlet

References

1. Clarence, E.L.B.; Alvin, T. Variable Stator Vanes. U.S. Patent No. 3237918, 1 March 1966.
2. Svetlana, K. Application and Improvement of Gas Turbine Blades Film Cooling. In Proceedings of the International Conference on Aerospace System Science and Engineering; Jing, Z., Ed.; Springer: Singapore, 2018; pp. 69–91.
3. Zhou, Z.; Li, H.; Wang, H.; Xie, G.; You, R. Film cooling of cylindrical holes on turbine blade suction side near leading edge. *Int. J. Heat Mass Transf.* **2019**, *141*, 669–679. [CrossRef]
4. Zhao, W.; Chi, Z.; Zang, S. Scaling criteria accuracy for turbine blade film cooling effectiveness at unmatched temperature ratio conditions. *Appl. Therm. Eng.* **2021**, *197*, 117363. [CrossRef]
5. Ravi, D.; Parammasivam, M. Enhancing film cooling effectiveness in a gas turbine end-wall with a passive semi cylindrical trench. *Therm. Sci.* **2019**, *23*, 2013–2023. [CrossRef]
6. Berkache, A.; Dizene, R. Numerical and experimental investigation of turbine blade film cooling. *Heat Mass Transf.* **2017**, *53*, 3443–3458. [CrossRef]
7. Continente, J.; Shehab, E.; Salonitis, K.; Tammineni, S.; Chinchapatnam, P. Cooling and Capability Analysis Methodology: Towards Development of a Cost Model for Turbine Blades Film Cooling Holes. In Proceedings of the 22nd ISPE Conference on Concurrent Engineering, Delft, The Netherlands, 20–23 July 2015; Volume: Advances in Transdisciplinary Engineering. pp. 386–395.
8. Moritz, N.; Kusterer, K.; Bohn, D.; Sugimoto, T.; Tanaka, R.; Taniguchi, T. Conjugate calculation of a film-cooled blade for improvement of the leading edge cooling configuration. *Propuls. Power Res.* **2013**, *2*, 1–9. [CrossRef]
9. Wang, Y. Study of Power Minimization of Co-Flow Jet Active Flow Control with Fixed Airfoil Geometry. Master's Thesis, University of Miami, Coral Gables, FL, USA, 2019.

10. Padilla, I.; Saavedra, J.; Paniagua, G.; Pinna, F. Analysis of the boundary layer stability to assess flow separation control capability in low-pressure turbines. In Proceedings of the 7th European Conference on Computational Fluid Dynamics, Glasgow, UK, 11–15 June 2018.

11. Wang, Y.; Saavedra, J.; Ferrer, E.; Paniagua, G.; Valero, E. Optimisation in Active Flow Separation Control in Low-Pressure Turbines. In Proceedings of the 8th European Conference for Aeronautics and Aerospace Science, Madrid, Spain, 1–4 July 2019.

12. Saavedra, J.; Paniagua, G. Transient Performance of Separated Flows: Characterization and Active Flow Control. In Proceedings of the ASME Turbo Expo 2018: Turbomachinery Technical Conference and Exposition, Volume 6: Ceramics; Controls, Diagnostics, and Instrumentation; Education; Manufacturing Materials and Metallurgy, Oslo, Norway, 11–15 June 2018.

13. Alimi, A.; Wünsch, O. Numerical Investigation of Steady and Harmonic Vortex Generator Jets Flow Separation Control. *Fluids* **2018**, *3*, 94. [CrossRef]

14. Lyko, C.; Dähnert, J.; Peitsch, D. Forcing of Separation Bubbles by Main Flow Unsteadiness or Pulsed Vortex Generating Jets: A Comparison. In Proceedings of the ASME Turbo Expo 2013: Turbine Technical Conference and Exposition, Volume 6C: Turbomachinery, San Antonio, TX, USA, 2–7 June 2013.

15. Galbraith, M.C.; Kasliwal, A.; Ghia, K.; Ghia, U. Numerical Simulation of a Low Pressure Turbine Blade Employing Active Flow Control. In Proceedings of the ASME 2006 2nd Joint U.S.-European Fluids Engineering Summer Meeting Collocated With the 14th International Conference on Nuclear Engineering, Volume 2, Miami, FL, USA, 17–20 July 2006; pp. 797–804.

16. Carbajosa, C.; Martinez-Cava, A.; Valero, E.; Paniagua, G. Efficiency of Pulsating Base Bleeding to Control Trailing Edge Flow Configurations. *Appl. Sci.* **2022**, *12*, 6760. [CrossRef]

17. Benton, S.I.; Bons, J.P.; Sondergaard, R. Secondary Flow Loss Reduction Through Blowing for a High-Lift Front-Loaded Low Pressure Turbine Cascade. *ASME J. Turbomach.* **2013**, *135*, 021020. [CrossRef]

18. Benton, S.I.; Bernardini, C.; Bons, J.P.; Sondergaard, R. Parametric Optimization of Unsteady Endwall Blowing on a Highly Loaded LPT. In Proceedings of the ASME Turbo Expo 2013: Turbine Technical Conference and Exposition, Volume 6A: Turbomachinery, San Antonio, TX, USA, 3–7 June 2013.

19. Rohr, C.; Yang, Z. A numerical study of active flow control for low pressure turbine blades. In Proceedings of the 4th International Symposium on Jet Propulsion and Power Engineering, Xi'an, China, 10–12 September 2012.

20. Nicoara, R.; Crunteanu, D.; Vilag, V. Numerical Study Of Axial Turbines Performance Enhancement Technique By Specific Fluid Injection. *U.P.B. Sci. Bull. Ser. D* **2022**, *83*, 69–80.

21. Ghaffari, P. Passive Tip-Injection in Shrouded Low Pressure Turbines. Doctor of Engineering Sciences Thesis, Technische Universität Wien, Vienna, Austria, 2016.

22. Nisizima, S.; Yoshizawa, A. Turbulent channel and Couette flows using an anisotropic k-epsilon model. *AIAA J.* **1987**, *25*, 414–420. [CrossRef]

23. Mishra, P.; Aharwal, K.R. A review on selection of turbulence model for CFD analysis of air flow within a cold storage. *IOP Conf. Ser. Mater. Sci. Eng.* **2018**, *402*, 012145. [CrossRef]

24. Tomovska, E. Turbulent Flow Modelling And Prediction Of Blade Profile Losses In Radial Outflow Turbine. Master Thesis, Lahti University of Technology, Lappeenranta, Finland, 2021.

Article

Numerical Simulations on the Performance of Two-Dimensional Serpentine Nozzle: Effect of Cone Mixer Angle and Aft-Deck

Hamada Mohmed Abdelmotalib Ahmed [1,*], Byung-Guk Ahn [2] and Jeekeun Lee [3]

1   Department of Mechanical Power and Energy Engineering, Faculty of Engineering, Minia University, Minia 61511, Egypt
2   School of Advanced Materials Engineering, College of Engineering, Jeonbuk National University, Jeonju-si 54896, Republic of Korea
3   Division of Mechanical System Engineering, Jeonbuk National University, Jeonju-si 54896, Republic of Korea
*   Correspondence: en_hamada83@yahoo.com; Tel.: +20-1002892756

Abstract: The current study addresses the effect of different designs of the exhaust mixer and aft-deck on the performance of a two-dimensional convergent nozzle represented by the internal and external flows and heat transfer process. The effect of different exhaust mixer cone angles of 10°, 15°, and 20°, and different aft-deck lengths of 140 mm, 280 mm, and 420 mm on the nozzle performance was investigated. To address the effect of an aft-deck, the flow behavior of a nozzle with an aft-deck was compared to that of a nozzle without an aft-deck. Then, the effect of different aft-deck lengths and different aft-decks with rectangular and trapezoid shapes was investigated. The results demonstrated that increasing the mixer cone angle resulted in decreasing the high-temperature core flow and increasing the low-temperature bypass flow. Increasing the mixer cone angle resulted in reducing the velocity inside the nozzle and at the exhausted jet, which can reduce the noise generated by the engine. Furthermore, increasing the mixer cone angle decreased the internal temperature of the nozzle and, along with the exhausted jet, decreased the infrared radiation. The results also illustrated that the presence of the aft-deck resulted in decreasing the pressure, temperature, and velocity inside the nozzle. The aft-deck also decreased the length and size of the potential core. The aft-deck length had no clear effect on the internal flow. However, increasing the aft-deck length resulted in a decrease in the exhaust gas temperature, which can decrease the infrared radiation. On another hand, using trapezoid and triangle aft-deck can enhance the performance of the nozzle by decreasing the velocity and temperature inside the nozzle and at the exhausted jet.

Keywords: aft-deck length; external jet; infrared radiation; two-dimensional nozzle

Citation: Ahmed, H.M.A.; Ahn, B.-G.; Lee, J. Numerical Simulations on the Performance of Two-Dimensional Serpentine Nozzle: Effect of Cone Mixer Angle and Aft-Deck. *Aerospace* **2023**, *10*, 76. https://doi.org/10.3390/aerospace 10010076

Academic Editors: Dan Zhao, Chenzhen Ji and Hexia Huang

Received: 5 December 2022
Revised: 24 December 2022
Accepted: 9 January 2023
Published: 11 January 2023

1. Introduction

An area of military application interest is stealth technology that needs to decrease infrared radiation and jet noise. The infrared radiation is mainly emitted by the walls of high-temperature including nozzle walls, the exit of the turbine, and the booster sleeve, in addition to hot gases generated from the exhaust system. About 90% of infrared radiation is provided by the exhaust nozzle and turbine section [1]. Therefore, the management and suppression of the infrared radiation of the exhaust system are very important to enhance the aircraft's stealth performance [2,3]. There are various techniques that can be used to minimize infrared radiation. The most vital methods that are used in almost all military aircraft include using a jacket in cold water around the exhaust walls and using a material with low emissivity for coating the nozzle's inner surface. However, using a jacket of cold-water results in extra weight problems on the aircraft, and using coating technology has a limitation on the capability of material heat resistance [4]. Moreover, a nozzle with a curved shape and rectangular or ellipse shape can be used to reduce both infrared radiation and jet noise. Serpentine nozzles are widely employed on different types of aircraft to increase survivability for their performance to suppress the infrared radiation signature. The curved shape of the serpentine nozzle can reduce the turbine exit's high temperature. The serpentine nozzle can be used to decrease infrared radiation by 70%. Consequently, a serpentine nozzle has been used in a series of UAVs and stealth bombers to suppress the IR radiation resulting from an engine exhaust system [1]. The two-dimensional rectangular exit of the nozzle can intensify the mixing process between the hot jet and atmospheric air, which can reduce the size and length of the high-temperature potential core. The performance of the serpentine nozzle can be further improved by adding an aft-deck at the nozzle exit [5].

In the open literature, there are few studies [1–23] that investigate the flow characteristics of the serpentine nozzle. The double serpentine nozzle flow characteristics were numerically investigated by Xiao et al. [1]. The influence of the length ratio that defined the first S length to the second S length on the nozzle performance was studied. The Flow characteristics, such as total and static pressures, wall shear stress, and streamlines of the Mach number, were investigated. Compared to the axisymmetric nozzle, a double serpentine nozzle has different static pressure distribution at the lower walls and upper walls. The influence of the ratio of length to the diameter, aspect, and offset ratios on the infrared radiation and flow characteristics of the serpentine nozzle was studied by Yong et al. [2]. The study indicated that both thrust coefficient and total pressure recovery were first improved by increasing the offset ratio, and both parameters rapidly decreased. The friction and viscous losses resulting from stream vortices dominated the aerodynamic performance of the nozzle. The infrared radiation can be decreased by a higher 50% double S-shaped nozzle compared to the circular nozzle. Increasing the aspect and offset ratios can suppress the plume radiation. In another study by Xiao et al. [3], a numerical study was performed to study the effect of specific design parameters at the first serpentine exit on the flow field. Among the six different turbulent models used in this study, a shear-stress transport (SST) k-ω model achieved more accurate results and was used to simulate a serpentine nozzle flow. Increasing the width ratio W1/D increased the friction losses because of increasing the wetted perimeter, while small values of the width ratio resulted in increasing secondary flow losses. High flow velocity was induced in the first duct for a small area of the first serpentine duct. High local losses resulted from the steep offset distance of the first serpentine nozzle. These results demonstrated that the first serpentine duct width was recommended to range from one to three, and its area should be as large as possible, while the offset distance should be small. The serpentine nozzle friction losses were high because of the steep passage slope and rapid turning. Wen et al. [4] investigated the influence of engine swirl on the infrared radiation of the serpentine nozzle. The study results indicated the ability of engine swirl to suppress the infrared signature. The hot streak can be produced by the engine swirl, which can affect the infrared signature on the walls. The total infrared radiation could be reduced by 31.21% in the vertical plane and by 13.84% in the horizontal plane, compared to the engine without a swirl. An experimental study was conducted by Rajkumar et al. [5] to study the influence of pressure ratio on the flow characteristics in a single serpentine nozzle. The flow behavior was analyzed based on wall static pressure measurements and exhausted jet flow visualization. The pressure coefficient variation was symmetric, along with various pressure ratios. The flow of the core increased because the geometric transition resulted in transferring to sonic and supersonic flow. The symmetry of flow is reduced at the extension section exit with the constant area. The presence of an aft-deck can prevent barrel shock formation. In the study by Hamada et al. [6], the effect of different types of annular mixers on the performance of a two-dimensional nozzle was investigated. An annular mixer with different lengths of 140 and 280 mm, different diameters of 320 and 340 mm, and different shapes were used. The results indicated that decreasing the mixer diameter increased the cold bypass flow and improved the mixing between the hot core flow and cold bypass flow, which resulted in decreasing the temperature inside the nozzle and at the external jet. The mixer with a cone shape decreased the temperature inside the nozzle and increased it at the external jet. The results demonstrated that the best nozzle performance could be obtained using a smaller cylindrical mixer with a shorter length and a smaller diameter. Sun et al. [7] studied the effect of a serpentine nozzle on the bypass ratio of the turbofan engine exhaust and compared it to that of an axisymmetric nozzle. The results indicated that the bedded duct at the serpentine nozzle generated a pressure gradient in the streamline's directions that resulted in a change in the flow through the bypass and core. Compared to the axisymmetric nozzle, the exhaust bypass ratio of the serpentine nozzle was higher by 8.1%. The impact of the shield ratio on the infrared radiation of the serpentine nozzle was studied by Wen et al. [8]. The infrared radiation of the serpentine and circular nozzles was numerically studied at various shield ratios. The results showed that, compared to the circular nozzle, the serpentine nozzle could reduce infrared radiation by 28.9. The complete nozzle shielding had no apparent advantages on the shielding ratio and infrared radiation, but it had a small influence on the serpentine nozzle outlet gases. The area ratio of double and single serpentine nozzles should be lower than 0.35 and 0.15, respectively, to achieve an effective reduction in the infrared radiation. The influence of aspect ratio on the flow characteristics of the double serpentine nozzle was studied by Sun et al. [9]. The aerodynamic performance based on internal flow and the external jet was numerically investigated for six different models of a nozzle with different aspect ratios of 3, 5, 7, 9, and 11. The variation in the lateral width leads to different pressure gradients and vortex distribution in the lateral direction. The length of the potential core was reduced by increasing the aspect ratio. The results demonstrated that an aspect ratio of five could achieve the best aerodynamic characteristics. The double serpentine nozzle flow characteristics for different inlet configurations were studied by Sun et al. [10]. The influence of strut setting angles and inlet swirl angles on the performance and flow field was also studied. The results indicated that the vortices found at the corner did not affect the tail con, struts, and bypass. There were differences in high-vorticity regimes of the core region due to the presence of a tail cone. Increasing strut setting angles and an inlet swirl angle resulted in reducing the high static temperature region. As the angles of the strut setting and inlet swirl increased, the serpentine nozzle decreased.

Most studies, including the above studies, focus on investigating the performance of the serpentine nozzle without considering the exhaust system, which exists in real turbofan engines. Furthermore, the utilization of the aft-deck can improve the nozzle performance by decreasing the external jet length and shielding the high-temperature sections, which can reduce IR radiation by 90% [16]. Therefore, this study investigates numerically the serpentine nozzle performance and considers the engine exhaust system and aft-deck. This study is an extension of a previous study [6] to achieve more understanding of the internal flow and exhausted jet characteristics of the nozzle. The effect of different mixer cone angles of 10°, 15°, and 20° on the serpentine nozzle operation was studied. Then, the effect of an aft-deck on the flow characteristics and heat transfer process was studied. First, the nozzle performance of a nozzle with an aft-deck was compared to a nozzle without an aft-deck. Second, the influence of different aft-deck lengths of 140 mm, 280 mm, and 420 mm on the nozzle performance was studied. The simulation study was conducted using Commercial STAR CCM+ software via a three-dimensional model of the serpentine nozzle.

2. The Exhaust System Geometry

The exhaust system of an aircraft represents the main source of IR radiation in a band of 3–5 μm due to the large quantity of heat generated by the engine turbine, which increases the temperature of turbine entry above 2000 K [4]. Therefore, infrared radiation treatment requires more attention during the engine exhaust system design process. The exhaust system of the aircraft has two sections: the exhaust mixer and the nozzle section. The mixer part has double ducts called a bypass and core, as well as a mixer and tail cone. The exiting hot gases flow through the core, while low-temperature fluids flow through the bypass. The nozzle, as illustrated in Figure 1, has a curved shape with a circular inlet and an internal diameter of 470 mm, which is converted to a rectangular exit with a major and minor axis of

728 × 146 mm and hydraulic diameter (D_h) of 266 mm. The transformation from a circular inlet shape into a rectangular exit provides a serpentine shape so that the serpentine nozzle can shield the high-temperature parts. The serpentine nozzle was attached to a conical mixer with a length of 140 mm and diameter (D_2) of 320 mm, and different cone angles (θ) of 10°, 15°, and 20°. The mixer is coupled with a central cone with a length of 280 mm and a diameter (D_3) of 135 mm. The nozzle exit is attached to an aft-deck with a width of 728 mm with different lengths of 140 mm, 280 mm, and 420 mm, and different shapes.

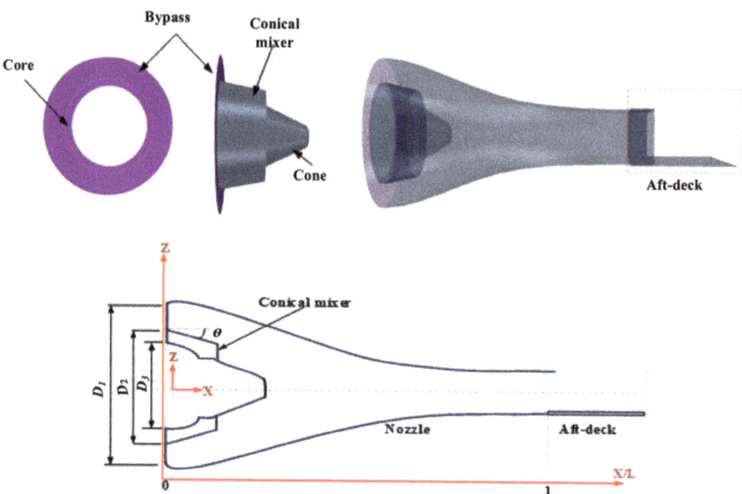

Figure 1. The CAD model and schematic of the serpentine nozzle with conical mixer under study.

3. Numerical Method

3.1. Numerical Model Setup

The three-dimensional model of the serpentine nozzle used in the simulation study is shown in Figure 2. A block with dimensions of 5 × 5 × 20 m was set as the computational flow domain to simulate the flow from the nozzle and the exhaust jet. The grid system shown in Figure 2 was used in the simulations, and the base size of the flow domain was 10 mm; for the nozzle wall, cone, and annular mixer, the mesh size was reduced to 0.5 mm. Two prism layers were used as the boundary cells with a total length of 0.5 mm. The bypass and core were considered pressure inlet boundary conditions with a uniform total temperature, pressure, and turbulence intensity. The nozzle exit was treated as a pressure outlet and subjected to atmospheric pressure and temperature. The walls of the nozzle were considered adiabatic walls, where the insulating material is typically used, and no-slip boundary conditions were considered. The SST (k-ω) turbulence model was used to model the flow inside the nozzle. For wall treatment, the all-y+ wall treatment was selected and provided the best results compared to other wall treatments provided by STAR CCM+ software. The all-y+ wall treatment is a hybrid treatment that uses a mixed wall function that emulates the high y+ wall treatment for coarse mesh and the low-y+ wall treatment for fine mesh. Consequently, the all-y+ wall treatment is relevant for a wide range of near-wall grid conditions. Table 1 summarizes the pressure and temperature values at different flow regions. The simulations were carried out using STAR CCM+ software, in which the equations were discretized with a finite volume method for each control volume. For spatial discretization, a second-order upwind scheme was used. The conservation equations of continuity, momentum, and energy were solved by using a coupled implicit flow solver in which the conservation equations are solved simultaneously as a vector of equations. The velocity field is obtained from the momentum equation, and the density is estimated from the equation of state. The flow inside the nozzle was treated as the ideal gas generated by mixing the high-temperature gases from the core with low-temperature gases from the bypass at high pressure, which results in different complicated features inside the nozzle. In this study, the flow-through core and bypass regions were considered air with the properties given in Table 2.

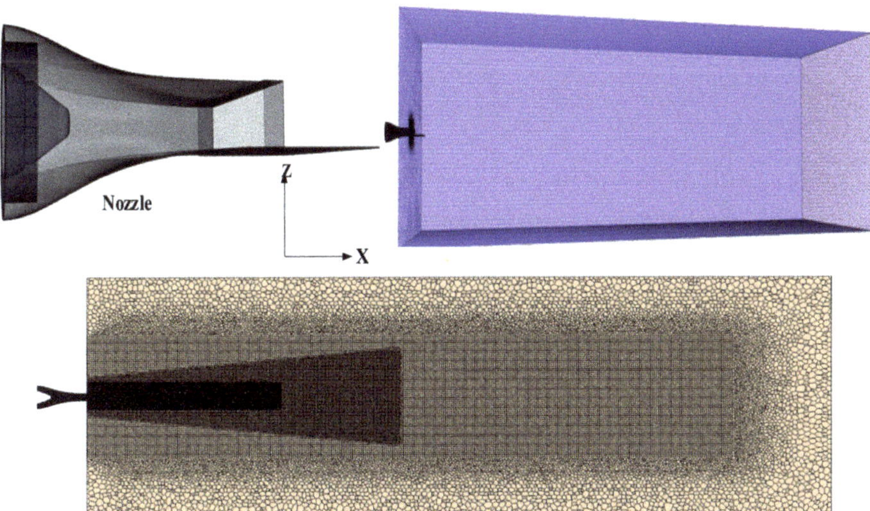

Figure 2. Grid system used in the simulations [6].

Table 1. Boundary conditions at different regions used in the simulations.

Region	Type	Pressure (Bar)	Temperature (K)	Turbulent Intensity
Core	Pressure inlet	0.45	800	0.01
Bypass	Pressure inlet	0.48	330	0.01
Outlet	Pressure outlet	1.013	288	0.01

Table 2. Properties of air used in the calculations.

Property	Value	Unit
Dynamic viscosity	1.85508×10^{-5}	Pa·s
Molecular weight	28.9664	kg/kmol
Specific heat	1003.62	J/kg·K
Thermal conductivity	0.0260305	W/m·K
Prandtl Number	0.9	-

The flow characteristics inside the nozzle were determined at different locations in the axial direction (X), which were non-dimensionalized with the length of the nozzle (L). The axial locations of 0, 0.12, 0.26, 0.5, 0.78, and 1 were selected to evaluate the different flow characteristics. The axial location $X/L = 0$ refers to the nozzle inlet, the axial location $X/L = 0.12$ refers to the annular mixer end, the axial location $X/L = 0.26$ refers to the cone outlet, and the axial location $X/L = 1$ refers to the nozzle exit [6]. The variation of the nozzle frontal area along its length is shown in Figure 3. The frontal area is defined as the area projected along the fluid flow path to a plane perpendicular to the direction of motion. As illustrated by the figure, the frontal area gradually increased until it reached the axial location of 0.26 and then gradually reduced until the axial location of 0.78 before the nozzle area was almost the nozzle outlet.

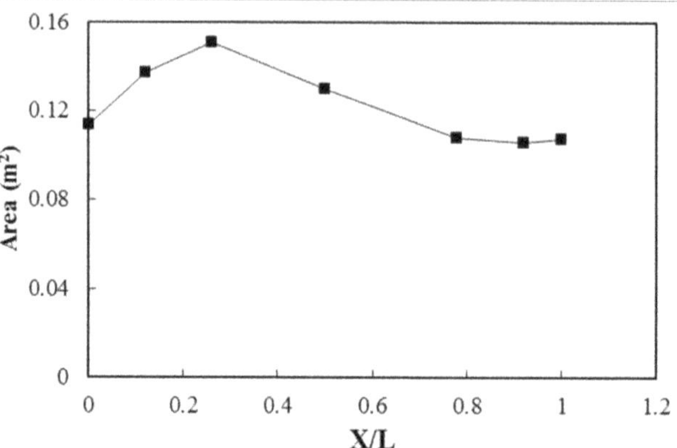

Figure 3. Variation of the frontal area along the nozzle length.

For the grid independence study, three grids were used in the simulations with a total number of cells of one million and 200,150 cells, five million and 420,193 cells, and twelve million and 360,253 cells for the course, medium, and fine grids, respectively. Figure 4 indicates the flow characteristics inside the nozzle at different sections along the nozzle length for the different grids. As shown in the figure, both medium and fine grids have almost the same results indicating that at a certain cell size, the results are not affected by the cell size. To save the solution time, the medium grid is used in the simulations.

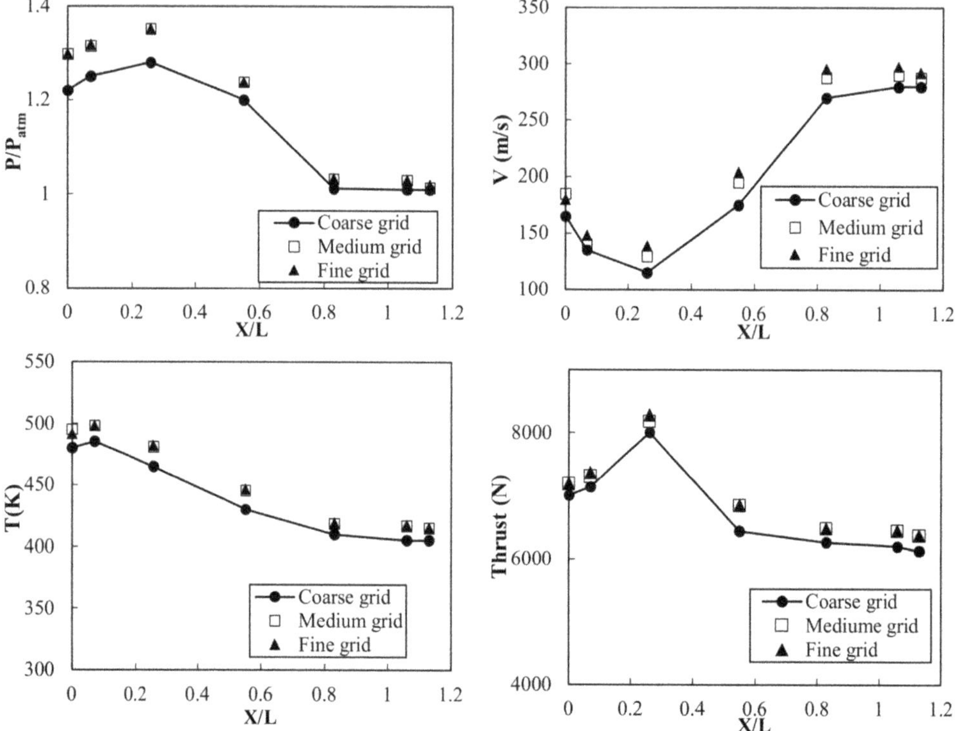

Figure 4. Flow characteristics inside the nozzle for different grids.

3.2. Model Validation

The simulation results were validated using the experimental results. The experimental procedure is mentioned in detail in the previous study [6]. The nozzle used in the experiments, as shown in Figure 5, was manufactured from a polylactic acid material using a technology of 3D printing and modeled using CATIA V5 software. The nozzle had an around inlet and rectangular exit attached with a rectangular aft-deck. To measure the wall static pressure, the nozzle thickness of 10 mm was used to attach the required pressure taps. The upper and lower wall pressures were measured using ten pressure taps connected to a microprocessor micro-manometer (FCO510) and LAB VIEW data acquisition software. The experiments were carried out at a temperature of 295 K and at an inlet air pressure of 0.25 bar. The experiments were repeated ten times, and the averaged values were considered. Figure 6 shows the profile of static pressure for the lower and upper walls of the nozzle. The figure indicates a good agreement between experimental and simulation results with a deviation of 5%. The method by Kline and Mcclintock [24] was selected for the uncertainty analysis.

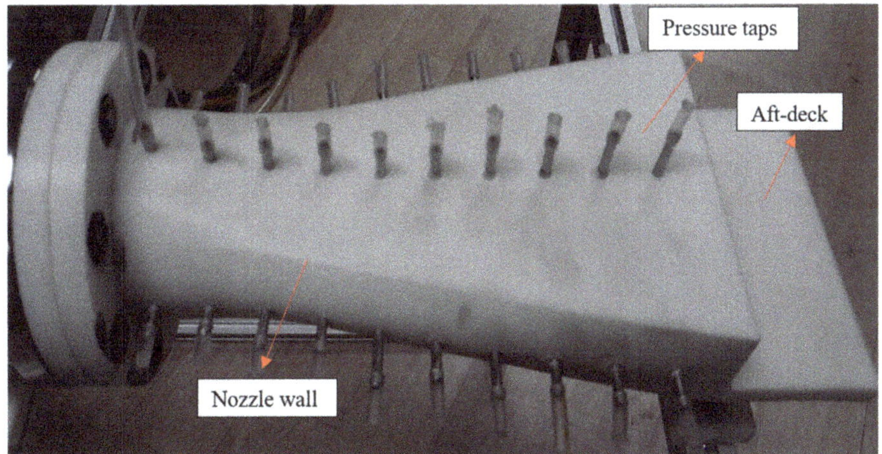

Figure 5. Photo of serpentine nozzle used in the experiments [6].

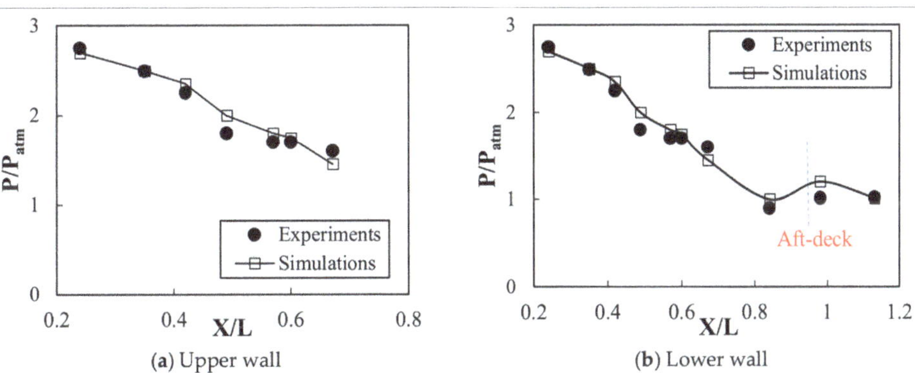

Figure 6. Axial distribution of the pressure at the upper and lower nozzle walls for experiments and simulations [6].

4. Results and Discussion

4.1. Effect of the Mixer Cone Angle

The conical mixer has two flow areas: the core inlet, where the hot gases enter, and the bypass flow inlet, where cold air enters. The flow through the bypass region depends on the flow through the core region and the static pressure distribution at the conical mixer end. The ratio between the core and bypass flow rates is an important parameter that affects the operation of the nozzle. The values of the core and bypass flow rates at different mixer cone angles are given in Table 3. Despite all used models being subjected to the same inlet conditions, both the core and bypass flow rates varied with the variation in the mixer cone angle. Increasing the mixer cone angle resulted in reducing core flow rates and increasing bypass flow rates. The variation in the core and bypass flow rates is due to the variation in the static pressure at the end of the conical mixer resulting from the core flow with a high velocity. Decreasing the cone angle resulted in decreasing the area available for core flow, and increasing the area available for bypass flow resulted in reducing core flow rates and increasing bypass flow rates. The effect of the variation of core and bypass flow rates with mixer angles will be explained based on internal flow and exhausted jet characteristics in the following sections.

Table 3. The flow rates at core and bypass regions for different mixer cone angles.

Cone Angle	Core Flowrate (kg/s)	Bypass Flowrate (kg/s)	Total Flowrate (kg/s)
10°	3.08	27	30.08
15°	2.16	28.35	30.46
20°	1.27	29.17	30.44

The variation in static pressure along the upper and lower nozzle walls at different mixer cone angles is shown in Figure 7. As shown by the figure, the pressure gradually decreased with the decreasing nozzle area. The pressure decreased with the increasing mixer cone angle because increasing the cone angle led to a greater expansion of flow inside the nozzle, which resulted in decreasing the pressure near the nozzle walls. The pressure along the upper and lower walls was almost identical due to the symmetry of the nozzle. Increasing the mixer cone angle reduced the pressure along the upper and lower walls.

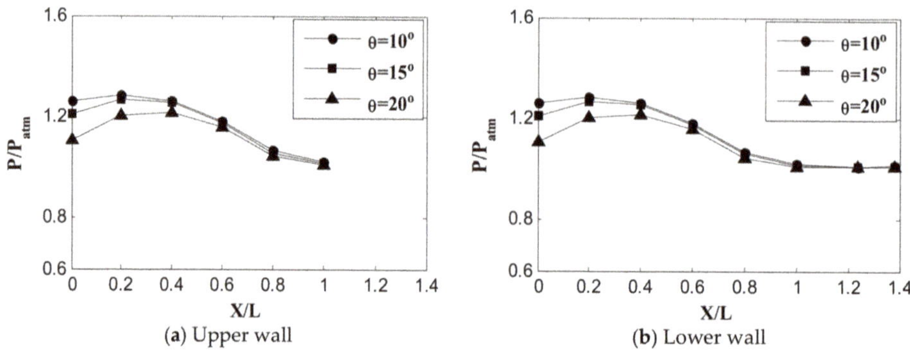

Figure 7. Pressure distribution on the upper and lower nozzle walls for different mixer cone angles.

Figure 8 indicates the velocity contour and distribution at different locations inside the nozzle at different mixer cone angles. The figure indicated that the velocity decreased from the nozzle inlet until the end of the cone at X/L of 0.26, then the velocity gradually increased up to the nozzle exit as the area of the nozzle decreased. The velocity contours show the exiting of the core region at the center of the nozzle with the highest velocity. The size and velocity of the core region decreased with the increasing mixer cone angle. Figure 8d shows the average velocity distribution at different locations along the nozzle length. The figure demonstrates that increasing the mixer cone angle decreased the velocity inside the nozzle. This may be due to the flow separation point moving upstream. Additionally, as indicated by the velocity contour, the size of the core region with the highest velocity decreased with the increasing mixer cone angle due to the reduction in the core flow rate, as given in Table 3. This indicated that the velocity inside the nozzle strongly depended on the core flow rates and hence the mixer cone angle despite the constant area of the nozzle.

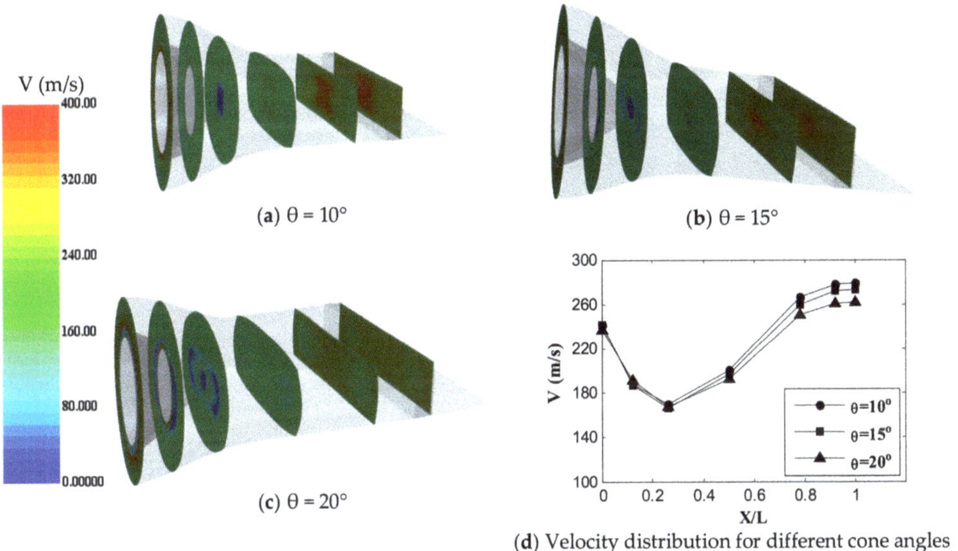

Figure 8. Velocity contour and distribution at different locations along the nozzle length for different mixer cone angles.

The turbulent kinetic energy (TKE) contours and distribution are shown in Figure 9. TKE is the most important parameter in the fluid flow because it is a measure of turbulence intensity and is directly related to momentum and heat. TKE can be used as an indication of mixing intensity. TKE was the highest at the end of the mixer cone, where the mixing between the lower-temperature bypass flow and higher-temperature core flow occurs. Figure 9d indicates the average values of TKE at different locations along the nozzle length. The figure indicates that the TKE was the highest at the largest mixer cone angle, demonstrating that increasing the mixer cone angle increased the turbulence motion inside the nozzle and hence improved the mixing process between the hot and cold streams, which resulted in decreasing the temperature.

Figure 10 indicates the distribution of thrust force at different locations for different cone angles. The thrust force proportion with the nozzle area increased gradually up to the mixer end as the nozzle area increased, then gradually decreased and became nearly constant near the nozzle exit, where the area was almost the same. The thrust force accelerates the flow inside the nozzle by decreasing the pressure and increasing the velocity so that the velocity gradually increases after the mixer ends up at the nozzle exit, as shown in Figure 8. The figure shows that the thrust decreased with the increasing cone angle. As the cone angle increased, the area available for flow decreased, resulting in a reduction in the flow rate and hence decreasing the thrust.

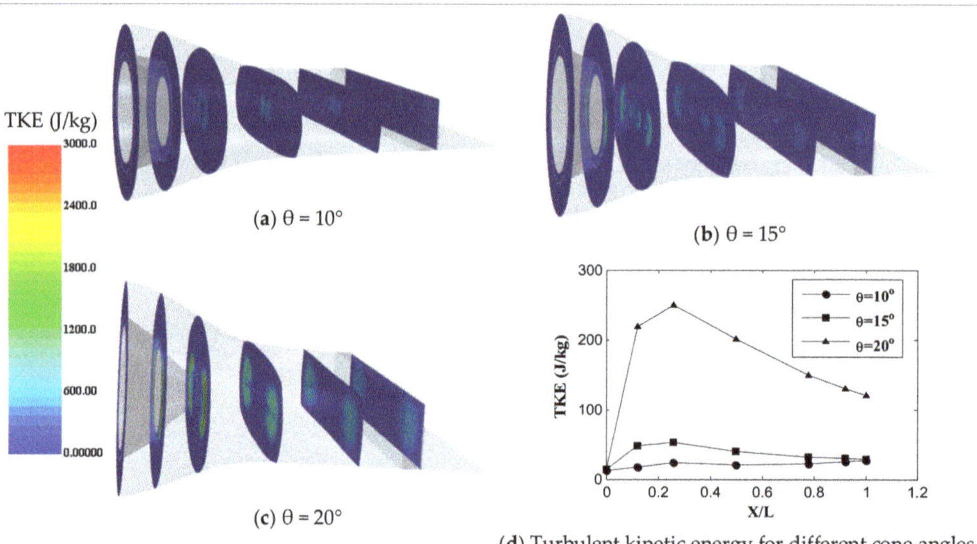

Figure 9. Turbulent kinetic energy contour and distribution at different locations along the nozzle length for different mixer cone angles.

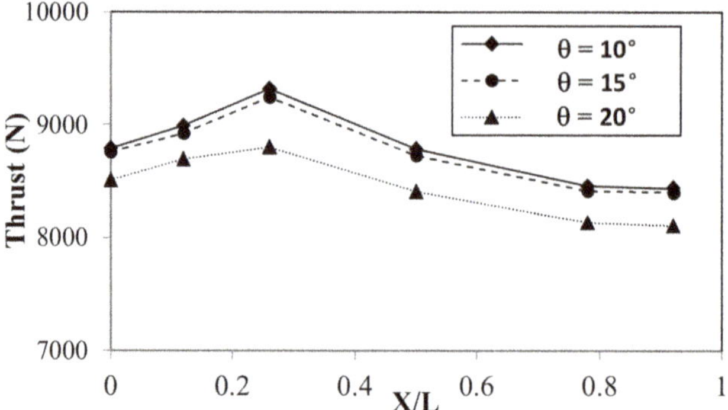

Figure 10. Thrust force distribution at different locations along the nozzle length for different mixer cone angles.

The distribution and temperature contours at different sections along the length of the nozzle for different cone angles of the mixer are illustrated in Figure 11. As shown by the temperature contours, there was a core flow region with high temperature surrounded and separated by a bypass flow with low temperature. As indicated by the figure, the temperature of the core region at the center of the nozzle decreased with the increasing mixer cone angle due to the reduction in the core flow rate. The average temperature of the selected sections, as shown in Figure 11d, decreased gradually from the end of the tail cone up to the nozzle exit because the nozzle geometry shields the temperature. The reduction in the temperature results from increasing the mixer cone angle, which is due to the reduction in a high-temperature core flow and increasing bypass cold flow, as indicated in Table 3. Additionally, improving the mixing process due to increasing the turbulence motion, as illustrated by TKE in Figure 9, was another reason.

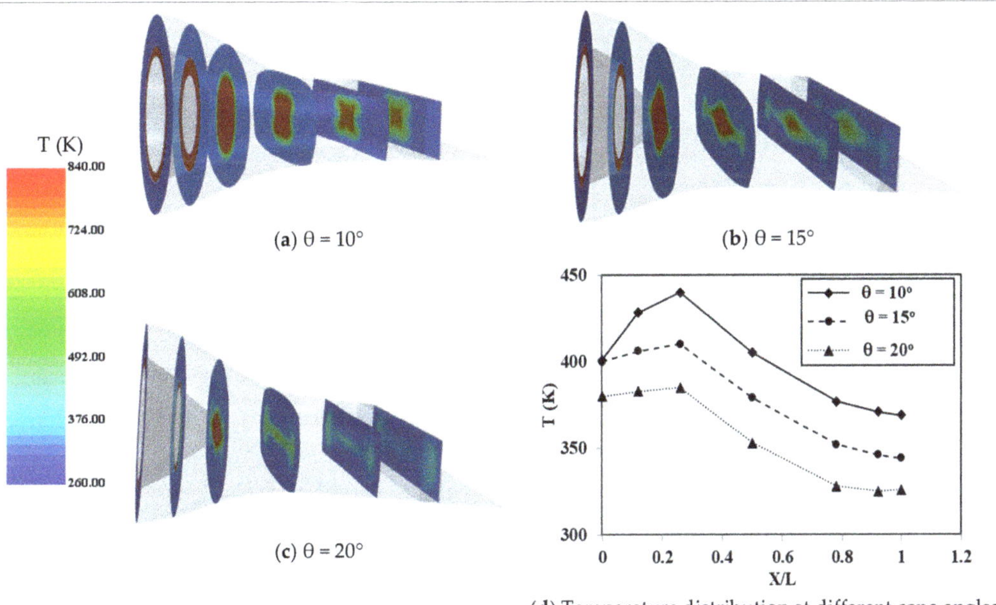

(**d**) Temperature distribution at different cone angles.

Figure 11. Temperature contour and distribution at different locations along the nozzle length for different mixer cone angles.

The exhausted jet, which is indicated by velocity contour and radial velocity distribution at different axial locations along the jet length at different mixer cone angles, is shown in Figure 12. As indicated by the velocity contour, the external jet is characterized by the highest velocity region at its centerline, which is called a potential core region. The potential core is known as a region in the jet where the centerline velocity is uniform and equal to the exit centerline velocity, as indicated by the velocity distribution. The jet's initial regime is known as the region of the establishment of flow and extends from the exit of the nozzle until the potential core apex. The other region is known as the region of stabled flow, which starts at the potential core end and is characterized by the gradual reduction in the centerline velocity. The velocity distribution illustrated in Figure 13 shows that the velocity increased at the nozzle exit, where the mixing between the ambient air and exhausting hot gases was weak. With increasing the distance from the nozzle exit, the surrounding air was entrained into the exhausted jet and resulting in an increase in the jet size in the axial direction and a reduction in the centerline velocity. The figure also indicated that the higher mixer cone angle had the lowest velocity at the nozzle exit at X/L = 1. This demonstrated that increasing the mixer cone angle could decrease the noise generated by the engine nozzle by decreasing the peak velocity at the nozzle exit.

Figure 14 indicates constrained streamlines at the vertical plane along with the computational domain for different mixer cone angles. As indicated by the figure, large-scale vortices are formed on the jet boundary so that vortices can entrain the surrounding air into the exhausted jet. The entrained ambient air with low momentum tries to gain momentum from the high-velocity exhausted jet that prevails in the centerline of the exhausted jet. This results in the transfer of entertain air into the jet centerline, leading to a reduction in the velocity and temperature at this region.

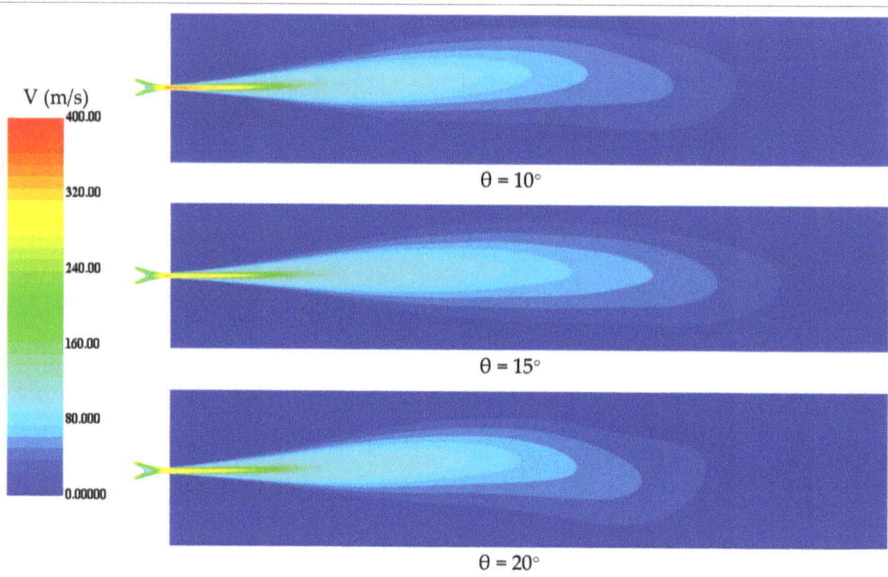

Figure 12. Velocity contour at the vertical plan downstream nozzle exit for different mixer cone angles.

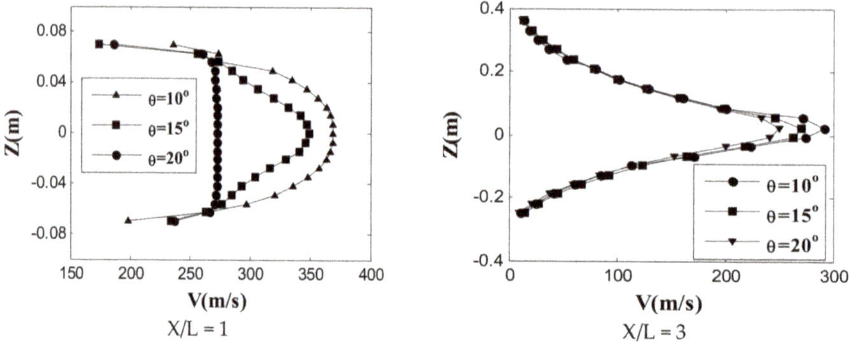

Figure 13. Velocity distribution at the centerline of the exhausted jet for different mixer cone angles.

The M distribution at the centerline of the exhausted jet for different mixer cone angles is shown in Figure 15. As shown by the figure, M increased from the nozzle exit up to $X/L = 1.3$, indicating the extension of the core region in the axial direction. After the core region, there was a steep reduction in M along the centerline of the jet. At the core region, the mixing was the lowest and resulted in increasing M, and with increasing the distance from the nozzle exit, the mixing process increased until reaching the jet centerline, which resulted in increasing the spread of the jet and vanishing the core region leading to a further decrease in M. The M distribution at the jet centerline decreased with the increasing mixer cone angle, indicating the improvement of the mixing process between the exhausted jet and entrained ambient air. The velocity contour and M distribution show that the length of the core region is almost the same for different mixer angles.

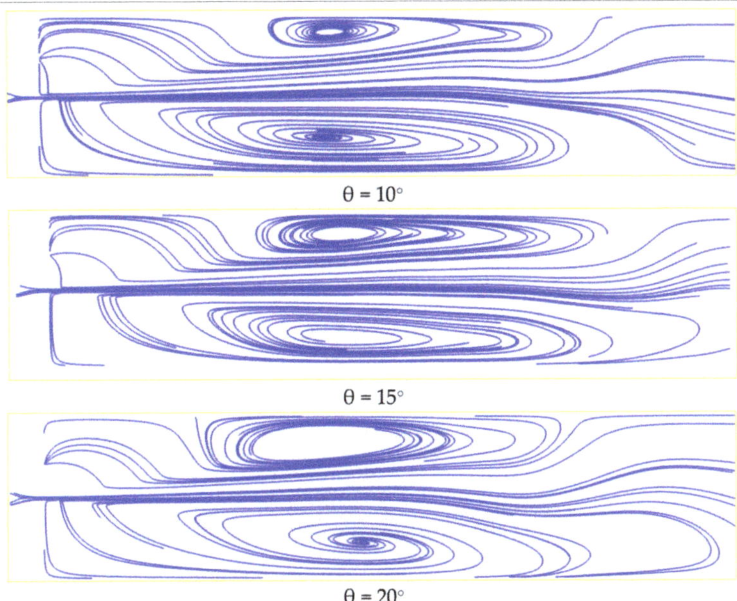

Figure 14. Constrain streamlines at the vertical plane for different mixer cone angles.

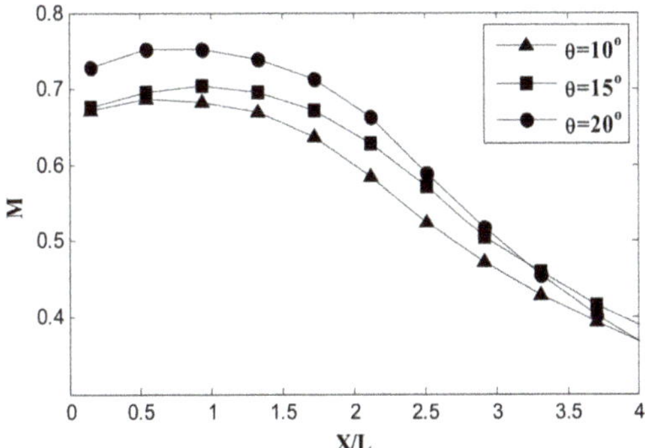

Figure 15. M distribution at the centerline of the exhausted jet for different mixer cone angles.

Figure 16 illustrates the temperature distribution along the centerline of the exhausted jet at different angles. The temperature increased near the nozzle exit along the potential core region, and then the temperature gradually reduced due to the spreading of the jet, thus improving the mixing between the exhausted gases and surrounding air. The figure indicated that the temperature distribution along the jet centerline decreased with the increasing mixer cone angle. As shown by the temperature contour in Figure 11, the size and temperature at the nozzle exit decreased with the increasing mixer cone angle, which resulted in the reduction of the centerline temperature. Moreover, as discussed in the previous sections, increasing the mixer cone angle improved the mixing between the hot gases and cold surrounding air, which resulted in reducing the temperature. The reduction in the temperature downstream of the nozzle exit reduced the infrared radiation.

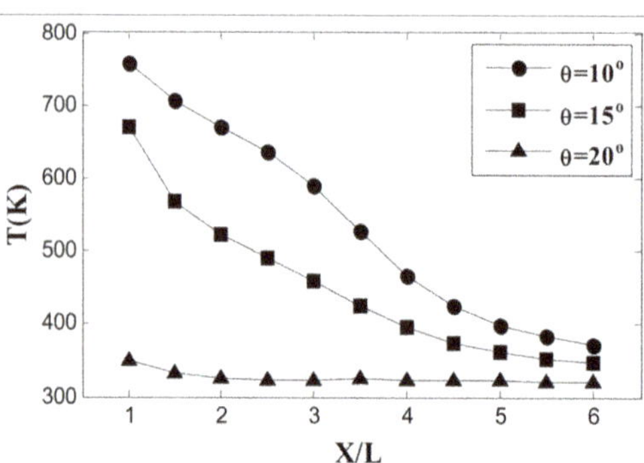

Figure 16. Temperature distribution at the centerline of the exhausted jet for different mixer cone angles.

4.2. Effect of Aft-Deck

In this section, the effect of the aft-deck on the nozzle performance was discussed. The nozzle performance with aft-deck was compared to that of the nozzle without aft-deck. Then, the effect of the aft-deck length and shape was studied. The internal flow at different sections of the serpentine nozzle without and with an aft-deck is shown in Figure 17. Figure 17a indicates the pressure distribution at different sections along the length of the nozzle. The pressure (P) was nondimensionalized with atmospheric pressure (Patm). For the two nozzles, the pressure increased gradually until X/L = 0.26 as the area increased, and then it decreased downstream until the distance of X/L = 0.78 as the area gradually decreased, and the flow expanded. The two nozzles have an almost constant area region from distance X/L = 0.78; in this region, the pressure was almost constant up to the nozzle exit. The nozzle without the aft-deck had an almost higher-pressure distribution than that of the nozzle with the aft-deck. As mentioned by Nageswara and Kushari [15], the presence of the aft-deck resulted in an additional nozzle length leading to an increase in the wetted surface and boundary layer thickness in this section, so the flow accelerated in the central part of the nozzle cross-sectional area to satisfy the mass conservation that resulted in a greater expansion of flow inside the nozzle and hence decreased the pressure inside the nozzle.

Figure 17b indicates the velocity magnitude at different locations inside the nozzle without and with an aft-deck. As shown by the figures, the velocity decreased gradually from the nozzle inlet until the distance of X/L = 0.26 and then increased gradually up to the nozzle exit exhibiting the opposite trend of pressure. As it is known, the area variation of the nozzle area along its length resulted in converting pressure energy into kinetic energy. The figure illustrated that the velocity of the nozzle with aft-deck was lower than that of the nozzle without the aft-deck. The lower velocity of the nozzle with an aft-deck may be attributed to the viscous interaction between the aft- deck and the jet, which decreased the velocity.

The distribution of Ma at various locations along the nozzle length without and with an aft-deck is shown in Figure 17c. As shown by the figure, the Ma of the nozzle with an aft-deck is lower than that of the nozzle without an aft-deck. The aft-deck at the exit of the nozzle restricted the entrainment of the surrounding fluid at the aft-deck side. This resulted in reducing the momentum gains from the fluid with a high momentum surrounding the jet axis, leading to a reduction in the spreading of the jet.

Figure 17d illustrates the distribution of the temperature at different locations along the length of the nozzle without and with an aft-deck. The temperature decreased gradually after a distance of X/L = 0.26, where the annular mixer intensified the mixing process between cold air from bypass and core hot gases, and resulted in a reduction in temperature. The temperature inside the nozzle with the aft-deck was almost lower than that of the nozzle without an aft-deck. As mentioned in the above sections, the presence of the aft-deck increases the flow expansion inside the nozzle and enhances the mixing process between the bypass cold air and core hot gases leading to a decrease in the temperature inside the nozzle.

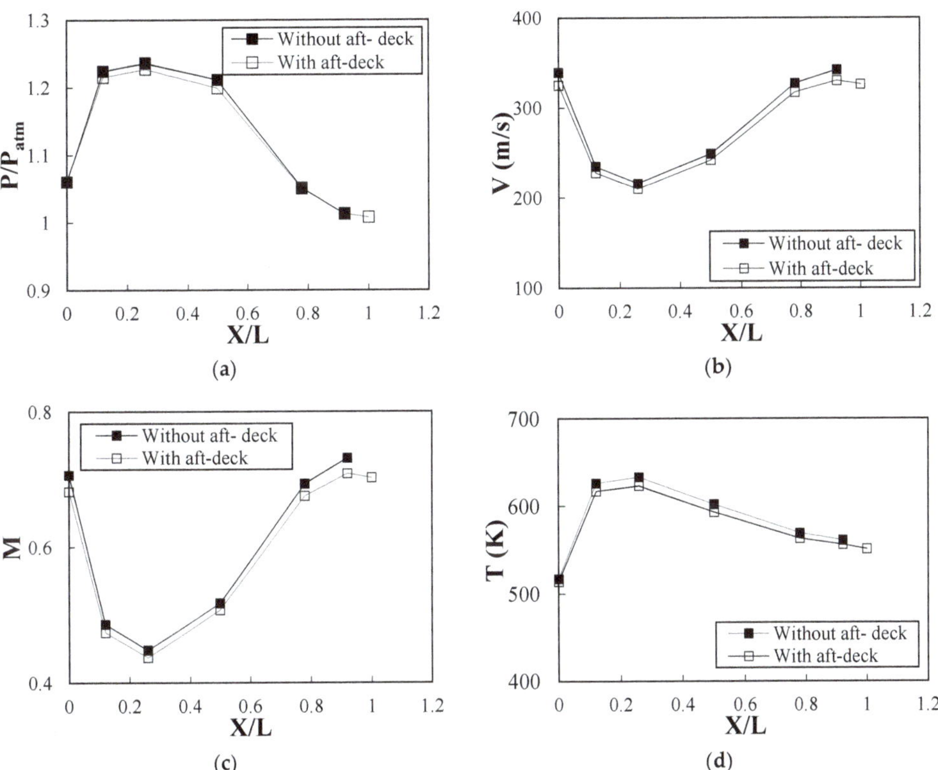

Figure 17. Internal flow characteristics at different sections along the length of the nozzle without and with an aft-deck. (**a**) Pressure distribution, (**b**) Velocity distribution, (**c**) M distribution, (**d**) Temperature distribution.

The velocity contour at the horizontal plane and radial velocity profiles along the jet length of the nozzle with and without an aft-deck is shown in Figures 18 and 19, respectively. As shown in the figure, at the exit of the nozzle (X/L = 1), the jet velocity was uniform and at the highest level, with a nearly top-hat profile because the mixing process between the jet and ambient air was weak. A thin shear layer was created at the exit of the nozzle because of the large velocity difference between the jet and ambient air, and this shear layer continuously grew downstream. With the increase in the shear layer, the entrainment of the ambient air into the jet increased, increasing the mixing process. As a consequence, the jet spreads outward in a radial direction, and the velocity of the jet gradually decreases downstream. The region at the nozzle exit along the central portion of the jet with an almost uniform mean velocity refers to the potential core region (represented as a dotted box). The potential core eventually disappears due to the spreading of the jet and the shear layer. Beyond the potential core region, the velocity profiles convert to a bell-shaped profile at an axial distance (X/L) from three to five. The aft-deck increased the length of the nozzle; hence, the wetted perimeter area at the exit of the nozzle increased, which resulted in a reduction of the size and length of the potential core of the jet, which is considered the highest temperature region, as illustrated by the velocity contour shown in Figure 18 that can reduce the infrared radiation. On the other hand, the reduction in the velocity at the nozzle exit can reduce the noise produced by the nozzle.

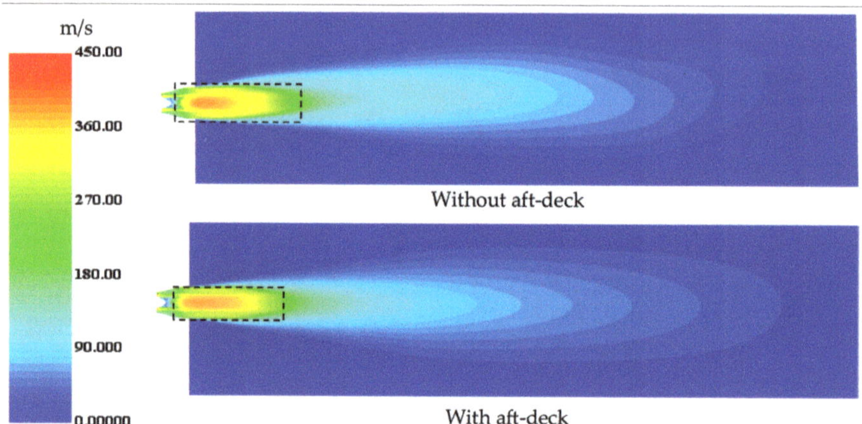

Figure 18. Velocity magnitude contours for nozzle without and with an aft-deck.

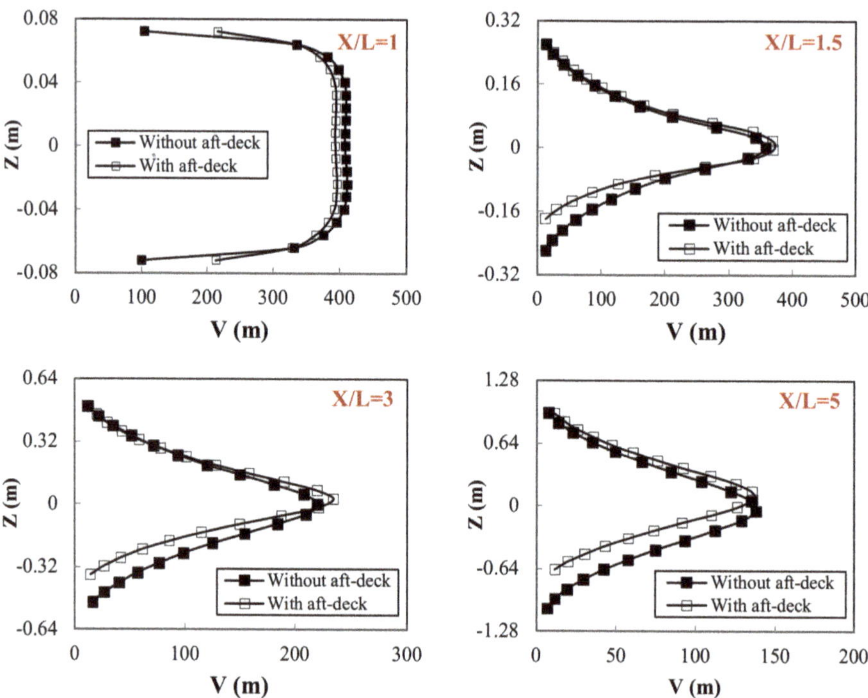

Figure 19. Radial profiles of velocity magnitude at different locations along jet length.

Figure 20 illustrates the temperature profiles at different locations (X/L) of 1, 1.5, 3, and 5 along the jet length for the nozzle with and without an aft-deck. As indicated by the figure, the temperature was the highest at the nozzle exit where the highest temperature core region is exited. With the increasing distance from the nozzle exit, the temperature decreased as the mixing between the exhausted gases and ambient air increased, and the exhausted jet expanded more. Furthermore, the figure demonstrated that the presence of the aft-deck decreased the temperature of the exhausted gas. As discussed in the above sections, the aft-deck reduced the length and size of the potential core with the highest temperature. Additionally, the aft-deck increased the wetted area at the nozzle exit, which improved the mixing process between exhausted gases and ambient air and increased the expansion of the exhausted gases, which resulted in a decreasing temperature.

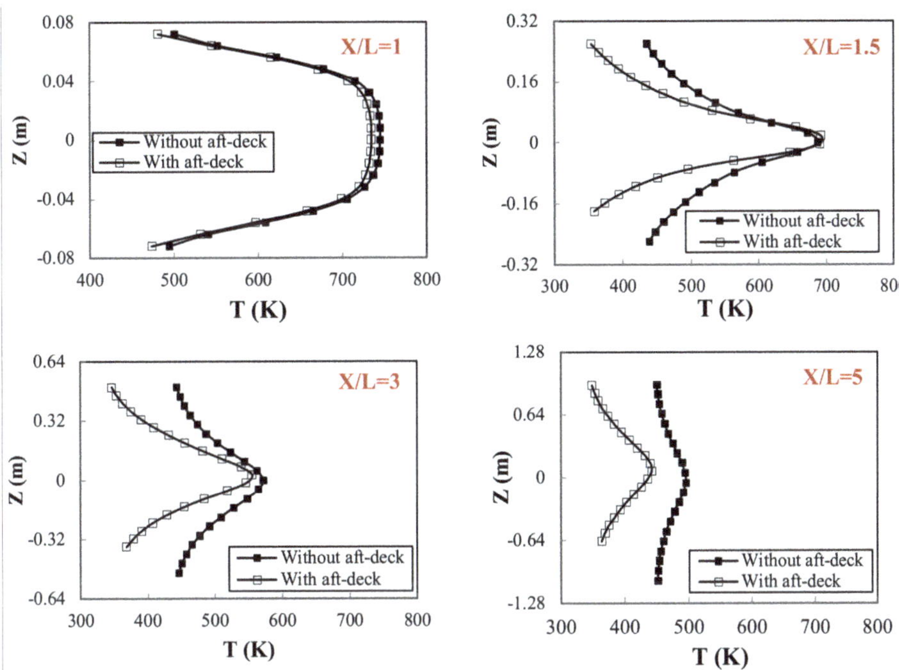

Figure 20. Temperature radial profiles at different locations along jet length. Effect of the length and shape of the aft-deck.

This section investigated the effect of the aft-deck length and shape on the flow characteristics. Three aft-deck lengths of 140, 280, and 420 mm and three different models of aft-deck with trapezoid, triangle, and rectangular shapes, as shown in Figure 21, were investigated. The length of all shapes was kept at the largest length of 420 mm.

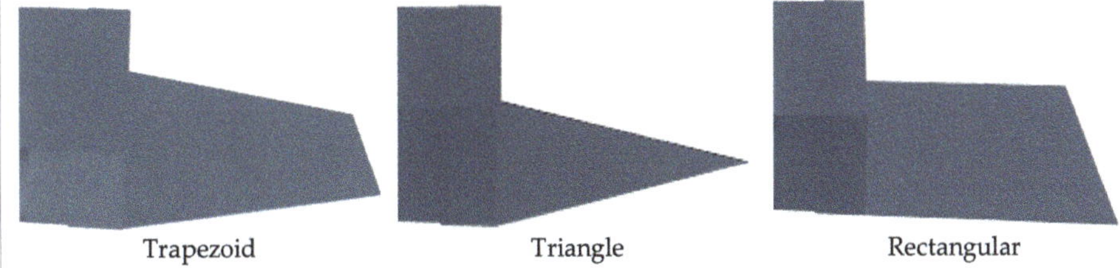

Figure 21. Different shapes of aft-deck used in the study.

Figure 22 indicates the average velocity magnitude and temperature distributions at different locations along the nozzle length for different lengths and shapes of the aft-deck. As shown in Figure 22a,b, the aft-deck had no effect on the velocity and temperature distributions inside the nozzle. Although increasing the aft-deck length increased the length of the nozzle and wetted area at the exit, this had no effect on the internal flow. On another hand, as indicated by Figure 22c,d, the velocity and temperature inside the nozzles with a trapezoid and triangle aft-deck was the same, and it was high for the nozzle with a rectangular aft- deck. This demonstrated that the aft-deck with a rectangular shape can increase the expansion of flow inside the nozzle, which resulted in decreasing the pressure and, hence, increasing the velocity. The increase in temperature for the nozzle with a rectangular aft-deck can be explained using the temperature contours indicated by Figure 23a–c. As illustrated by the figure, the size and temperature of the hottest core region for the nozzle with a trapezoid and triangle aft-deck were smaller than that of the nozzle with a rectangular aft-deck. Therefore, the average temperature at each location, indicated by the temperature contour for the triangle and trapezoid aft-deck shapes, was lower than that of a rectangular aft-deck. It can be concluded that the aft-deck with trapezoid and triangle shapes had the same effect on the internal flow characteristics. The trapezoid and triangle aft-deck shapes could increase the flow expansion inside the nozzle, which can reduce the velocity, and the noise generated by the nozzle, as well as decrease the wall temperature by decreasing the high-temperature core flow, which results in decreased infrared radiation. The figure demonstrated that the length of the aft-deck had no effect on the internal flow characteristics, but its shape could affect it.

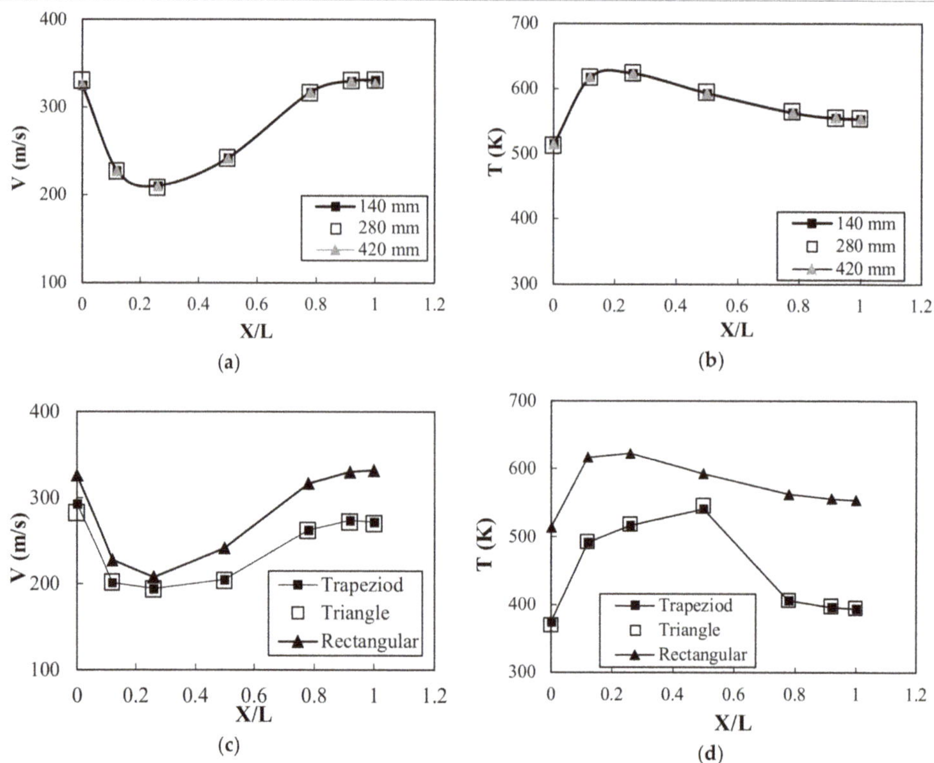

Figure 22. The velocity and temperature profiles inside the nozzle at different axial positions: (**a**,**b**) For different aft-deck lengths, (**c**,**d**) For different aft-deck shapes.

Figures 24 and 25 illustrate the velocity contour in the horizontal plane and the radial distribution of velocity at different aft-deck lengths. The two figures demonstrate that an aft-deck had no obvious effect on the jet velocity. However, the velocity contours shown in Figure 24 indicate that an aft-deck can increase the spreading of the jet. This can be explained by using a new parameter that can characterize the external jet called the half jet width. The velocity distribution shown in Figure 22 indicated that the maximum mean velocity was located at the jet centerline. The distance from the jet centerline to that at which the velocity is equal to a half maximum velocity is known as the width of the half-jet (*b*), and this characterizes the jet growth. The jet half-width (*b*) normalized by the nozzle exit hydraulic diameter (D_h) at different axial lengths for different aft-deck lengths is shown in Figure 26. As indicated by the figure, increasing the axial distance from the exit of the nozzle increased the jet's half-width because the jet gradually spread. Figures 24 and 26 demonstrate that increasing the aft-deck length increased the half-jet width, which resulted in the increased spreading of the jet.

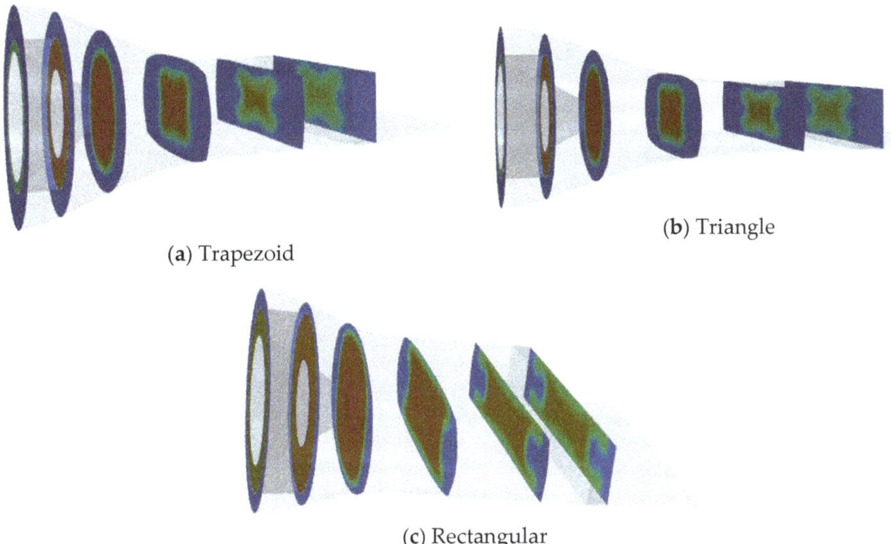

(**a**) Trapezoid

(**b**) Triangle

(**c**) Rectangular

Figure 23. Temperature contour inside the nozzle at different axial locations for different aft-deck shapes.

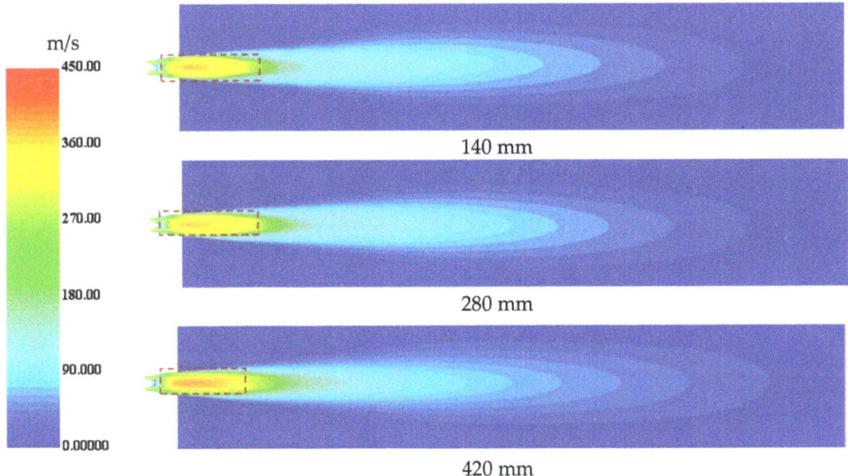

Figure 24. Velocity contour of the nozzle at different aft-deck lengths.

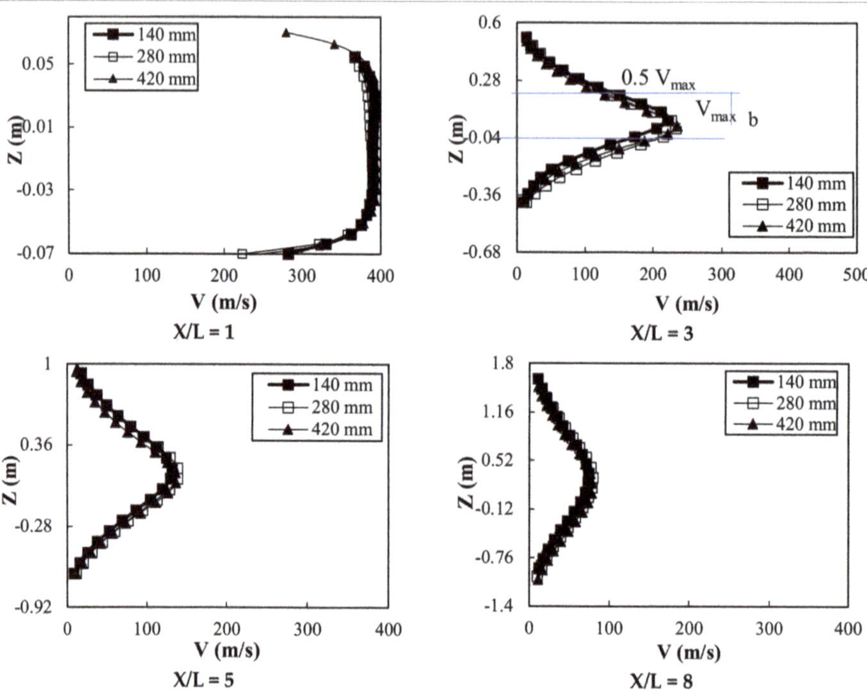

Figure 25. Radial velocity distribution at different axial lengths along the jet centerline for different aft-deck lengths.

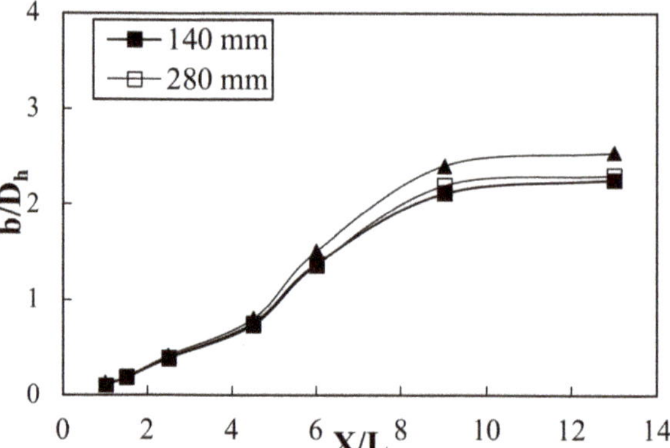

Figure 26. Jet half-length distributions at different axial lengths for different aft-deck lengths.

The temperature profiles at different locations at the jet centerline and at different aft-deck lengths are shown in Figure 27. The figure indicates that increasing the length of the aft-deck resulted in decreasing the temperature from the nozzle exit by up to 10 m. Increasing the aft-deck length leads to an increase in the wetted area and, hence, increases the expansion of the exhausted jet, as shown in Figure 24. This resulted in intensive mixing between the jet and ambient air, which resulted in reducing the temperature of the exhausted jet. Furthermore, with increasing the distance from the nozzle exit, the temperature reduced as the jet expanded more, as shown in velocity contours.

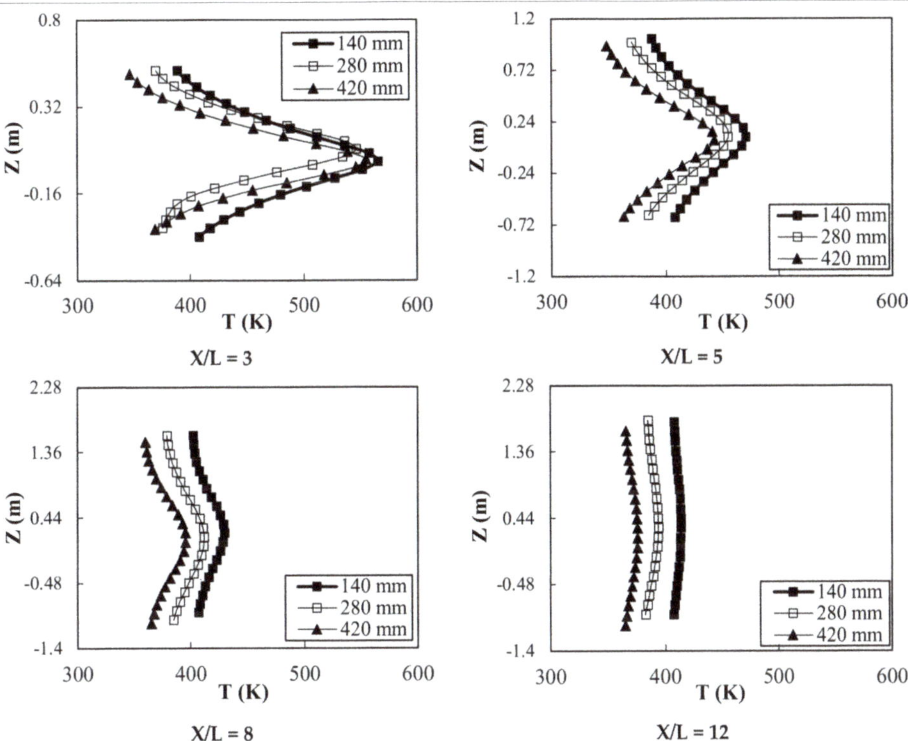

Figure 27. Temperature distribution at different locations along the jet centerline for different aft-deck lengths.

Figure 28a,b illustrate the velocity contour and velocity distribution, respectively, downstream at the nozzle exit. Figure 28a shows the exhausted jet indicated by velocity contours at a vertical plane for different aft-deck shapes. The velocity contour indicates that the potential core of the exhausted jet for the nozzle with a rectangular aft-deck was almost higher than that of the nozzle with a trapezoid or triangle aft-deck. The velocity distribution along the centerline of the exhausted jet is shown in Figure 28b. The velocity at the nozzle exit was the highest and gradually decreased with the increasing distance from the nozzle exit. Due to the momentum exchange, the surrounding air was entrained with the exhausted jet. As a consequence, the flow of the jet gradually increased in the axial direction, which led to the expansion of the jet volume and, hence, a decrease in the velocity. As shown by the velocity distribution, the velocity at the nozzle exit of the nozzle with a trapezoid and a triangle aft-deck was lower than that of the nozzle with a rectangular aft-deck shape; this also demonstrated that using both the triangle and trapezoid aft-deck could reduce the noise generated by the engine and improve the performance of the nozzle.

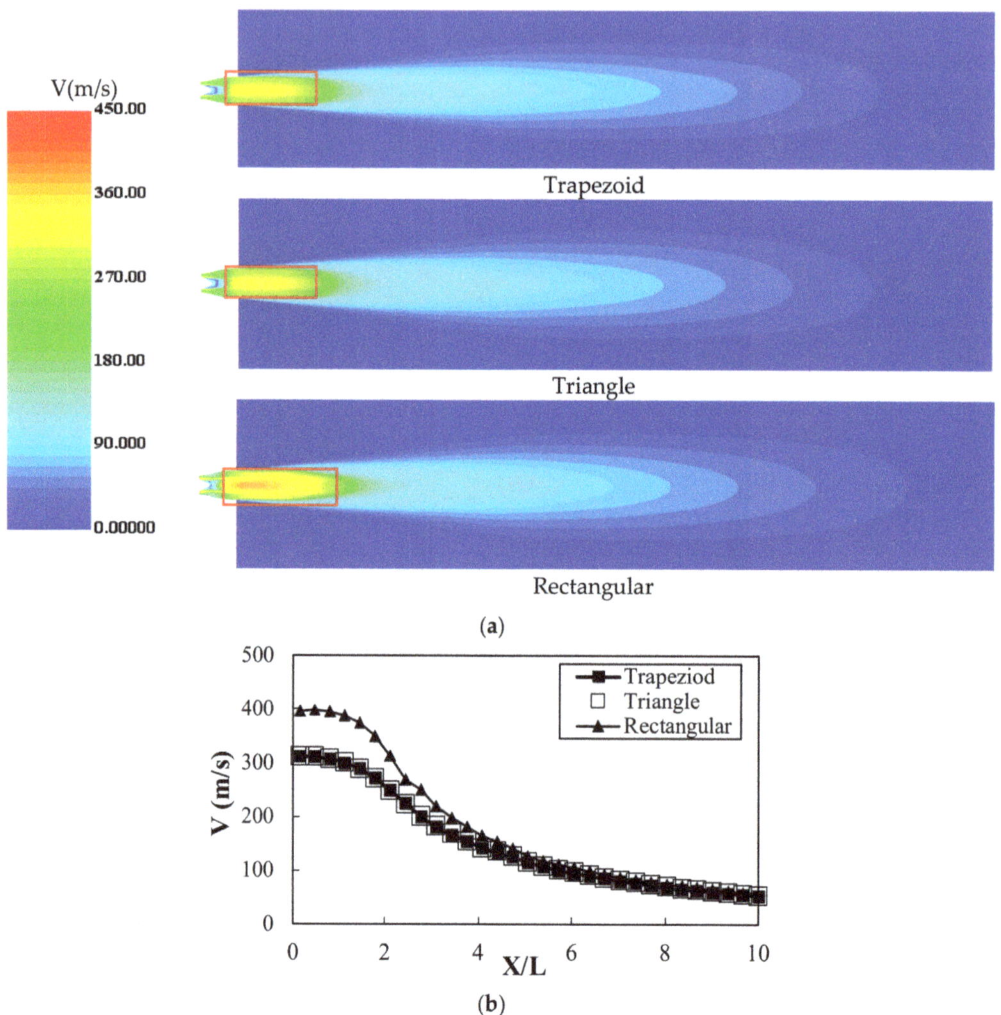

Figure 28. Velocity contour and (**a**) velocity distribution, (**b**) at the exhausted jet for different aft-deck shapes.

The temperature distribution along the centerline of the exhausted jet for different shapes of the aft-deck is shown in Figure 29. The figure shows that the temperature was the highest at the nozzle exit where the potential core region exited, and then the temperature gradually reduced with the increasing distance from the nozzle exit due to the mixing between the exhausted jet and ambient air. The figure demonstrates that using a nozzle with a trapezoid or triangle aft-deck can reduce the temperature at the nozzle exit compared to using a nozzle with a rectangular aft-deck. This can reduce the infrared radiation and hence improve the nozzle performance. Both the trapezoid and triangle aft-deck increased the projected area available for the exhausted jet to mix with the surrounding air, which enhanced the mixing process and led to the decreasing size and length of the potential core region, as shown in Figure 28.

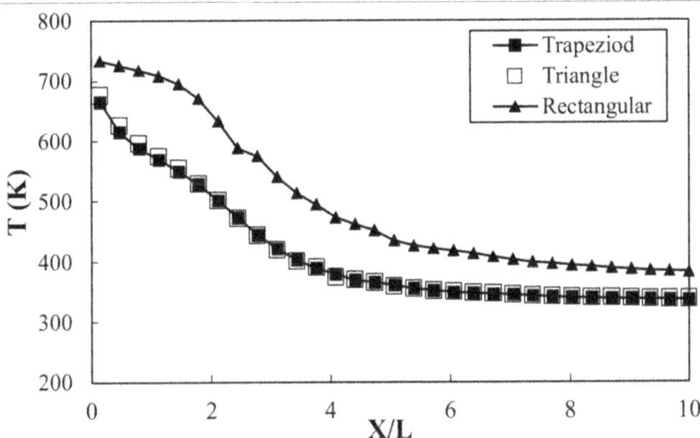

Figure 29. Temperature distribution at the centerline of the exhausted jet for different aft-deck shapes.

5. Conclusions

The serpentine nozzle is of strategic interest to researchers and military forces because it can shield the high-temperature components of the engine. Most studies in the open literature only investigate the serpentine nozzle performance; however, in the actual aircraft, the serpentine nozzle is attached to the engine exhaust system. The engine exhaust system consists of a double duct, mixer, and tail cone. Furthermore, the serpentine nozzle performance can be further improved by adding an aft-deck to the end of the nozzle. Therefore, the main goal of this study was to investigate the performance of a serpentine nozzle considering the engine exhaust system and aft-deck and using the computational fluid dynamics technique. The main conclusions can be summarized in the following.

- A new design of a mixer with a cone shape was used, and the effect of different mixer cone angles of $10°$, $15°$, and $20°$ was studied.
- The importance of the aft-deck was indicated by comparing the performance of nozzles with and without an aft-deck. Then, the influence of aft-deck lengths and shapes was also investigated.
- Increasing the mixer cone angle resulted in a decrease in the high-temperature core flow and an increase in the low-temperature bypass flow. The velocity inside the nozzle and at the exhausted jet decreased with the increasing mixer cone angle and could reduce the noise generated from engine exhaust systems.
- Increasing the mixer cone angle resulted in a decrease in the internal temperature of the nozzle and at the exit jet, which could minimize the infrared radiation.
- Compared to the nozzle without an aft-deck, the presence of an aft-deck resulted in a decrease in the pressure, temperature, and velocity inside the nozzle. The aft- deck also reduced the temperature of exhausted gases, which helped decrease infrared radiation. Furthermore, the aft-deck can decrease the velocity at the nozzle exit, which can decrease the noise generated from the nozzle.
- The aft-deck length did not affect the internal flow characteristics. However, increasing the aft-deck length can decrease the temperature of the external jet, which can reduce infrared radiation.
- The shape of the aft-deck affects the internal flow and exhaust jet characteristics. Using a trapezoid and triangle aft-deck improved the nozzle performance compared to a rectangular aft-deck. Using the aft-deck with trapezoid and triangle shapes reduced the velocity and temperature inside the nozzle, and the exhausted jet could reduce the noise and infrared radiation.

Author Contributions: H.M.A.A.: Conceptualization, Methodology, analysis, Investigation; B.-G.A.: resources, review and editing; J.L.: funding acquisition, project administration, supervision, review and editing. All authors have read and agreed to the published version of the manuscript.

Funding: This research received no external funding.

Data Availability Statement: The data that support the findings of this study are available from the corresponding author upon reasonable request.

Conflicts of Interest: The authors have no conflict to disclose.

References

1. Sun, X.-l.; Wang, Z.-x.; Shi, J.-w.; Li, Z.; Liu, Z.-w. Experimental and computational investigation of double serpentine nozzle. *Proc. Inst. Mech. Eng. Part G J. Aerosp. Eng.* **2014**, *229*, 2035–2050. [CrossRef]
2. Shan, Y.; Zhou, X.; Tan, X.; Zhang, J.; Wu, Y. Parametric Design Method and Performance Analysis of Double S-Shaped Nozzles. *Int. J. Aerosp. Eng.* **2019**, *2019*, 4694837. [CrossRef]
3. Sun, X.-l.; Wang, Z.-x.; Li, Z.; Liu, Z.-w.; Shi, J.-w. Influences of Design Parameters on a Double Serpentine Convergent Nozzle. *J. Eng. Gas Turbines Power* **2016**, *138*, 072301. [CrossRef]
4. Cheng, W.; Wang, Z.; Zhou, L.; Shi, J.; Sun, X. Infrared signature of serpentine nozzle with engine swirl. *Aerosp. Sci. Technol.* **2019**, *86*, 794–804. [CrossRef]
5. Rajkumar, P.; Chandra, S.T.; Abhijit, K.; Bhavik, M.; Biju, U. Flow Characterization for a Shallow Single Serpentine Nozzle with Aft deck. *J. Propuls. Power* **2017**, *33*, 1–10. [CrossRef]
6. Hamada, M.A.; Changwook, L.; Yechan, S.; Jeekeun, L. A computational study of two-dimensional serpentine nozzle performance with different annular mixer configurations. *Int. J. Mech. Sci.* **2021**, *208*, 106690.
7. Sun, P.; Zhou, L.; Wang, Z.; Sun, X. Effect of Serpentine Nozzle on Bypass Ratio of Turbofan Engine Exhaust. In Proceedings of the American Institute of Aeronautics and Astronautics 2018 Joint Propulsion Conference, Cincinnati, OH, USA, 9–11 July 2018. [CrossRef]
8. Chen, W.; Wang, Z.; Zhou, L.; Sun, X.; Shi, J. Influences of shield ratio on the infrared signature of serpentine nozzle. *Aerosp. Sci. Technol.* **2017**, *71*, 299–311. [CrossRef]
9. Sun, X.; Wang, Z.; Zhou, L.; Shi, J.; Cheng, W. Internal flow and external jet characteristics of double serpentine nozzle with different aspect ratio. *Proc. Inst. Mech. Part G J. Aerosp. Eng.* **2017**, *231*, 1–15. [CrossRef]
10. Sun, X.; Wang, Z.; Zhou, L.; Shi, J.; Cheng, W. Flow characteristics of double serpentine convergent nozzle with different inlet configuration. *ASME J. Eng. Gas Turbines Power* **2018**, *140*, 082602. [CrossRef]
11. Zhang, Y.C.; Wang, Z.X.; Shi, J.W.; Fang, L.; Kong, D.Y. Analysis on flow and infrared radiation characteristics of double S-nozzle. *J. Aerosp. Power* **2013**, *28*, 2468–2474. [CrossRef]
12. Yu, M.F.; Ji, H.H.; Li, N. Numerical analysis on infrared radiation characteristics of two-dimensional S-nozzle with small offset. *J. Aerosp. Power* **2015**, *30*, 2080–2087. [CrossRef]
13. Wang, D.; Ji, H.H.; Hang, W. A serpentine 2-D exhaust system with high thrust and low infrared signature. *J. Eng. Thermophys.* **2017**, *38*, 1944–1951.
14. Parviz, B.; James, J.M. Underexpanded Jet Development from a Rectangular Nozzle with Aft-Deck. *AIAA J.* **2015**, *53*, 1287–1298. [CrossRef]
15. Wei, Y.B.; Ai, J.Q. Parameter design method of double juxtaposition 3-D asymmetric several S-shaped nozzles. *J. Aerosp. Power* **2015**, *30*, 271–280. [CrossRef]
16. Nageswara, R.A.; Kushari, A. Underexpanded Supersonic Jets from Elliptical Nozzle with Aft-deck. *J. Propuls. Power* **2019**, *36*, 1–15. [CrossRef]
17. Luo, M.D.; Ji, H.H.; Huang, W. Numerical evaluation on infrared radiant intensity of exhaust system of turbine engine without afterburning. *J. Aerosp. Power* **2007**, *22*, 1609–1616.
18. Xie, Y.; Zhong, C.; Ruan, D.F.; Liu, K.; Zheng, B. Effect of core flow inlet swirl angle on performance of lobed mixing exhaust system. *J. Mech.* **2016**, *32*, 325–337. [CrossRef]
19. Li, Y.; Yang, Q. Influence on radar cross-section of S-shaped two-dimensional convergent nozzles with different outlet width-height radios. *J. Aerosp Power* **2014**, *29*, 645–651. [CrossRef]
20. Zhang, J.; Xie, Z. *Three-Dimensional Computational Study for Flow Fields within Forced Lobe Mixer*; ISABE Paper 2003-1104; ISABE: Cleveland, OH, USA, 2003.
21. Shan, Y.; Zhang, J.Z. Numerical investigation of flow mixture enhancement and infrared radiation shield by lobed forced mixer. *Appl. Therm. Eng.* **2009**, *29*, 3687–3695. [CrossRef]
22. Liu, C.; Ji, H. Numerical simulation on infrared radiation characteristics of serpentine 2-D nozzle. *J. Aerosp Power* **2013**, *28*, 482–1488. [CrossRef]

23.	Liu, C.; Ji, H. Numerical simulation on infrared radiant characteristics of 2D S-nozzle. *J. Eng.* **2010**, *31*, 1567–1570.

24.	Kline, S.J.; McClintock, F.A. Describing uncertainties in single sample experiments. *Mech. Eng.* **1953**, *75*, 3–8.

Review

A Review of Working Fluids and Flow State Effects on Thermal Performance of Micro-Channel Oscillating Heat Pipe for Aerospace Heat Dissipation

Xiaohuan Zhao [1], Limin Su [1], Jiang Jiang [1], Wenyu Deng [1] and Dan Zhao [2,*]

[1]   Energy and Electricity Research Center, International Energy College, Zhuhai Campus, Jinan University, Zhuhai 519070, China

[2]   Department of Mechanical Engineering, University of Canterbury, Private Bag 4800, Christchurch 8140, New Zealand

*   Correspondence: dan.zhao@canterbury.ac.nz

Abstract: A MCOHP (micro-channel oscillating heat pipe) can provide lightweight and efficient temperature control capabilities for aerospace spacecraft with a high power and small size. The research about the heat flow effects on the thermal performance of MCOHPs is both necessary and essential for aerospace heat dissipation. In this paper, the heat flow effects on the thermal performance of MCOHPs are summarized and studied. The flow thermal performance enhancement changes of MCOHPs are given, which are caused by the heat flow work fluids of nano-fluids, gases, single liquids, mixed liquids, surfactants, and self-humidifying fluids. The use of graphene nano-fluids as the heat flow work medium can reduce the thermal resistance by 83.6%, which can enhance the maximum thermal conductivity by 105%. The influences of gravity and flow characteristics are also discussed. The heat flow pattern changes with the work stage, which affects the flow mode and the heat and mass transfer efficiency of OHP. The effective thermal conductivity varies from 4.8 kW/(m·K) to 70 kW/(m·K) when different gases are selected as the working fluid in OHP. The study of heat flow effects on the thermal performance of MCOHPs is conducive to exploring in-depth aerospace applications.

Keywords: aerospace heat dissipation; micro-channel oscillating heat pipe; heat flow; thermal performance; heat flow pattern

Citation: Zhao, X.; Su, L.; Jiang, J.; Deng, W.; Zhao, D. A Review of Working Fluids and Flow State Effects on Thermal Performance of Micro-Channel Oscillating Heat Pipe for Aerospace Heat Dissipation. *Aerospace* **2023**, *10*, 179. https://doi.org/10.3390/aerospace10020179

Academic Editors: Tiegang Fang and Phil Ligrani

Received: 29 November 2022
Revised: 7 February 2023
Accepted: 9 February 2023
Published: 14 February 2023

1. Introduction

The working environment of aerospace equipment is special, including cold and heat exchange [1,2], heat transfer [3,4], energy mode conversion [5,6], and energy management [7,8]. The heat dissipation that is caused by heat conduction [9,10], heat convection [11,12], and radiation [13,14] during the operation of aerospace equipment should not be underestimated. Oscillating heat pipe (OHP) is the preferred heat dissipation technology for aerospace equipment, battery, and electronic equipment, due to its superior heat transfer performance, simple structure, and miniaturization [15,16]. The changes of the array layout, heating mode, working fluid, pipe wall material, and working fluid flow in the pipe are the primary factors that affect the OHP heat transfer performance [17]. The OHP has been studied using a variety of working fluids and nano-fluids have emerged as a research fad [18]. OHPs with a 3D structure have been mentioned in recent years [19], which can provide excellent performance in some space heat transfer scenarios [20]. The OHP turns [21], and the heating settings and pulsing heating techniques also have an impact on the thermal performance [22,23].

Based on the outcomes of the visualization, Senjaya et al. [24] developed a model for the creation and expansion of tube diameter bubbles. The liquid slug velocity was less than 0.2 m/s, which promotes bubbles formation. Ando et al. [25] tested the OHP with a check valve, which is located near the condensation section. The valve intermittently supplied liquid plugs to the evaporation section to start the pipe successfully. Wang et al. [26] investigated the intricate hydrodynamic processes and presented a new, closed OHP with periodic expansion and contraction condensation. The thermal efficiency was raised by 45% and the contraction condensation increased the oscillation frequency of the vapor plug/liquid slug. Shi et al. [27] performed an experiment investigation on a closed-loop OHP with a 0.5% mass-concentrated micro-encapsulated phase change material suspension. The vertical installation performed better for heat transfer under the effect of gravity. Lim et al. [28] adopted a randomly arranged pipeline layout OHP and studied the internal working fluid oscillation. The liquid slugs oscillated with large amplitude in every channel and the thermal performance was 32% higher than MCOHPs with a uniform channel layout. Liu et al. [29] made 15 turns of the anti-gravity OHP and carried out the research with the heat recovery rate of 1.66 times than that of pure copper tubes, which proved that gravity had a positive effect on the internal hydrodynamics.

The heat transmission capabilities of liquid metal and water as the working fluids of OHPs were compared by Hao et al. [30]. The heat transfer performance of the OHP is increased by 13% when the liquid metal is the working fluid instead of pure water. Schwarz et al. [31] proposed two design methods of floral OHP and star-shaped OHP and conducted experiments. The floral OHP increased the latent heat transfer, while the star-shaped OHP improved the convective heat transfer. The floral OHP design reduced the thermal resistance by 7% in the horizontal position and 12% in the vertical position. Kwon et al. [32] reported the influence of double-diameter tubes on single-turn OHPs. The circular flow was promoted by double-diameter tubes, which reduced the thermal resistance of the OHP by 45%. Liu et al. [33] designed and manufactured a new type of flat-plate OHP (FPOHP) with double serpentine channels. FPOHPs can successfully start at all tilt angles from 0° to 90°. The thermal conductivity of FPOHPs is 5.8 times than that of ordinary OHPs. Arai et al. [34] designed three kinds of polycarbonate OHPs of actual flow channel structures with an additive manufacture method and conducted research. By comparison of the effective thermal conductivity of different flow channel sizes at the same filling ratio, the effective thermal conductivity of a 0.8 mm square flow channel was about seven times that of a 2 mm square flow channel.

The working fluids and flow state have effects on the thermal performance of the micro-channel oscillating heat pipe, which are confirmed by all the above studies. In this paper, the effects of working fluids and the flow state on OHPs factors on the heat transfer performance are studied. The heat dissipation mechanism and technical characteristics of OHP are introduced. The effects of various filling working fluids including metal nano-fluids, filling working fluids, non-metallic nano-fluids, mixed nano-fluids, gas, organic solvents, mixed liquids, surfactants, and self-rewetting fluids (SRWFs) on heat dissipation are listed in detail. The influence of gravity on the flow, the flow pattern characteristics, and the two-phase oscillating flow are compared.

2. Heat Dissipation Mechanism and Characteristics of the MCOHP

2.1. Heat Dissipation Mechanism of the OHP

The working fluid is filled into a vacuum pipe with a certain proportion, which can be blended into various shapes and divided into an evaporation section, adiabatic section, and a condensation section [35]. The working fluid absorbs heat in the evaporation section and releases heat in the condensation section, which can transfer the heat generated at the evaporation section to the condensation section to achieve heat dissipation [36]. The mass forms randomly distributed air and liquid plugs under the effect of temperature difference between the cold and hot ends and the surface tension [37,38]. Due to the pressure difference between the evaporation section and the condensation section, the working fluid is driven to flow to the condensation section. The working fluid flows back to the evaporation section under the gravity action after the heat release [39]. The working process of the OHP is given in Figure 1.

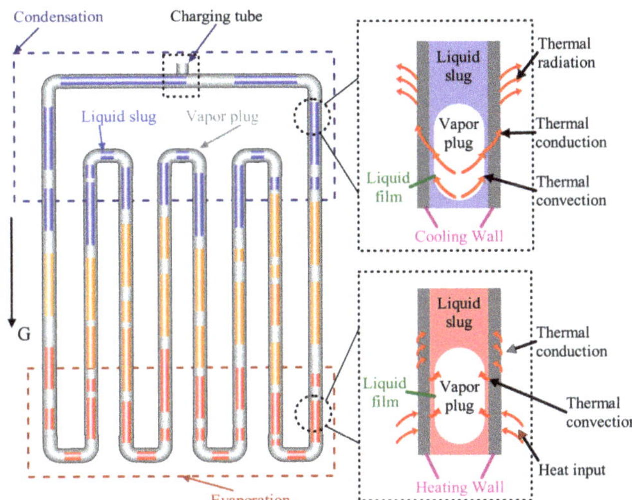

Figure 1. Working process of OHP.

There are many complex physical phenomenon and heat transfer processes in the internal operation of the OHP, including heat convection, latent heat transfer, pressure difference, temperature difference, inertial force, friction, gravity, and other factors [40], which are complex coupling processes [41,42]. When the OHP is placed vertically and the heater is located at the bottom, the thermal resistance representing the heat transfer performance of the OHP can be obtained from Equation (1) [43], after reaching a pseudo-steady state under each heating power condition.

$$R_{th} = \frac{\overline{T}_{evap} - \overline{T}_{cond}}{Q_{in}} = \frac{\frac{1}{t_a}\frac{1}{W}\int_0^{t_a}\int_0^W T_{Si}(x,L,t)dxdt - \frac{1}{t_a}\frac{1}{W}\int_0^{t_a}\int_0^W T_{Si}(x,0,t)dxdt}{Q_{in}} \tag{1}$$

where R_{th} is the thermal resistance, K/W. $\overline{T}_{evap}$ and $\overline{T}_{cond}$ are the average temperature of evaporation and condensation, respectively, K. Q_{in} is the input power, W. t_a is the time interval for time averaging, s. W and L are the width and length of the OHP, respectively, m. T_{Si} is the temperature of the silicon substrate, K. x is the horizontal coordinate, m. t is the time, s.

It is assumed that the temperature distribution of the liquid film is linearly related to the thickness of the liquid film. The total heat transferred from the heating wall to the liquid film and the heat transferred from the liquid film to the cooling wall is calculated by Equation (2) [44].

$$\begin{cases} Q_{w,H} = (T_w - T_{sur,H})\pi dx_H \lambda_l / \delta \\ Q_{w,C} = (T_{sur,C} - T_w)\pi dx_C \lambda_l / \delta \end{cases} \tag{2}$$

where $Q_{w,H}$ and $Q_{w,C}$ are the total heats transferred from the heating wall to the liquid film and from the liquid film to the cooling wall, W. T_w is the wall temperature, $T_{sur,H}$ and $T_{sur,C}$ are the temperature of liquid film during heating and that of cooling, respectively, K. λ_l is the thermal conductivity of the liquid, W/m·K. δ is the liquid film thickness, m.

In OHPs, the liquid phase is regarded as an incompressible flow and the gas phase as an ideal gas. The temperature and pressure in each vapor plug are uniform. The temperature changes for liquid plugs and vapor plugs can be calculated by Equation (3) [45].

$$\begin{cases} c_{p,l}\rho_l A_{cr}\frac{\partial T_l}{\partial t} = h_{w,l}(T_w - T_l)S + \lambda_l A_{cr}\frac{\partial^2 T_l}{\partial l^2} \\ \frac{dT_v}{dt} = \left(Q_{v,sen} + Q_{v,lat} - P_v\frac{dV_v}{dt}\right)/c_{v,v}m_v \end{cases} \tag{3}$$

where $c_{p,l}$ is the specific heat of the liquid, J/(kg·K). ρ_l is the density of the liquid, kg/m^3. A_{cr} is the cross-sectional area of the flow path, m^2. T_w and T_l are the temperature of channel wall and liquid plug, respectively, K. $h_{w,l}$ is the heat transfer coefficient between the channel wall and the liquid plug, W/(m^2·K). S is the perimeter of the liquid plug, m. T_v is the temperature of the vapor plug, K. $Q_{v,sen}$ and $Q_{v,lat}$ are the total amount of sensible heat and latent heat, respectively, W. P_v is the pressure of vapor plug, Pa. V_v is the volume of the vapor plug, m^3. $c_{v,v}$ is the specific heat at constant volume, J/(kg·K). m_v is the mass of the vapor plug, kg.

2.2. Technical Characteristics of OHP

2.2.1. Excellent Heat Transfer Performance

Yu et al. [46] designed the 18-channel 3D-OHP with liquid metal as the working fluid, as listed in Figure 2a. The thermal resistance was as low as 0.0351 °C/W and the heat transfer performance was improved by 20.57% compared with the pure ammonia OHP. Thompson et al. [47] measured the thermal resistance data (0.08 °C/W) of a plate OHP with staggered micro-channels when the working fluid, heating mode, and heat pipe matched. Ji et al. [48] manufactured the high-temperature OHP for liquid metal. The minimum thermal resistance was 0.08 °C/W with a filling rate of 45% and a heating power of 3168 W. Ji et al. [49] tested the high-temperature vibration OHP of a liquid metal through different proportions of NaK. The minimum thermal resistance was 0.071 °C/W with 3528 W input power and a 90° inclination angle. Czajkowski et al. [50] studied the patterned OHP with a special rotation system. The thermal resistance was 0.05 °C/W when the rotation speed was higher than 200 rpm. Qu et al. [51] discovered the OHP of spherical Al$_3$O$_3$. The maximum thermal resistance of the OHP decreased by 0.14 °C/W with the filling rate of 70%. The thermal resistance of water was lower than that of ethanol [52]. Tokuda et al. [53] tested the double-loop closed OHP made of Incoloy 800 HT with sodium and the effective thermal conductivity was 2.6×10^3 to 2.3×10^4 W/(m·K). Zhao et al. [54] constructed a sizable OHP experiment to investigate the variables of thermal conductivity. The OHP had a significant thermal load with an effective thermal conductivity of 5676 W·m^{-1}·C^{-1} when the filling rate was 40%. Lin et al. [55] took aluminum materials to make interconnected rectangular channels of the OHP for heat dissipation of high-power LED, as given in Figure 2b. The thermal resistance was 0.18 °C/W with the heating power of 110 W to achieve a good heat dissipation.

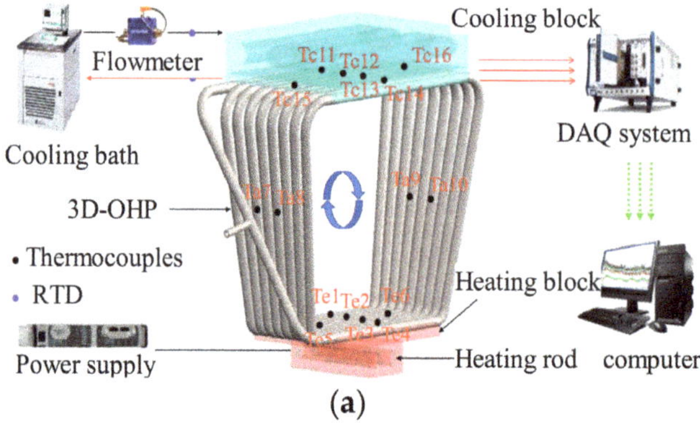

Figure 2. *Cont.*

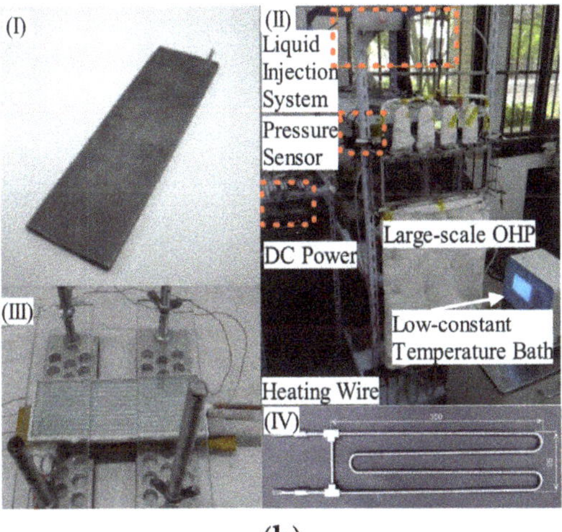

(b)

Figure 2. Two-dimensional and three-dimensional OHP devices. (**a**) An 18-channel 3D-OHP experimental device; (**b**) 2D OHP unit. (I) Thompson et al., (2011), [47]. (II) Zhao et al., (2017), [54]. (III) Lin et al., (2011), [55]. (IV) Tokuda et al., (2022), [53].

2.2.2. Simple Structure of the OHP with a Small Volume

The most prominent characteristics of OHPs are its miniaturization abilities and simple structure, as listed in Figure 3a. An OHP can oscillate by itself without the liquid suction core and the assistance of other equipment. Qian et al. [56] applied loop OHP to study the heat dissipation of grinding wheels. Monroe et al. [57] used a four-loop OHP to test the fluid stirring of magnets to collect energy with aluminum blocks to assist heating and cooling. Zhao et al. [58] carried out the experiment with three turns of the OHPs to achieve the thermal properties of the coupling phase change materials. Qu et al. [59] made three kinds of copper OHPs (a 2D-OHP, three-layer 3D-OHP, and four-layer 3D-OHP) and studied the coupled heat transfer of the phase change materials. Jin et al. [60] used high-temperature quartz glass to make transparent OHPs with high solar light transmittance, which can realize the experiment research of solar energy-absorbing nano-fluids. The highest thermal conductivity can be achieved when the OHP is filled with 3.0 wt.% nano-fluids. Alqahtani et al. [61] explored the influence of the bending degree of OHPs on heat transfer. There is no significant effect on the thermal performance when the bending angle increases. Iwata et al. [62] developed 10 laps of OHPs. A flexible and highly conductive tropical belt was formed, which can be used as a cooling device in a spacecraft. Wei et al. [63] clamped between commercial battery packs to simulate the thermal power generated by two adjacent battery modules, as shown in Figure 3b. The OHP was filled with the dual fluid mixed ethanol–water and the size was consistent with the length of the battery pack. The evaporation section was heated by the battery and the condensation section used fans to dissipate heat, which had a good battery cooling effect and provided a new idea for the battery cooling of electric vehicles.

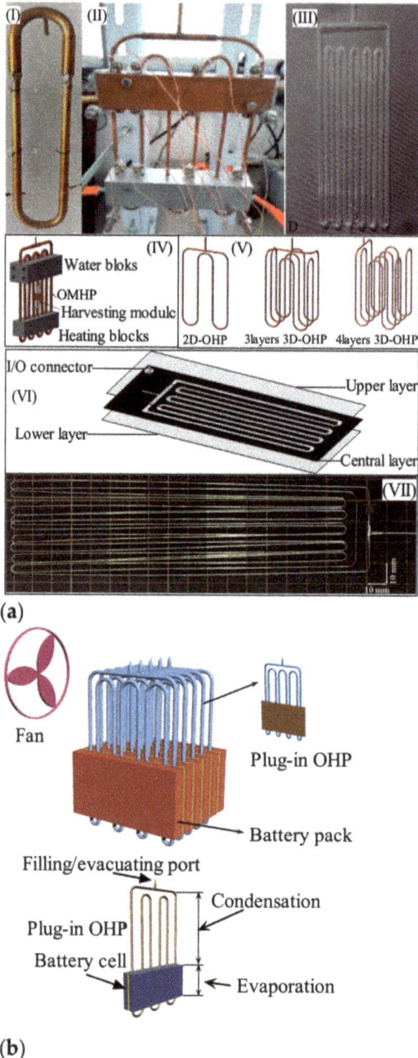

Figure 3. OHP array and distribution structure for cooling. (**a**) Multi-loop OHP array and distribution structure for spacecraft cooling; (**b**) OHP distribution for battery cooling. (I) Qian et al., (2019), [56]. (II) Zhao et al., (2016), [58]. (III). Jin et al., (2019), [60]. (IV). Monroe et al., (2018), [57]. (V). Qu et al., (2019), [59]. (VI). Alqahtani et al., (2022), [61]. (VII). Iwata et al., (2021), [62].

3. Effect of Various Filling Working Fluids of OHPs on Heat Dissipation

The common working fluids used in OHPs are nano-fluids, gases, single liquids, mixed liquids, surfactants, and SWRF.

3.1. Metal Nano-Fluid

A metal nano-fluid refers to a new type of heat transfer medium with uniform, stable, and high thermal conductivity, which is prepared by dispersing metal nano-powders into the base liquid [64]. At an appropriate concentration, nano-fluids have better thermal conductivity [65,66] and higher heat transfer limits [67] than traditional working fluids. The common nano-particles are metals (Al, Ag, Cu, Fe, etc.) and metal oxides (Al_2O_3,

Fe$_3$O$_4$, TiO$_2$, etc.). Table 1 demonstrates the metal nano-fluids effect on the heat transfer performance of OHPs. The heat transfer performance of the OHP can be significantly improved by the appropriate particle size [68,69], fluid concentration [70,71], filling rate [72], tilt angle [73], and heating power [74] of metal nano-particles. Furthermore, for metal nano-fluid OHPs, an applied magnetic field helps to reach the start-up faster at low heat input conditions [75,76].

Table 1. Thermal properties of metal nano-fluids.

Metal Nano-Category	Particle Size	Concentration	Liquid Filling Rate	Inclination Angle/°	Heating Power/W	Reduction of Thermal Resistance
Al$_2$O$_3$ [68]	–	0, 0.1, and 0.5 wt.%	50%	0, 90	10~80	15.8%
Al$_2$O$_3$ [69]	56 nm	0~1.2 wt.%	50%	90	20~140	25.7%
Ag [70]	50 nm	50, 200, and 600 ppm	–	–	314, 385, and 488	30%
Al$_2$O$_3$ [71]	10~30	0.5, 1, and 3 wt.%	20%, 40%, 60%, and 80%	10, 40, 70, and 90	20, 30, and 40	Improved thermal performance by 56.3%
ɤ-Fe$_2$O$_3$ [72]	20	2 vol.%	50%	90	0~160	12%
Fe$_2$O$_3$ [73]	20	2 vol.%	50%	0~90	10~90	24.1%
Fe$_3$O$_4$ [74]	5~20	90, 270, and 450 ppm	70%	90	20, 55, 90, 125, and 160	27.6%
Fe$_3$O$_4$ [75]	25	0.2 wt.%	50%	90	0~200	11%
NiFe$_2$O$_4$ [76]	25	1.5, 3 wt.%	–	90	200, 300, and 400	30.4%

Karthikeyan et al. [77] conducted an experimental study on a COHP with colloidal nano-fluids of Cu (average nano-particle size is 100 nm) and Ag (average nano-particle size is 60 nm). Compared with deionized water, the Ag nano-fluid increased the OHP heat transfer limit by 33.3% and the evaporation wall temperature was lower. The shape of the nano-particles also had an impact on the thermal resistance. Kim et al. [78] found that the OHP thermal resistance of Al$_2$O$_3$/acetone nano-fluids with spherical, brick, and cylindrical nano-particles decreased 33%, 29%, and 16%, respectively. The nano-particles effect on the thermal performance of the OHP was revealed by Jafarmadar et al. [79] with Al$_2$O$_3$, CuO, and Ag. The flow, heat transfer, and entropy generation of the OHP in the case of pure water were checked. The entropy produced by Ag was the highest. The volume concentration of nano-particles was 0.5~1%, which can minimize the generation of entropy and proper thermal operation. Goshayeshi et al. [80] studied the influence of nano-fluids on the flow and thermal properties of OHPs with Fe$_2$O$_3$/kerosene, as Figure 4 displayed. The five-flow modes were obtained of the evaporation section when the filling rate of the Fe$_2$O$_3$ nano-fluids was 50% (average nano-particle size of 20 nm with the concentration of 5 vol.%). With the increase in heat (10~80 W), the bubble flow, slug flow, foam flow, annular steak flow, and annular flow will gradually appear in the evaporation section. When the liquid plug speed was ≤ 0.15 m/s, bubbles with a diameter equal to the inner diameter of the pipe were generated. Subsequently, [81] compared Fe$_3$O$_4$/water and the effect of -Fe$_2$O$_3$/kerosene nano-fluids on the heat transfer performance. Fe$_3$O$_4$/water and -Fe$_2$O$_3$/kerosene nano-fluids reduced the thermal resistance by 30.8% and 16.7%. Gandomkar et al. [82] studied the glass and copper OHP of ferromagnetic fluid under different magnetic fields through visual experiments. The place with the magnetic field had a smaller thermal resistance and the best thermal performance of the copper OHP. The performance without a magnetic field was the best for the glass OHP. Monroe et al. [83] examined the performance of solenoid-assisted OHPs for CoFe$_2$O$_4$ nano-fluids. A ring magnet was used to magnetize and the CoFe$_2$O$_4$ nano-fluids improved the heat transfer of heat pipes by 58%.

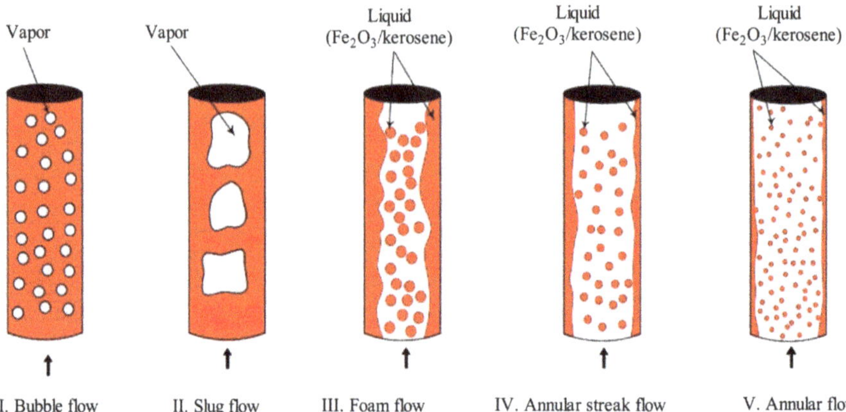

Figure 4. Flow pattern of Fe_2O_3/kerosene metal nano-fluid of OHP [80]. Goshayeshi et al., (2016).

3.2. Non-Metallic Nano-Fluid

Non-metallic nano-fluids are SiC, CNT, graphene, $CaCO_3$, and other compounds. Table 2 highlights the influence of non-metallic nano-fluids on the heat transfer performance of OHPs. In the OHP, heat transfer occurs due to repeated pressure fluctuations, with higher heat transfer occurring with more repetitions of pressure fluctuations. To provide a higher frequency of pressure and an average pressure inside the OHP, Tanshen et al. [84] used an aqueous solution of 0.2 wt.% of multi-walled carbon nano-tubes (MWCNTs) to experimentally investigate the thermal resistance and pressure fluctuations inside the OHP. Sadeghinezhad et al. [85] found through experimental studies that the deposition of graphene formed a coating on the surface of the sintered core in the evaporator section. This coating increased the surface wettability and thus improved the thermal performance of the heat pipe. Kim and Bang [86] discovered that the capillary limit of heat pipes containing graphene oxide/water nano-fluids was higher than that of the aqueous heat pipes. This is because the nano-particle coating changes the effective capillary radius and the bending moon surface, leading to an increase in the maximum fluid flow rate through the core structure. On the other hand, Wu et al. [87] showed in their study that the variation of thermal load has a greater effect on the thermal performance of the OHP than the variation of concentration. Beyond this, the addition of nano-particles to the working fluid can significantly enhance the heat transfer characteristics of the OHP and further improve the heat dissipation capacity of the OHP [88]. Zhou et al. [89] indicated that the addition of a graphene nano-sheet nano-fluid to distilled water can alleviate drying and improve the heat transfer performance of OHPs. Nazari et al. [90] reported that the addition of graphene oxide flakes improved the thermal conductivity and viscosity of the base fluid. Furthermore, the high concentration of nano-fluids reduces the thermal properties of OHPs compared to pure water, which is attributed to the increase in the dynamic viscosity of the nano-fluid. To prepare graphene nano-fluids with excellent stability, Xu et al. [91] as well as Zhou et al. [92] chose to use ethanol–water mixtures as the base fluid. In addition to this, the addition of appropriate graphene oxide nano-particles improved the OHPs initiation performance [93]. Zhang et al. [94] revealed that the addition of nano-particles promoted the phase transition of the work fluid in the OHP on the one hand, while increasing the transient velocity and driving force of the work fluid on the other hand. These are conducive to the reflux of condensate, and they can effectively avoid the dry-out phenomenon.

Table 2. Thermal properties of non-metallic nano-fluids.

Non-Metallic Nano-Fluids	Concentration	Liquid Filling Rate/%	Inclination Angle/°	Input Power/W	Reduction of Thermal Resistance
MWCNTs [84]	0.05, 0.1, 0.2, and 0.3 wt.%	60	90	50 ~400	About 36.2%
Graphene [85]	0.025, 0.05, 0.075, and 0.1 wt.%	–	0~90	20~120	48.4%
Graphene oxide (GO) [86]	0.01 and 0.03 vol.%	100	90	50 ~400	Maximum heat transfer enhancement 25%
C60 [87]	0.1, 0.2, and 0.3 wt.%	50	50	10~60	36%
Hydroxylation MWNTs [88]	0.1~1 wt.%	50	90	–	34%
Graphene Nano-sheets [89]	1.2, 2, 5.7, 9.1, 13.8, and 16.7 vol.%	45, 55, 62, 70, and 90	90	10 ~100	83.6%
Graphene oxide [90]	0.25, 0.5, 1, and 1.5 g/L	50	90	10~70	42%
Oligographene (FLG) [91]	0.1, 0.3, 0.5, 0.75, and 1 mg/mL	55	90	20~60	25.16%
Carbon nano-tubes (CNTs) [92]	0.05, 0.1, 0.2, 0.3, and 0.5 wt.%	35	90	8~56	About 66.6%
Graphene oxide [93]	0.02~0.1 wt.%	20, 50, and 80	90	10~30	54.34%
SiO$_2$ [94]	0.5, 1, 1.5, and 2 wt.%	50	90	10~50	40.1%

Sadeghinezhad et al. [95] studied the thermal properties of copper sintered heat pipes with graphene nano-fluids at different dip angles of 0°~70° and liquid filling rates of 30~60%. The maximum thermal conductivity of the graphene nano-fluids heat pipe (5 vol.%) was increased by 105% and the thermal resistance was reduced by 26.4%. Khajehpour et al. [96] discovered the performance of the L-shaped OHP with SiO$_2$ nano-fluids with different nano-particle sizes (11~14 nm and 60~70 nm). The experiment thermal resistance increased with the nano-particle size. For SiO$_2$ nano-fluids at 11~14 nm (0.5 wt.%), the maximum reduction of thermal resistance was about 24% at the vertical position under a heat load of 10 W and a liquid filling rate of 100%. Li et al. [97] studied the thermal performance of OHPs with aqueous ethylene glycol-based graphene nano-fluids. The minimum thermal resistance of 0.36 K/W was achieved at a thermal load of 85 W and a liquid filling rate of 35% for 2 g/L graphene nano-fluids. Choi [98] tested the thermal performance of thermosyphon heat pipes with cellulose nano-fluids, which increased the boiling heat transfer coefficient by about 71.74%.

3.3. Mixed Nano-Fluid

Both metallic and non-metallic nano-fluids are prepared by suspending a single nano-particle in a base solution to obtain a stable suspension. Mixed nano-fluids are made up of two or more different nano-particles [99]. Zufar et al. [100] studied the thermal performance of Al$_2$O$_3$-CuO/water mixed nano-fluids (0.1 wt.%) and SiO$_2$-CuO/water mixed nano-fluids (0.1 wt.%) under different heat inputs (10–100 W) with the liquid filling rates of 50–60%. A minimum thermal resistance of 0.27 °C/W can be obtained with SiO$_2$-CuO mixed nano-fluid. The thermal resistance of Al$_2$O$_3$-CuO and SiO$_2$-CuO mixed nano-fluids were reduced by 57% and 34%, respectively [101]. Moghadasi et al. [102] conducted a 3D numerical study on the laminar flow and heat transfer of Al$_2$O$_3$-CuO/water mixed nano-fluids in a U-shaped bend in porous media. The temperature and velocity contour of different volume fractions with water at the base fluid and mixed nano-fluids are given in Figure 5a,b, where φ is the volume fraction and r_p is the porosity. Nano-fluids are applied in the presence of porous media as fluids accumulate near the walls and enhance the heat transfer. When the volume fraction changed from 1% to 5%, the velocity distribution improved, and the temperature gradient increased. Xu et al. [103] studied the performance of thermosyphon OHPs mixed with Al$_2$O$_3$-TiO$_2$/water (0.2 vol.%). The conditions under different filling rates (30~70%) and coolant flow rates (0.4~0.56 L/min) were compared with 25% (Al_2O_3) and 75% (TiO_2) mixed nano-fluids, which achieved a thermal resistance reduction of 26.8% and increased the thermal efficiency by 10.6%. Mukherjee et al. [104] configured the SiO_2-ZnO/water mixed nano-fluids of different mass fractions (0.025~0.10 wt.%) and Reynolds numbers (7743~23,228). The thermal conductivity at 60 °C can be increased by up to 30% with a mass fraction of 0.10 wt.%. Veeramachaneni et al. [105] fabricated a rectangular flat loop OHP for electron cooling applications with Cu-graphene/water mixed nano-fluids (0.1~0.2 vol.%). For a mixed nano-fluid with a volume concentration of 0.02%, the capillary limit increased by 36.97% and the wall temperature of the evaporation section decreased by 9.8%. The mixed nano-fluid with a copper/graphene ratio of 30:70 can obtain a minimum thermal resistance of 0.1 K/W.

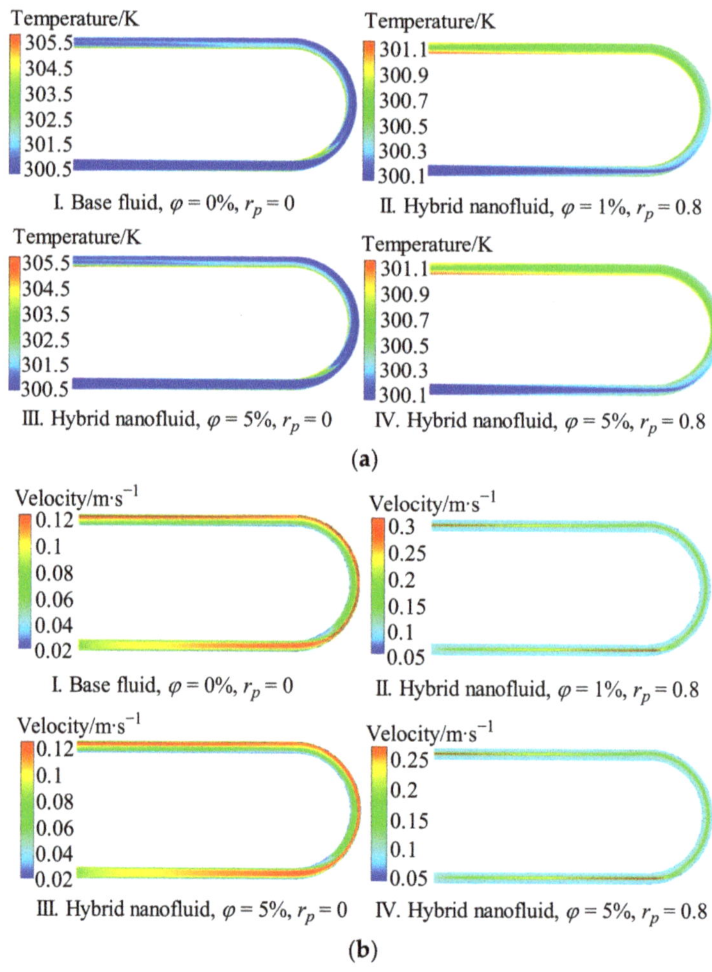

Figure 5. Temperature and velocity contours of mixed nano-fluids [102]. Moghadasi et al., (2020). (**a**) Temperature contours with mixed nano-fluids of different volume fractions. (**b**) Velocity contours using mixed nano-fluids of different volume fractions.

3.4. Gas Working Fluid

The gas can be applied as the working fluid of OHPs if the temperature of the working environment is too low. The researchers explored various gases of neon, argon, nitrogen [106,107], helium [108], and hydrogen [109,110]. Liang et al. [111,112] experimentally tested OHPs with neon, and the maximum effective thermal conductivity was 22.18 kW/(m·K) at an optimal filling rate of 24.5%. Barba et al. [113] found that the expansion and contraction of gases play an important role in the circulation of working fluids. The circulation was hindered when the filling rate was too low and the OHP could not be started normally. Sun et al. [114] simulated OHPs with hydrogen as the working fluid. The influence of hydrogen on the latent heat transfer was 45~51%, which is proportional to the volume fraction of gas. Li et al. [115] researched nitrogen as the working fluid and the thermal conductivity of the bottom heated was about 16 k W/(m·K), which was about 32 times that of pure copper. Xu et al. [116] conducted experimental studies on low-temperature OHPs filled with helium. The effective thermal conductivity was 4.8~13 kW/(m·K) with an inclination angle of 30° and a liquid filling rate of 70.8%. Fonseca et al. [117] took the experiments between 77 K and 80 K with nitrogen. The temperature difference between the OHP sections was small and the maximum thermal conductivity was 70 kW/(m·K), with a liquid filling rate of 20% [118]. The thermal performance of more than 2000 helium working fluid OHPs under the filling rate of 20~90% was tested. The liquid filling rate was 69.5% and the maximum effective thermal conductivity was 50 kW/(m·K). The non-condensable gas of the OHP reduced the evaporation amount to slow down the circulation of the working fluid, which can weaken the oscillation and reduce the heat transfer performance [119,120]. The OHP with a heat flux constant and wall temperature constant is displayed in Figure 6a. The effect of non-condensable gases on OHPs is demonstrated in Figure 6b. (Q is the heat flux, T is the temperature, and ω is the mass concentration). The higher the temperature of the evaporation section, the less influence of the non-condensable gas. Chen et al. [121] conducted a series of experiments to study the thermal performance of ethane OHPs (EOHPs) in the medium and low temperature regions ($-90\sim0$ °C). The liquid filling rate of the best performance of EOHPs was not affected by the operating temperature and heat input, which was always maintained at about 30%. The lowest corresponding thermal resistance was 0.02 °C/W at the inclination angle of 30° and the temperature of -80 °C. At a high heat input of 30~50 W, the latent heat of vaporization was the main characteristic that determined the thermal performance of the EOHP.

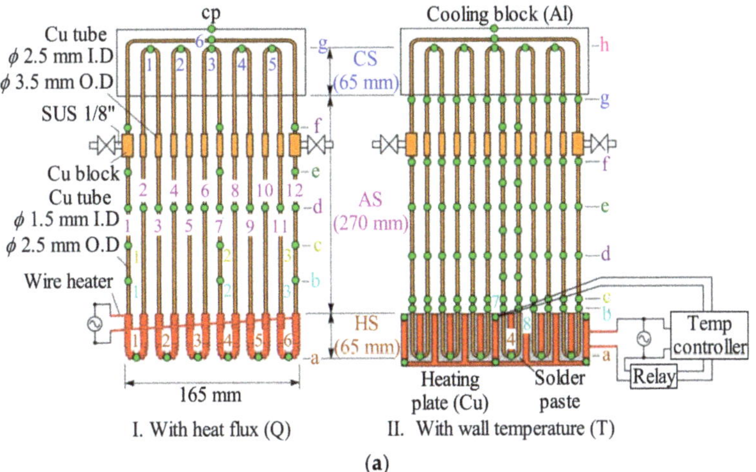

Figure 6. *Cont.*

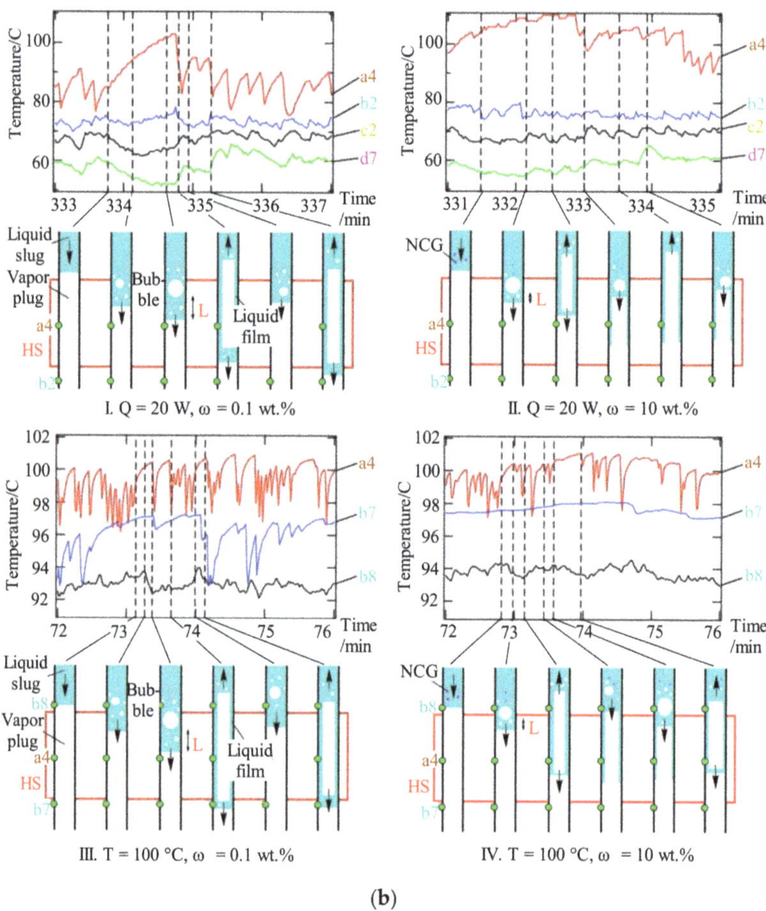

Figure 6. Effect of non-condensable gas on heat pipe [120]. Senjaya et al., (2014). (**a**) EOHP with heat flux constant and wall temperature constant. (**b**) Measurement point temperature status and flow image of OHP.

3.5. Organic Solvent

Table 3 summarizes some trends in the heat transfer performance of OHPs with different organic solvent liquids as the working fluids. From the information in Table 3, the working fluids, filling rate, input power, lowest thermal resistance, and lowest thermal resistance obtained from the nine papers are compared. At a low heat input, the heat transfer depends strongly on whether the oscillations are triggered or whether the oscillatory flow is triggered fast. In contrast, the effect of viscosity on the heat input and the effect of the latent heat of vaporization increases at a high heat input. It can be concluded from Table 3 that the choice of the working medium and filling ratio should be determined according to the actual situation. In terms of taking full advantage of low thermal resistance, 50% is a good filling ratio for ethanol, which should have a low heat input. In the case of a large filling ratio and large heat input, methanol is a better choice. Acetone has a good fill ratio of 50–70% with high heat input. The best filling ratio for Ionic liquids is about 44% with high heat input. The LiCl solution performs well at a 62% filling rate. R1233zd (E) with a filling rate of about 50% at moderate heat input is also an option.

Table 3. Heat transfer performance with different organic solvent liquids as working fluids.

Working Fluids	Filling Rate/%	Input Power/W	Lowest Thermal Resistance/°C·W^{-1}	Lowest Thermal Resistance Obtained
Ethanol [122]	0, 25, 37.5, 50, 62.5, 75, and 100	–	0.95	50%
Methanol [123]	20~95	5~100	0.2	95% and 100 W
Ethanol [124]	50	15~50	0.6244	50 W
Acetone [125]	50 ± 5% and 70 ± 5%	60~300	0.092	70% and 260 W
Lonic liquids [126]	65 ± 5	50~250	0.15	44.4% and 250 W
Acetone [127]	50	10~200	0.14	200 W
LiCl solution [128]	45, 55, 62, 70, 80, and 90	10~100	0.9	62% and 10 wt.%
Acetone [129]	0~100	10~120	0.39	60% and 100 W
R1233zd(E) [130]	40~70	0~200	0.1184	50% and 70 W

Takawale et al. [131] studied the performance of FPOHPs and capillary tube OHPs (CTOHPs) under different heat inputs (20 W~180 W) and liquid filling rates (40%, 60%, and 80%). After ethanol was filled into the OHP, the thermal resistance of the FPOHP and CTOHP decreased by 83% and 35%, respectively. Bastakoti et al. [132] tested the heat transfer performance of OHPs with methanol, ethanol, cetyltrimethylammonium chloride (CTAC), and deionized water as the working fluids. The heat pipe charged into the CTAC had the lowest thermal resistance of 0.30 K/W. The thermal resistance of the OHP with deionized water, methanol, ethanol, and acetone as the working fluid tended to increase after the heating power reached 65 W identically [133]. Bae et al. [134] established a numerical model of the OHP and simulated the change in the liquid film thickness. The numerical model was based on a 1D piston flow hypothesis. Figure 7a is a schematic diagram of two vertical heat pipes. Figure 7b is the plug flow on the z-axis. Figure 7c is a piston flow. Figure 7d is the liquid plug. Figure 7e is the plug (bubble). Figure 7f is the liquid film. Figure 7g is the pipe wall. The simulation results had an error of less than 20% compared with the experimental data, which proved that the oscillation prediction of fluids needs to consider membrane dynamics.

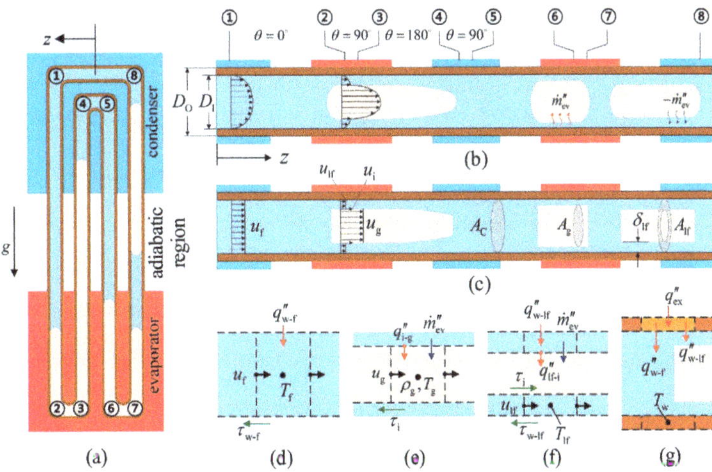

Figure 7. Schematics of vertical heat pipe and control volume analysis [134]. Bae et al., (2017). (a) Schematic. (b) Plug flow on z-axis. (c) One-dimensional plug flow. (d) Liquid slug. (e) Vapor plug. (f) Liquid film. (g) Wall.

Sun et al. [135] studied the hydro-thermodynamic behavior of the ethanol-based bubble distribution, bubble motion, and temperature of the working fluid. The proportion of small-sized bubbles increased with the improvement in the liquid filling rate and heating power. The proportion of medium-sized and large-sized bubbles decreased when the oscillation frequency and amplitude of bubbles increased. The high boiling point fluid working fluid heat pipes was studied by Mahapatra et al. [136] using the Buckingham's pi theorem to perform a dimensionless analysis of the heat transfer performance. High boiling point working fluids alleviated locally high heat flux densities. Xue et al. [137] conducted a novel full-visualization experiment on ammonia water OHPs with a high-speed camera. As the heating power increased, the flow pattern changed from a stopper flow to a ring flow and the proportion of evaporation heat increased from 7.7% to 32.4%. Liu et al. [138] analyzed the starting performance of the OHP based on system identification theory. The working fluids with small dynamic viscosity, small specific heat, and large saturation pressure gradients favored the start-up of OHPs. Hao et al. [139] studied the effect of polytetrafluoroethylene with ionized water, ethanol, and acetone as the working fluids on the heat transfer performance. When acetone was used as the working fluid, the liquid plug oscillation amplitude and speed were the highest and the thermal resistance was 30~63% lower than that with water as the working fluid.

3.6. Mix Liquids

Different pure working fluids have their own advantages under working conditions. Non-azeotropic mixtures have the characteristics of phase change and temperature fluctuation, which can make the heat source and working fluid well-matched [140]. The mixture plays the superior characteristics of each different components, which cause the OHP to achieve better start-up and heat transfer performances [141]. Zhu et al. [142] concluded that the thermal resistance of the OHP was filled with a ketone-pure–water mixture. When the filling rate was high, the thermal resistance of the OHP filled with pure water and acetone was 45.8% and 38.7% lower than that of the ketone-pure–water mixture. The mixture had better resistance to dryness at a low liquid filling rate [143]. Shi et al. [144] studied the OHP with ethanol–water, ethanol–methanol, and ethanol–acetone as working fluids with different mixing ratios. When the filling rate was increased to 62%, the heat transfer performance of OHPs with pure working fluids was better than that with mixed working fluids. When the filling rate reached 70%, the thermal resistance of the different working fluids tended to be approximated with the increase in the heating power. When the filling rate was low, the methanol working fluid can inhibit the drying up of the OHP [145]. Xu et al. [146] tested the effect of HFE-7100 and the lowest thermal resistance was 0.1634 °C/W with the mixing ratio of the working fluid was 1:2. Chang et al. [147] obtained the internal pressure of methanol-deionized water OHPs with different mass ratios. When the mass ratio of binary working fluid methanol and deionized water was 1:5, the starting performance of the OHP was the best with the temperature of 80 °C, the thermal resistance of 0.114 °C/W, and the heat flux density of 1.47 W/cm^2.

Markal and Varol [148] studied the effects of the volume mixing ratio, inclination angle, and fill ratio on the OHP thermal performance of ethanol (E)–pentane (P) mixtures under different heat inputs. The ethanol–pentane mixture exhibited lower thermal resistance and had the best thermal performance with the filling rate FR of 30%, the dip angle IA of 90°, and the mixing ratio E:P of 1:3. Under the same filling rate and inclination conditions, the ternary mixture of deionized water (W), methanol (M), and pentane (P) had a better thermal performance when the mixing ratio was 1:2:3, as shown in Figure 8. The best thermal performance occurred when the filling rate was 50% [149]. Compared with the two binary mixtures of ethanol–pentane and methanol–pentane, the overall performance of the ternary mixture was low. Markal and Varol [150] also compared the effects of pentane-methanol, methanol–hexane, and water–methanol–pentane mixtures on OHP thermal properties. The immiscible pentane–methanol (P:M = 1:1) mixture had better thermal properties than the mixture of hexane–pentane.

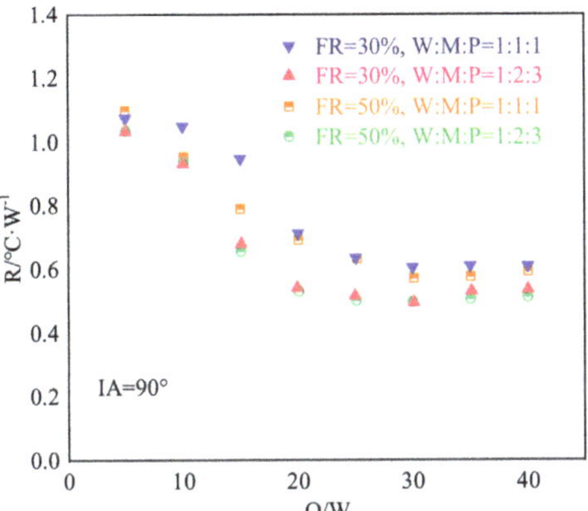

Figure 8. Change in thermal resistance with heat load for each mixture under different mixing ratios [149]. Markal et al., (2021).

3.7. Surfactants and Self-Rewetting Fluids

When the working fluid flows in the MCOHP, the flow of the working fluid will be affected by resistance due to the presence of surface tension. The surface tension of the working fluid can be reduced if surfactants are added. The capillary resistance can be reduced and the heat transfer performance of the OHP can be improved [151,152]. Hao et al. [153] conducted a series of experiments to study the effects of super-hydrophilic and hydrophilic surfaces on the segment plug motion of OHPs. The influence of surface wetting characteristics on the gas–liquid interface at the end of the plug is shown in Figure 9. The length of the film in super-hydrophilic OHPs is significantly increased. Compared with the copper OHP, the thermal resistance of the super-hydrophilic and hydrophilic OHP were reduced by about 5~15% and 15~25%, respectively. Xing et al. [154] obtained OHPs with a cetyltrimethylammonium bromide (CTAB) solution as the working fluid, which can reduce the surface tension of the solution and the contact angle. The thermal resistance of 0.25 wt.% of the CTAB solution is reduced by 48.5% with the filling rate of 50%. The addition of surfactants can increase the critical heat flux density of the heat pipe by 1.26 times with the enhancement of pressure fluctuations [155]. Compared with deionized hydraulic working fluid heat pipes, the thermal resistance of the cetyltrimethylammonium chloride working fluid was reduced by 4.78% [156], which prevents drying up. Bao et al. [157] experimentally proved that the thermal resistance of OHPs that had surfactants as the working fluid was reduced by a maximum of 27.8%.

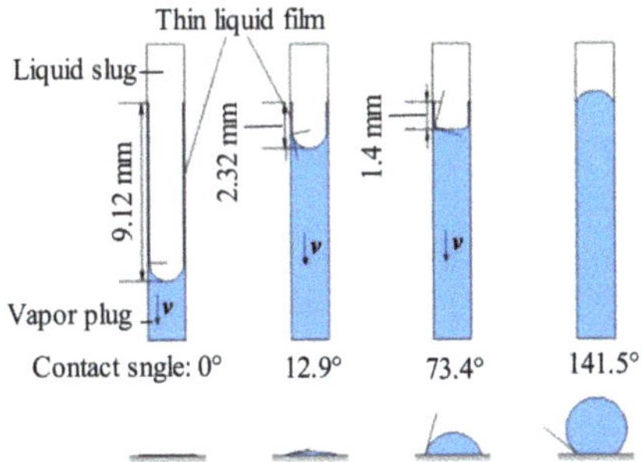

I. Superhydrophilic II. Hydrophilic III. Copper IV. Hydrophobic

Figure 9. Schematic of the liquid–gas interface in super-hydrophilic, hydrophilic, copper, and hydrophobic OHPs [153]. Hao et al., (2014).

Abe et al. [158] proposed the concept of SRWFs by the physical properties of dilute aqueous solutions with high carbon alcohols. A SRWF enhances the heat transfer performance and heat transfer limit of OHPs with an incensement of the surface tension and a reduction of the contact angle [159]. Hu et al. [160] used a heptanol–aqueous solution to study the enhancement effect of SRWFs. The characteristics of SRWFs caused the working fluid to be spontaneously wetted in the overheated part of the tube. Wu et al. [161] applied butanol at a concentration of 6% as a working fluid for performance testing. The critical heat load was 650 W, and the total thermal resistance was 0.25 °C/W with a reduction of 60%. The SRWF nano-fluids exhibited an excellent heat transfer performance over the entire heat load range, with a maximum enhancement rate of approximately 15% [162]. The influence of a SRWF nano-fluid base prepared by mixing graphene oxide dispersion with an n-butanol–aqueous solution on OHPs had been studied [163]. The following percentages, 0.07 wt.% and 0.7 wt.% were the optimal concentrations of graphene oxide and n-butanol. The heat transfer performances were increased by 16% and 12% compared with the SRWF and nano-fluid. Savino et al. [164] performed a microgravity heat pipe experiment. They researched working fluids including aqueous alcohol solutions, multi-component brine and nanoparticle suspensions. It was shown that self-rewetting brines and self-rewetting nanofluid brines have good thermal properties. Wang et al. [165] conducted numerical studies on CLOHP with different wettability (contact angles of 5°, 33°, 147°, and 175°). Figure 10 shows the volume fraction of the liquid and vapor distribution at a heat load 20 W. Compared with superhydrophobic surfaces, CLOHPs on super-hydrophilic surfaces had a 10.8% reduction in thermal resistance at an input heat load of 20 W.

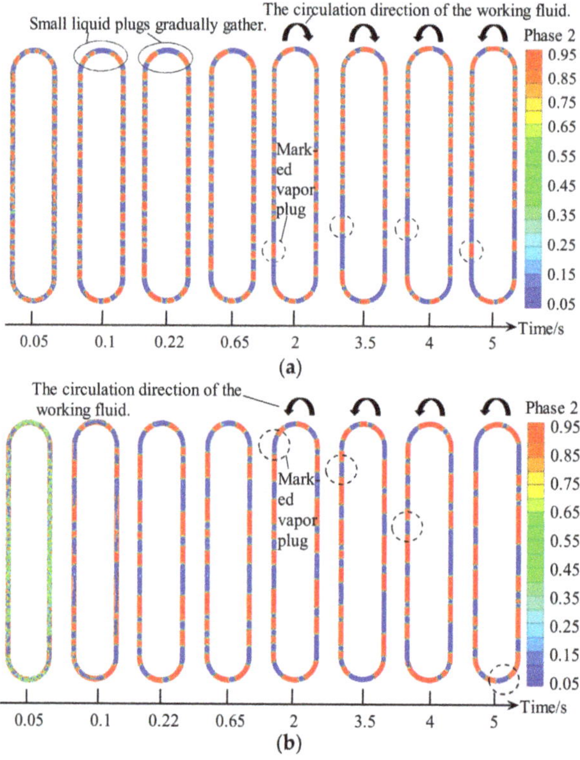

Figure 10. Volume fraction of liquid and vapor at heat load 20 W [165]. Wang et al., (2020). (**a**) Volume fraction of liquid and vapor at heat load 20 W and contact angle 147°. (**b**) Volume fraction of liquid and vapor at heat load 20 W and contact angle 33°.

4. Effect of In-Tube Flow State on Heat Dissipation Properties

The heating method, effect of gravity, characteristics of the flow pattern, and the oscillatory characteristics have influences on the heat dissipation properties.

4.1. Different Heating Method of Evaporation Section

An OHP has a very flexible use, which is reflected in its ability to heat in different positions; for example, its heating methods can be pulse heating, alternating heating, and continuous heating, and it can have a number of evaporation and condensation sections. Lin et al. [166] used vertical bottom heating to study the heat transfer performance of OHPs. The pulse heating method had a lower temperature difference than the continuous heating method [167]. Zhao et al. [168] found that the advantages of thermal resistance were pulse heating, alternating heating, and continuous heating. Taft et al. [169] compared the heat transfer performance of OHPs under DC and pulse modulation input modes. Chu et al. [170] used asymmetric heating to study a 3D helix OHP. The non-uniform heating method of multiple heat sources has also been studied in series with a two-channel flat OHP [171], as given in Figure 11a.

Mangini et al. [172] studied the heat transfer performance of hybrid OHPs in the super/microgravity environment and the non-uniform heating configuration promoted the net circulation of the fluid in the preferential direction, which improved the thermal performance relative to uniform heating. Peng et al. [173] took the bottom heating method to perform a numerical simulation study on the completely non-linear thermo-mechanical finite element model OHP. Qu et al. [174] used vertical and horizontal heating methods to study 3D-OHP. Yasuda et al. [175] studied flat-plate OHPs made of aluminum alloy by bottom heating and top heating. Lim et al. [176] used local heating to study the flat MOHP. As listed in Figure 11b, the heating method will directly affect the internal flow.

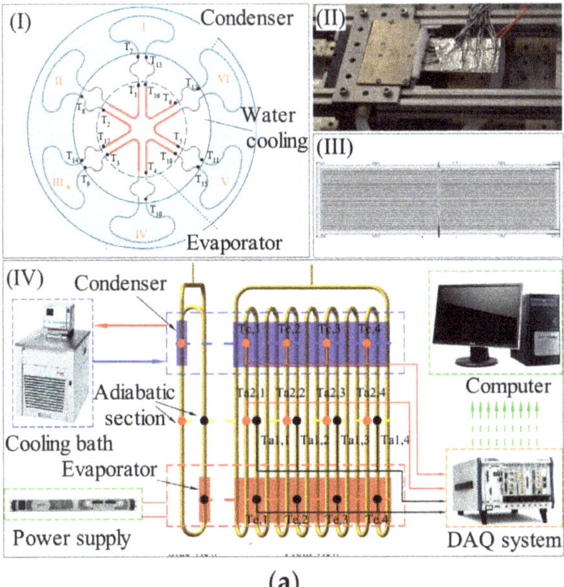

(a)

Figure 11. *Cont.*

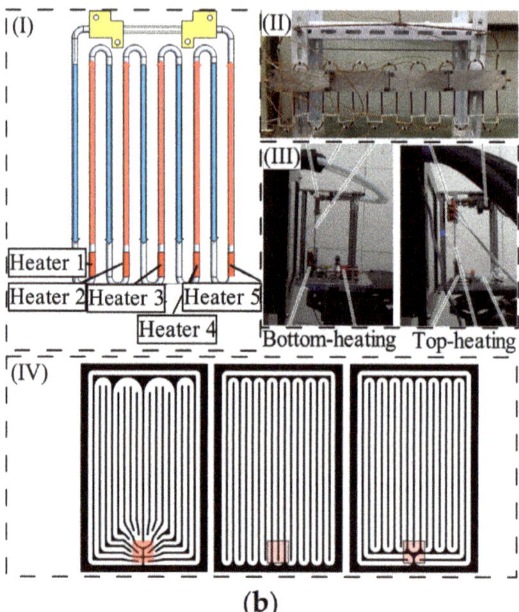

(b)

Figure 11. OHP for different heating methods. (**a**) OHP pulse heating and asymmetric heating method; (**b**) OHP for non-uniform heating and local heating. (I) [168]. Zhao et al., (2019); (II) [169]. Taft et al., (2017); (III) [171]. Chen et al., (2021); (IV) [170]. Chu et al., (2022); (**b**) OHP for non-uniform heating and local heating: (I) [172]. Mangini et al., (2017); (II) [174]. Qu et al., (2017); (III) [175]. Yasuda et al., (2022). (IV) [176]. Lim et al., (2021).

4.2. Flow State

E et al. [177] established a CLOHP model using the VOF (volume of fluid) method as the solution scheme to numerically simulate the liquid vapor in the two-phase conversion process. The distribution and fluctuation relationship of the pressure and vapor flow mode during the start-up under different vacuum degrees was determined. The interphase mass transfer due to evaporation and condensation in the VOF method can be applied to energy jump conditions, the Tanasawa model, and the Lee model [178]; the relevant equations are given in Table 4.

Table 4. Numerical model of interphase mass transfer.

Content	Remarks
Energy jump conditions [179] $$\dot{m} = -\frac{\left(L_v + (c_{p,l} - c_{p,v})(T_{sat} - T_{int})\right)}{[-k_{lv}\nabla T \cdot \vec{N}]_\Gamma} \quad (4)$$	$\dot{m}$ is the phase change local mass flow rate, kg/(m³·s). L_v is latent heat. $c_{p,v}$ is constant pressure specific heat of vapor, J/(kg·K). T_{sat} is the saturation temperature associated with the considered pressure, K. T_{int} is the local interface temperature, K. k_{lv} is thermal conductivity, W/(m·K). $\vec{N}$ is the normal vector pointing in the direction of the gas phase at the Γ of the interface.
Tanasawa model [180] $$\dot{m} = \frac{2\gamma}{2-\gamma}\left(\frac{M}{2\pi R_g}\right)^{1/2}\frac{\rho_v L_v(T_{if} - T_v)}{T_v^{3/2}} \quad (5)$$	γ is the adjustment factor. M is the molecular weight. R_g is a general gas constant. is 8.314 J/(mol K). ρ_v is the density of vapor, kg/m³. T_{if} is the interface temperature, K.
Lee model [181] $$\begin{cases} \dot{m}_{lv} = \alpha_l\rho_l \frac{c_{p,l}T_{sat}}{L_v} \cdot \frac{T_l - T_{sat}}{T_{sat}}, T_l > T_{sat} \\ \dot{m}_{vl} = \alpha_v\rho_v \frac{c_{p,v}T_{sat}}{L_v} \cdot \frac{T_{sat} - T_v}{T_{sat}}, T_{sat} > T_v \end{cases} \quad (6)$$	$\dot{m}_{lv}$ is the mass transfer of each time step in the evaporation process, kg/(m³·s). $\dot{m}_{vl}$ is the mass transfer of each time step in the condensation process, kg/(m³·s). α_l and α_v are the volume fraction of liquids and vapors.

The heat exchange units established by Nuntaphan et al. [182] can be used to evaluate the efficiency of the heat exchanger and the heat transfer coefficient of the air side. Qian et al. [183] proposed a novel heat transfer prediction model based on an extreme gradient boost algorithm and studied the design and cooling method of OHP prototypes during optimized processing. Sun et al. [184] established a model to study the oscillatory motion characteristics of liquid plug and vapor plug/bubble in OHPs. Nemati et al. [185] used numerical models to study the heat transfer mechanism of OHPs and predict heat transfer capacity, which simulated the oscillation behavior of the liquid plug, considering the thickness of the liquid film in the evaporation and the decrease in the liquid film thickness caused by evaporation. Daimaru et al. [186] proposed a numerical simulation method for OHPs with a check valve and new modeling features, including the pipe wall energy equation. The check valve model included pressure loss, the detailed boiling algorithm, and the pressure loss of a bending surface transformation. The temperature error of the heating section was less than 1.7 °C. Adachi et al. [187] developed transient models of fluid conditions to reproduce transmission lines. Odagiri et al. [188] combined thermo-fluid behavior in channels with thermal diffusion in OHP casing segments, which was in good agreement with the multi-branch OHP experiment results with a channel diameter of 1 mm and several turns of 42 turns.

4.3. Gravity Effect

Gravity has an important effect on the flow and circulation of the working fluid of OHPs. The gravity prevents the working fluid from flowing to the evaporation section in the top heating mode. The gravity promotes the flow of the working fluid to the evaporation section in the bottom heating mode. For micro-gravity research, the European Space Agency used parabolic flights of aircrafts to create different gravity environments. The variation of the tilt angle changed the flow pattern inside the OHP, resulting in different performance levels [189]. Mameli et al. [190] studied the OHP under different gravities. The change in gravity had a greater effect on the heat transfer performance of the OHP in the case of vertical heating compared with horizontal heating. The FC-72 working fluid was investigating the effect of gravity on the heat transfer performance at different heating powers [191]. The thermal resistance of vertical heating was lower than that of horizontal heating under the influence of gravity. The vertical heating was not as stable as the horizontal heating during operation. Ayel et al. [192] reported that closed-loop FPOHPs can respond to gravity changes more quickly and reach a steady state. Mangini et al. [193] tested the sudden loss of buoyancy and activated the oscillating segment plug/stuff flow state in micro-gravity, which had the lowest starting power. Cecere et al. [194] discovered that FPOHPs with SRWF (butanol–water) were easier to keep working under micro-gravity and a low heating power. Xing et al. [195] examined the effect of gravity on OHPs with a surfactant solution as the working fluid. The influence of gravity on the OHPs of the CTAB solution was relatively small. When the heating power was higher, the heat transfer performance of the CTAB solution OHP was stronger and the thermal resistance was reduced by 51%. Pagliarini et al. [196] trained the OHP used for the International Space Station in microgravity, where the two states of the working condition included intermittent flow (episodic fluid motion occurring in some channels) and full activation (steady fluid movement throughout the adiabatic section). The fully activated state is given in Figure 12. No significant variation between channel behaviors is observed with stable oscillations, high heat flux amplitude, and oscillation frequency. The heat flux amplitude increased almost linearly, 1500 W/m^2 at 202 W, with the power input from 1100 W to 202 W.

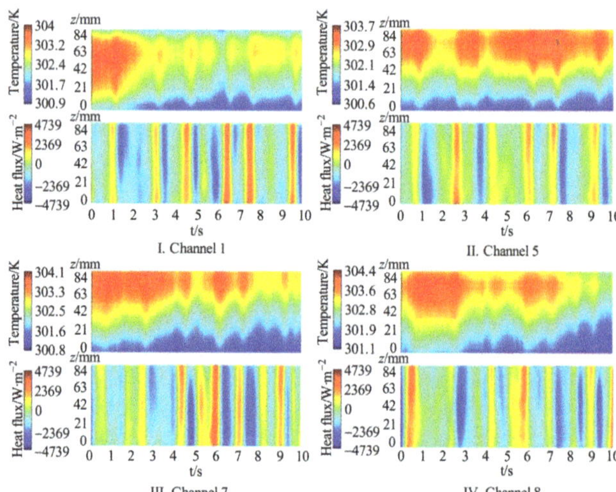

Figure 12. Wall temperature and heat fluxes in different channels under 202 W power input with stable microgravity conditions [196]. Pagliarini et al., (2021).

4.4. Characteristics of Flow Pattern in Tube

Awareness of flow boiling and two-phase instability are important parts of understanding the complex phenomena and developing OHP technology, to explore the physical mechanisms controlling the complex unsteady flow boiling heat transfer and two-phase phenomena [197], which provide insight into the heat and mass transfer relationships in OHPs. With the change in the OHP working stage, the flow pattern of the working fluid in OHPs also changes, which can directly affect the flow mode of the working fluid and the heat and mass transfer efficiency of OHPs. Yuan et al. [198] established a flow model of the liquid plug based on the Lagrange method. Liquid plug oscillation amplitude and angular frequency depend on the geometry of the OHP and the liquid filling rate. When the flow type is slug flow, the sensible heat transfer can account for more than 80% of the total heat transfer. Karthikeyan et al. [199] employed high-resolution infrared thermography to measure the flow characteristics inside OHPs. The flow is the working fluid without internal oscillation, intermittent oscillation, or continuous local oscillation. With the increase in heating power, the thermal resistance decreased from 1.90 K/W to 0.24 K/W. Spinato et al. [200] used time-strip image processing techniques to study the two-phase flow of the OHP. Low amplitude/high amplitude oscillations, cyclic oscillations, backflow, and steady cycles were observed. The nucleation and rapid growth of bubbles in the U-bend of the evaporation section lead to the transition of the working fluid from the circulating to the oscillating state. Xian et al. [201] visualized the flow behavior in an OHP duct with pulsed heating. The liquid film was thinner in pulsed heating than in continuous heating. Under the condition of a short period of pulse heating, the proportion of bubble flows increased. The flow pattern is the same as for continuous heating. Pouryoussefi et al. [202] worked with numerical methods to simulate the chaotic behavior of fluids, as listed in Figure 13. Volume fraction contours are provided for different time points for two different operating conditions (red is vapor and blue is liquid). Figure 13a is the volume fraction diagram of the OHP under a time series of 0.8 s, 2.5 s, 3.8 s, 5.5 s, 13 s, and 18 s, with the evaporation temperature $T_h = 145\,°C$, the condensation temperature $T_c = 35\,°C$, and the liquid filling rate of 30%. Figure 13 b presents the volume fraction plots for the time sequences of 2.2 s, 3.8 s, 5.8 s, 10 s, 16.5 s, and 18.6 s, with $T_h = 150\,°C$, $T_c = 35\,°C$, and a 60% filling rate. The relevant dimension increases by promoting the filling rate and evaporation temperature.

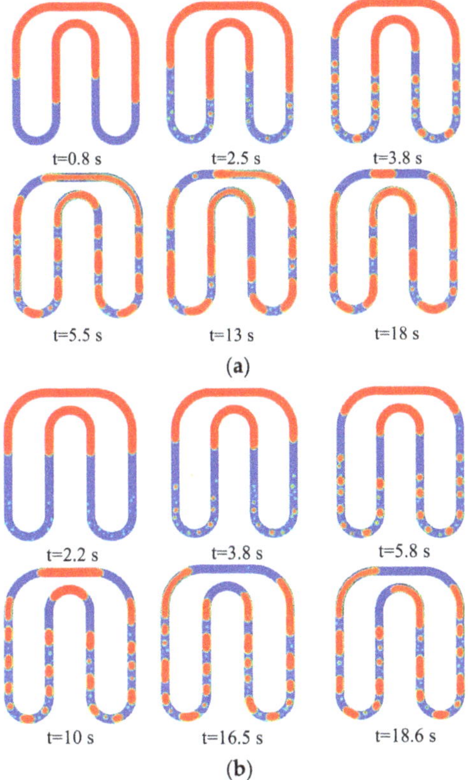

Figure 13. Volume fraction contours after formation of fluid flow [202]. Pouryoussefi et al., (2016). (a) Th = 145 °C, Tc = 35 °C, and 30% liquid filling rate; (b) Th = 150 °C, Tc = 35 °C and 60% liquid filling rate.

Feldmann et al. [203] simulated turbulence in a pipe at different Womersley numbers and found that disturbance energy is required to trigger the non-linear transition process in the subcritical state. Pouryoussefi et al. [204] modified the chaotic flow in a heat pipe based on the VOF method. There was an upper limit to the accuracy of the simulation as the fluid filling rate and heating power increased. The optimal filling rate and minimum thermal resistance were measured to be 60% and 1.62 °C/W. Mangini et al. [205] demonstrated the reliability of the infrared visualization of two-phase flows with a maximum error of ±1.5 °C in combination with a high-speed camera capable of detecting the wetting and drying of liquid films. Xia et al. [206] studied the properties of unsteady flows in parallel micro-channels. Continuous two-phase unsteady boiling often occurred when the flow rate and heat flow density were greater than 607.6 kg/m² and 30 W/cm². This phenomenon can be suppressed by increasing the forced convection heat transfer of 100% or increasing the flow boiling heat transfer of 50%. Yoon et al. [207] investigated the oscillation frequency of the liquid plug with a MOHP. The heat input can change the vapor temperature and affect the oscillation frequency. The oscillation frequency increased with the increase in the heat input. The liquid plug with a longer total length of the heat pipe had a lower oscillation frequency. Ling et al. [208] found that the heat pipe temperature fluctuated greatly during the stable circulation of the working fluid. Noh et al. [209] used a numerical model to simulate the heat transfer between the tube wall and the liquid slug/vapor plug and proposed guidelines for designing heat pipes. Figure 14a is the liquid slug and vapor plug distributions for a two-turn heat pipe with an input power of 50 W (t = 145 s) and a twenty-turn OHP with an input power of 48 W (t = 80 s). Figure 14b illustrates the heat flux distribution for the two heat pipes and Figure 14c presents the wall temperature distribution for the two OHPs. After the pseudo-steady state, the liquid slug in the two-loop heat pipe oscillates and shows a net circulating flow.

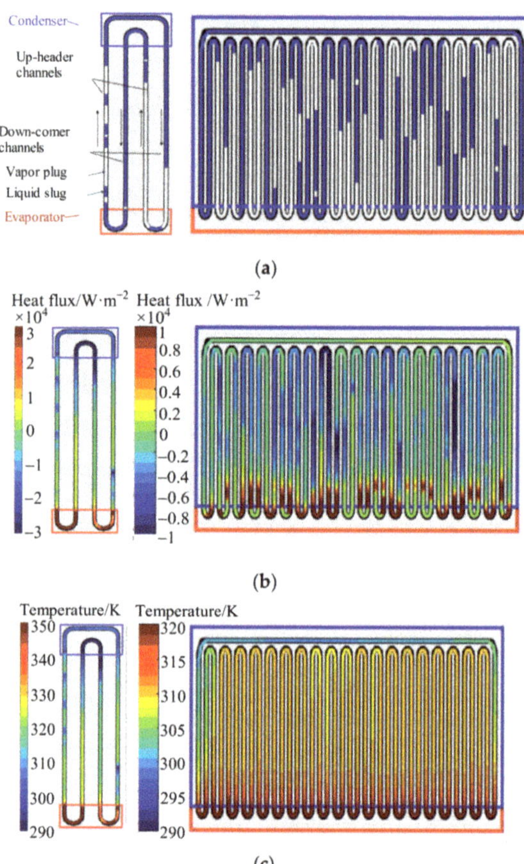

Figure 14. Heat pipe visualization data graph [209]. Noh et al., (2020). (**a**) Distribution of liquid slug and vapor plug. (**b**) Heat flux distribution. (**c**) Wall temperature distribution.

Ahmad et al. [210] investigated the heat transfer performance as a function of the flow pattern using OHPs with ethanol. The thermal resistance at a 50% filling rate was as low as 1.6 °C/W. Vo et al. [211] performed visualization experiments on the OHP and found that the cyclic motion dominated the motion of the working fluid. A 3D computational fluid dynamics model of the OHP was developed and the k-ε turbulence model was applicable to the heat pipe simulation. Schwarz et al. [212] discovered through visualization experiments that the thermal resistance of the fluid inside the heat pipe during start-up is constant at 0.43 °C/W when there is no flow (25 W to 75 W). The thermal resistance dropped to 0.34 °C/W (200 W)~0.36 °C/W (100 W) at an average velocity of 240 mm/s.

4.5. Oscillatory Characteristics

The pulsed heat pipe is a kind of non-equilibrium, passive two-phase heat transfer device with complicated internal transmission process. The heat transfer performance of OHPs depends on the oscillation degree and stability of the self-sustained oscillating two-phase flow, which leads to its unique heat transfer characteristics [213]. Yoon et al. [214] observed that the internal oscillation frequency of the OHP was between 40 Hz and 50 Hz.

Spinato et al. [215] examined the flow behavior inside the OHP using time-slot image processing technology. The flow pattern changed from oscillating to circulating state while the main frequency changed from 1.2 Hz to 0.6 Hz. Dilawar et al. [216] researched the oscillating two-phase flow in a micro-channel based on a numerical model. The pressure loss of the oscillating two-phase flow at the bend of the tube reduced the oscillation amplitude and weakened the heat transfer performance. Kato et al. [217] investigated the OHP consisting of a single straight pipe and open pipe. The heat transfer performance was enhanced with an increased amplitude of the oscillatory flow. Liquid exchange due to the oscillatory motion significantly enhanced the heat transfer with an effective thermal conductivity of up to 40 kW/(m·K). Miura et al. [218] studied liquid column oscillations with a forced oscillator. The evaporation of liquid film produced by the oscillation of working fluid is the main process of latent heat transfer. Both evaporation and condensation occur on the liquid film on the wall of micro-channel. Daimaru et al. [219] processed oscillation data based on a fast Fourier transform and mutual analysis. The vapor plug received or applied energy according to the direction of propagation. Das et al. [220] suggested a theoretical model considering the two-phase oscillatory equilibrium, which was calculated in good agreement with the experiment results. Jung et al. [221] evaluated the effect of the oscillation amplitude of a miniature OHP on the heat transfer performance. When the input power was 16 W, the time-resolved distribution of the heat flux and the corresponding flow visualization after the miniature OHP reached the pseudo-steady state are given in Figure 15. The total heat transfer rates of the evaporation and condensation sections were 14.4 W and 14.2 W, while the average value of latent heat ratio increased from 54.8% to 81.9%. When the input power increased from 7 W to 16 W, the oscillation amplitude increased from 3.4 mm to 8.3 mm, with a 13.5% reduction in thermal resistance. The oscillating motion of the liquid slug and vapor plug in the OHP promoted the heat transfer between the evaporation section and the condensing section. The oscillation amplitude and frequency are important parameters for estimating the heat flow density in OHPs [222].

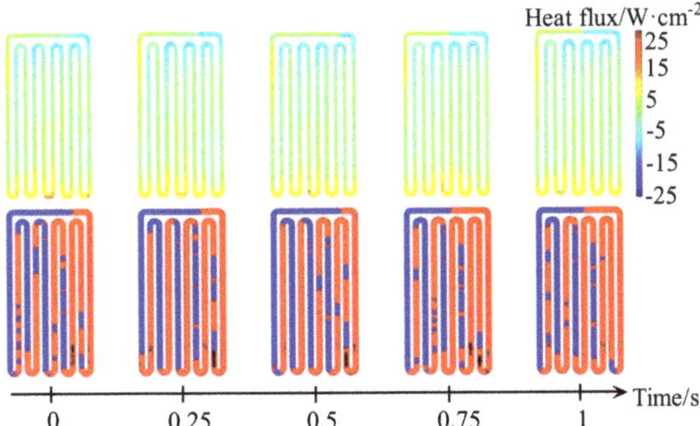

Figure 15. Time-resolved distribution and flow visualization of heat flux with input power of 16 W [221]. Jung et al., (2021).

Pai et al. [223] introduced a non-linear thermal model based on a U-shaped three-plug OHP. When the ratio of liquid slug mass or vapor plug length to the tube cross-sectional area decreased, the initial air pressure, liquid filling rate, gravity, and the oscillation frequency increased. As the temperature difference or heat transfer coefficient between the evaporation and condensation sections increased, the liquid filling rate or the initial temperature decreased. Perna et al. [224] tested the main frequency of pressure signal oscillation in the evaporation and condensation sections under micro-gravity. The main frequency of the pressure signal was in the range of 0.6~0.9 Hz, which increased with the heating power. Simplifying the entire OHP into a single unit can facilitate the study of various parameters of oscillation behavior [225]. Rao et al. [226,227] investigated self-sustained thermally driven oscillating in micro-channels. The motion of the meniscus generated a liquid film on the tube wall. The thickness of the liquid film and length at a given time determined the overall dynamics of the meniscus. Fourgeaud et al. [228] examined the variation of the liquid film thickness on the tube wall based on a single branch OHP. The film thickness was larger compared with the wedge-shaped film in a capillary OHP.

5. Bibliometric Study and Analysis

The OHP was first proposed in the 1990s and it has aroused wide attention in the academic circle. Scholars around the world have published a lot of research in this field. In order to understand the data from the relevant literature, we analyzed the literature on the oscillating heat pipe from the Web of Science Core Collection (WOSCC) during the 20 years from 2003 to 2022, through the research method of bibliometrics and through the data visualization analysis. Specific data are given as follows.

5.1. Publication Year and Number of Publications

In the WOSCC statistics, a total of 680 oscillating heat pipe literatures were published from 2003 to 2022. The number of papers published each year reflect trends in a particular field of research. As shown in Figure 16, the number of published articles in this field peaked at 65 in 2017. In the past two years, the number of published documents has fluctuated, but the overall trend is a slow rise.

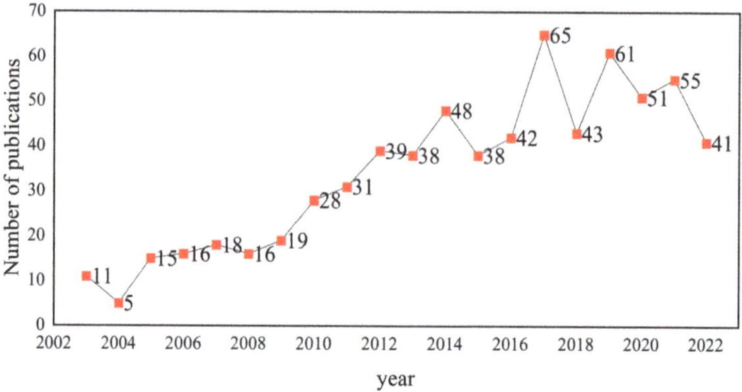

Figure 16. The growth trend of relevant literatures from 2003 to 2022.

5.2. Keyword Distribution

A keywords statistics of the above 680 literatures were carried out. A total of 41 keywords appeared more than 15 times. The top three keywords with the highest frequency are oscillating heat pipe, thermal performance, and flow, and the times are 135, 106, and 89, respectively. Figure 17 shows the connections among various words. The lines represent the number of times that these two keywords appear at the same time in a piece of literature. The thicker the lines are, the higher the number of times they appear together. The number of co-occurrences between these words is very high. Among them, the co-occurrence frequency of thermal performance and oscillating heat pipe is the highest, while the co-occurrence frequency of oscillating heat pipe and flow rank second. It can be seen that the OHP is closely related in the study of thermodynamics and fluid mechanics.

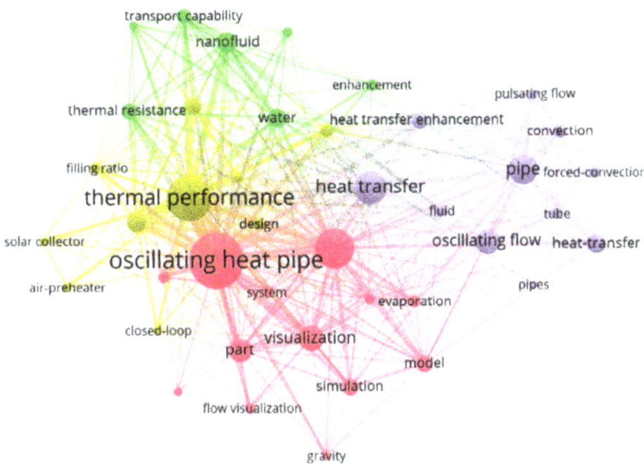

Figure 17. Analysis of each keyword.

6. Current Research Shortcomings and Prospects for Future Works

6.1. Shortcomings of the Current Studies

(1)　Effects of channel layout

Studies of OHPs have focused on the effect of the fluid volume on the heat transfer performance, which neglected the synergistic effects of the layout optimization and internal flow [229].

(2)　Limitation of materials

The material of the OHPs tube wall affects the heat transfer performance and the layout plasticity of the OHPs [230]. The overwhelming majority of the tube wall materials in the current study are conventional materials of copper, aluminum, and stainless steel [231,232]. The thermal conductivity of copper is relatively high, and some composite metal tube wall materials may have better performances in OHPs.

(3)　Insufficient understanding of working fluid properties

The performance of the working fluid is the core material for OHPs, which directly affects the heat transfer [233]. Most of the working fluids studies have focused on water, ethanol, etc. [234,235]. For mixed nano-fluids as well as non-azeotropic mixture liquids, the effect of the concentration and mixing ratio on the optimal thermal performance of OHPs should be explored in depth. In addition, there is a lack of studies on the time-dependent properties of nano-fluid OHPs.

(4)　Inadequate recognition of the heat transfer mechanism

The operation mechanism of OHPs is complex. It is not beneficial to optimize the performance of OHPs with a single thermodynamic theory to explain and derive the mechanism [236]. The interaction and influence between vapor and liquid phases have not been deeply studied, which provides a lack of accurate predictive numerical models [237].

(5)　Limitations of the actual application environment

The superior heat transfer performance of OHPs should be focused on the practical use aspects. The performance of OHPs [238,239] in different use environments (e.g., rotating, centrifugal, antigravity, and horizontal) can vary significantly from operating under laboratory conditions. The actual application scenario environment will limit the operation under different working conditions [240].

6.2. Prospects for Future Research

(1)   Further study of mixed nano-fluids

The nano-fluids have great potential for effective heat transfer as OHP working fluids. Through the collaborative combination of different types of nano-particles, the mixed nano-fluids can achieve higher advanced thermophysical properties and stability of single nano-fluids [241]. The influence of the mixed nano-fluid on the thermal performance of OHPs can be further studied by the optimization, analysis, and improvement of the thermal efficiency of the mixed nano-fluid [242,243].

(2)   Study of non-azeotropic mixtures

Theoretical analysis shows that the non-azeotropic immiscible binary mixture can expand the operating temperature range of OHPs [244]. The non-azeotropic immiscible binary mixture can solve the problems of starting temperatures and the drying under the high heat flux of OHPs [245,246], which can improve the heat transfer performance and heat transfer limit of OHPs [247,248]. Future studies can be performed on non-azeotropic mixtures OHPs with different configurations by the analysis of the correlation among profile mixing ratio, filling ratio, and power input.

(3)   The improvement in the numerical model

VOF is a widely used method in order to study the complex coupling of pressure and temperature of OHPs, which can capture the phase distribution and interface dynamics [249, 250]. Lee's model is used to explain the phase transition principle. Numerical model improvements can be made to better cover the microscopic and macroscopic levels [251]. Dreiling et al. [252] proposed a closed-interface tracking CFD-VOF method. The effect of factors of the film around the bubble and the curvature of the interface, turbulence, or mass transfer strength parameters on the OHP can be systematically investigated. More stable and effective turbulence models that are applicable for OHPs need to be developed in the prospective.

(4)   The combination of artificial intelligence technology

Artificial intelligence techniques, in terms of efficiency and intelligence, have also been applied in the field of OHPs [253]. Many cases of research with deep learning algorithms optimize the parameters of OHPs [254]. With the development of AI technology, the computational work of optimization algorithms has become more comfortable. Wen et al. [255] modeled acetone OHP thermal resistance with human neural networks, and the R-squared of the models proposed with MLP and GMDH were 0.989 and 0.965, which were able to predict and simulate acetone-filled OHP thermal resistance. Jokar et al. [256] proposed OHP simulation and optimization with the genetic algorithm approach and the obtained the optimum filling rate of 38.25%.

7. Conclusions

This paper describes the various filling working fluids and in-tube flow states on the heat dissipation properties of OHPs. The effect of metal nano-fluids, non-metallic nano-fluids, mixed nano-fluids, gas working fluids, organic solvents, mixed liquids, and SRWF fluids are given. The different heating methods of the evaporation section, flow state, gravity effect, flow pattern, and oscillatory characteristics of the tube are illustrated. The following are the main conclusions.

(1)   With the addition of non-metallic nano-fluids in OHPs, the thermal resistance decreases from 24% to 83.6% with the change in the type, size, and concentration of nano-particles. The maximum heat pipe thermal conductivity was enhanced by 105% with the graphene nano-fluids. OHPs with gas as the working fluid can be used in the field of low temperature cooling. The effective thermal conductivity varies from 4.8 kW/(m·K) to 70 kW/(m·K) when different gases are selected as the working fluid in OHPs.

(2)   Compared with the pure working fluid, the thermal resistance of OHPs can be reduced by 68.9% with the right mixture type, filling rate, and mixing ratio. The surfactant and SRWF can be added to reduce the surface tension of the working fluid and the thermal resistance of the OHP can be reduced by 4.78% to 60%.

(3)   The change in gravity has a significant effect on the heat transfer performance of OHPs with vertical heating. A sensible heat transfer can account for more than 80% of the total heat transfer when the internal flow type of OHPs is the slug flow. The heat transfer performance is enhanced with the increase in the oscillatory flow, amplitude and the effective thermal conductivity can reach 40 kW/(m·K). The input power is increased from 7 W to 16 W, the oscillation amplitude is increased from 3.4 mm to 8.3 mm, and the thermal resistance is reduced by 13.5%.

Author Contributions: X.Z.: conceptualization, writing and check.; L.S.: data processing, writing and check; J.J.: editing, correction; W.D.: drawing, English proofreading; D.Z.: conceptualization, editing, and English revision. All authors have read and agreed to the published version of the manuscript.

Funding: This work is supported by the Young Scientists Fund of the National Natural Science Foundation of China under the research grant of 52106151 and Guangdong Basic and Applied Basic Research Foundation of 2023A1515011918.

Data Availability Statement: Not applicable.

Conflicts of Interest: The authors declare that they have no conflict of interests regarding the publication of this paper.

Nomenclature

R_{th}	the thermal resistance, K/W
$\overline{T}_{evap}$	the average temperature of the evaporation, K
$\overline{T}_{cond}$	the average temperature of the condensation, K
Q_{in}	the input power, W
t_a	the time interval for time averaging, s
W	the width of OHP, m
L	the length of OHP, m
T_{Si}	the temperature of the silicon substrate, K
x	the horizontal coordinate, m
t	the time, s
$Q_{w,H}$	the total heat transferred from the heating wall to the liquid film, W
$Q_{w,C}$	the total heat transferred from the liquid film to the cooling wall, W
T_w	the wall temperature, K
$T_{sur,H}$	the temperature of liquid film during heating K
$T_{sur,C}$	the temperature of cooling, respectively, K
λ_l	the thermal conductivity of the liquid, W/(m·K)
δ	the liquid film thickness, m
$c_{p,l}$	the specific heat of the liquid, J/(kg·K)
ρ_l	the density of the liquid, kg/m^3
A_{cr}	the cross-sectional area of the flow path, m^2
T_w	the temperature of the channel wall, K
T_l	the temperature of liquid plug, K
$h_{w,l}$	the heat transfer coefficient between channel wall and liquid plug, W/(m^2·K)
S	the perimeter of the liquid plug, m
T_v	the temperature of the vapor plug, K
$Q_{v,sen}$	the total amount of sensible heat, W
$Q_{v,lat}$	the total amount of latent heat, W
P_v	the pressure of vapor plug, Pa
V_v	the volume of the vapor plug, m^3

$c_{v,v}$	the specific heat at constant volume, J/(kg·K)
m_v	the mass of the vapor plug, kg
$\dot{m}$	the phase change local mass flow rate, kg/(m³·s)
L_v	the latent heat
$c_{p,v}$	the constant pressure specific heat of vapor, J/(kg·K)
T_{sat}	the saturation temperature associated with the considered pressure, K
T_{int}	the local interface temperature, K
k_{lv}	the thermal conductivity, W/(m·K)
$\vec{N}$	the normal vector pointing in the direction of the gas phase at the Γ of the interface
γ	the adjustment factor
M	the molecular weight
R_g	the general gas constant, J/(mol·K)
ρ_v	the density of vapor, kg/m³
T_{if}	the interface temperature, K
$\dot{m}_{lv}$	the mass transfer of each time step in the evaporation process, kg/(m³·s)
$\dot{m}_{vl}$	the mass transfer of each time step in the condensation process, kg/(m³·s)
α_l	the volume fraction of liquid
α_v	the volume fraction of vapor
MCOHP	Micro-channel oscillating heat pipe
OHP	Oscillating heat pipe
TS	Tube diameter
CVOHP	Oscillating heat pipe with check valves
CLOHP	Closed-loop oscillating heat pipe
FPOHP	Flat-plate oscillating heat pipe
SRWF	Self-rewetting fluid
MWCNTs	Multi walled carbon nano-tubes
EOHP	Ethane oscillating heat pipe
CTOHP	Capillary tube oscillating heat pipe
CTAC	Cetyltrimethylammonium chloride
CTAB	Cetyltrimethylammonium bromide
VOF	Volume of fluid

References

1. Zhao, D.; Guan, Y. Characterizing Modal Exponential Behaviors of Self-excited Transverse and Longitudinal Combustion Instabilities. *Phys. Fluids* **2022**, *34*, 024109. [CrossRef]
2. Zhao, D.; Ji, C. Non-orthogonality analysis of thermoacoustic system with a premixed V-shaped flame. *Energy Convers. Manag.* **2014**, *85*, 102–111.
3. Li, S.; Zhao, D.; Li, J. Combustion Instabilities in a Bifurcating Tube: Open- and Closed-loop Measurements. *AIAA J.* **2014**, *52*, 2513–2523. [CrossRef]
4. Zhao, D.; Li, J. Prediction of stability behaviors of longitudinal and circumferential eigenmodes in a choked thermoacoustic combustor. *Aerosp. Sci. Technol.* **2015**, *46*, 12–21.
5. Zhao, D.; Guan, Y.; Reinecke, A. Characterizing Hydrogen-fuelled Pulsating Combustion on Thermodynamic Properties of a Combustor. *Commun. Phys.* **2019**, *2*, 44. [CrossRef]
6. Zhao, X.; Jiaqiang, E.; Liao, G.; Zhang, F.; Chen, J.; Deng, Y. Numerical simulation study on soot continuous regeneration combustion model of diesel particulate filter under exhaust gas heavy load. *Fuel* **2021**, *290*, 119795. [CrossRef]
7. Zhao, X.; Jiaqiang, E.; Zhang, Z.; Chen, J.; Liao, G.; Zhang, F.; Leng, E.; Han, D.; Hu, W. A review on heat enhancement in thermal energy conversion and management using Field Synergy Principle. *Appl. Energy* **2020**, *257*, 113995. [CrossRef]
8. Ji, C.; Zhao, D. Two-dimensional lattice Boltzmann investigation of sound absorption of perforated orifices with different geometric shapes. *Aerosp. Sci. Technol.* **2014**, *39*, 40–47. [CrossRef]
9. Zhao, D.; Reyhanoglu, M. Feedback control of acoustic disturbance transient growth in triggering thermoacoustic instability. *J. Sound Vib.* **2014**, *333*, 3639–3656. [CrossRef]
10. Zhao, D.; Gutmark, E.; Reneicke, A. Mitigating self-excited flame pulsating and thermoacoustic oscillations using perforated liners. *Sci. Bull.* **2019**, *64*, 941–952. [CrossRef]
11. Zhao, D.; Li, S.; Yang, W.; Zhang, Z. Numerical investigation of the effect of distributed heat sources on heat-to-sound conversion in a T-shaped thermoacoustic system. *Appl. Energy* **2015**, *144*, 204–213. [CrossRef]
12. Zhang, Z.; Zhao, D.; Ni, S.; Sun, Y.Z.; Wang, B.; Chen, Y.; Li, G.; Li, S. Experimental characterizing combustion emissions and thermodynamic properties of a thermoacoustic swirl combustor. *Appl. Energy* **2019**, *235*, 463–472. [CrossRef]

13. Zhao, D. Transient growth of flow disturbances in triggering a Rijke tube combustion instability. *Combust. Flame* **2012**, *159*, 2126–2137. [CrossRef]
14. Thompson, S.M.; Ma, H.B.; Winholtz, R.A.; Wilson, C. Experimental investigation of miniature three-dimensional flat-plate oscillating heat pipe. *J. Heat Transf.* **2009**, *131*, 043210. [CrossRef]
15. Zhao, D.; Qin, J.; Zheng, L.; Cao, M. Amorphous vanadium oxide/molybdenum oxide hybrid with three-dimensional ordered hierarchically porous structure as a high-performance Li-ion battery anode. *Chem. Mater.* **2016**, *28*, 4180–4190. [CrossRef]
16. Im, Y.H.; Lee, J.Y.; Ahn, T.I.; Youn, Y.J. Operational characteristics of oscillating heat pipe charged with R-134a for heat recovery at low temperature. *Int. J. Heat Mass Transf.* **2022**, *196*, 123231. [CrossRef]
17. Charoensawan, P.; Wilaipon, P.; Seehawong, N. Flat plate solar water heater with closed-loop oscillating heat pipes. *Therm. Sci.* **2021**, *25 Pt A*, 3607–3614. [CrossRef]
18. Dehshali, M.E.; Nazari, M.A.; Shafii, M.B. Thermal performance of rotating closed-loop pulsating heat pipes: Experimental investigation and semi-empirical correlation. *Int. J. Therm. Sci.* **2018**, *123*, 14–26. [CrossRef]
19. Rittidech, S.; Sangiamsuk, S. Internal flow patterns on heat transfer performance of a closed-loop oscillating heat pipe with check valves. *Exp. Heat Transf.* **2012**, *25*, 48–57. [CrossRef]
20. Gully, P.; Bonnet, F.; Nikolayev, V.S.; Luchier, N.; Tran, T.Q. Evaluation of the vapor thermodynamic state in PHP. *Heat Pipe Sci. Technol. Int. J.* **2014**, *5*, 369–376. [CrossRef]
21. Barrak, A.S.; Saleh, A.A.M.; Naji, Z.H. An experimental study of using water, methanol, and binary fluids in oscillating heat pipe heat exchanger. *Eng. Sci. Technol. Int. J.* **2020**, *23*, 357–364. [CrossRef]
22. Shafii, M.B.; Arabnejad, S.; Saboohi, Y.; Jamshidi, H. Experimental investigation of pulsating heat pipes and a proposed correlation. *Heat Transf. Eng.* **2010**, *31*, 854–861. [CrossRef]
23. Lin, H.; Tian, P.; Luo, C.; Wang, H.; Zhang, J.; Yang, J.; Peng, H. Luminescent nanofluids of organometal halide perovskite nanocrystals in silicone oils with ultrastability. *ACS Appl. Mater. Interfaces* **2018**, *10*, 27244–27251. [CrossRef] [PubMed]
24. Senjaya, R.; Inoue, T. Bubble generation in oscillating heat pipe. *Appl. Therm. Eng.* **2013**, *60*, 251–255. [CrossRef]
25. Ando, M.; Okamoto, A.; Nagai, H. Start-up and heat transfer characteristics of oscillating heat pipe with different check valve layouts. *Appl. Therm. Eng.* **2021**, *196*, 117286. [CrossRef]
26. Wang, W.W.; Wang, L.; Cai, Y.; Yang, G.B.; Zhao, F.Y.; Liu, D.; Yu, Q.H. Thermo-hydrodynamic model and parametric optimization of a novel miniature closed oscillating heat pipe with periodic expansion-constriction condensations. *Int. J. Heat Mass Transf.* **2020**, *152*, 119460. [CrossRef]
27. Shi, W.; Chen, H.; Pan, L.; Wang, Q. Starting and running performance of a pulsating heat pipe with micro encapsulated phase change material suspension. *Appl. Therm. Eng.* **2022**, *212*, 118626. [CrossRef]
28. Lim, J.; Kim, S.J. Effect of a channel layout on the thermal performance of a flat plate micro pulsating heat pipe under the local heating condition. *Int. J. Heat Mass Transf.* **2019**, *137*, 1232–1240. [CrossRef]
29. Liu, X.; Han, X.; Wang, Z.; Hao, G.; Zhang, Z.; Chen, Y. Application of an anti-gravity oscillating heat pipe on enhancement of waste heat recovery. *Energy Convers. Manag.* **2020**, *205*, 112404. [CrossRef]
30. Hao, T.; Ma, H.; Ma, X. Experimental investigation of oscillating heat pipe with hybrid fluids of liquid metal and water. *J. Heat Transf.* **2019**, *141*, 071802. [CrossRef]
31. Schwarz, F.; Messmer, P.; Lodermeyer, A.; Danov, V.; Fleßner, C.; Becker, S.; Hellinger, R. Analysis of improved pulsating heat pipe designs for hot spot applications. *Int. J. Heat Mass Transf.* **2022**, *196*, 123294. [CrossRef]
32. Kwon, G.H.; Kim, S.J. Operational characteristics of pulsating heat pipes with a dual-diameter tube. *Int. J. Heat Mass Transf.* **2014**, *75*, 184–195. [CrossRef]
33. Liu, X.; Chen, X.; Zhang, Z.; Chen, Y. Thermal performance of a novel dual-serpentine-channel flat-plate oscillating heat pipe used for multiple heat sources and sinks. *Int. J. Heat Mass Transf.* **2020**, *161*, 120293. [CrossRef]
34. Arai, T.; Kawaji, M. Thermal performance and flow characteristics in additive manufactured polycarbonate pulsating heat pipes with Novec 7000. *Appl. Therm. Eng.* **2021**, *197*, 117273. [CrossRef]
35. Okazaki, S.; Fuke, H.; Ogawa, H. Performance of circular Oscillating Heat Pipe for highly adaptable heat transfer layout. *Appl. Therm. Eng.* **2021**, *198*, 117497. [CrossRef]
36. Kim, S.; Zhang, Y.; Choi, J. Effects of fluctuations of heating and cooling section temperatures on performance of a pulsating heat pipe. *Appl. Therm. Eng.* **2013**, *58*, 42–51. [CrossRef]
37. Barba, M.; Bruce, R.; Bouchet, F.; Bonelli, A.; Baudouy, B. Effect of the thermo-physical properties of the working fluid on the performance of a 1-m long cryogenic horizontal pulsating heat pipe. *Int. J. Heat Mass Transf.* **2022**, *187*, 122458. [CrossRef]
38. Monroe, J.G.; Ibrahim, O.T.; Thompson, S.M. Effect of harvesting module design on the thermal performance and voltage generation of a thermoelectric oscillating heat pipe. *Appl. Therm. Eng.* **2022**, *201*, 117651. [CrossRef]
39. Bastakoti, D.; Zhang, H.; Li, D.; Cai, W.; Li, F. An overview on the developing trend of pulsating heat pipe and its performance. *Appl. Therm. Eng.* **2018**, *141*, 305–332. [CrossRef]
40. Gürsel, G.; Frijns, A.J.H.; Homburg, F.G.A.; Van Steenhoven, A.A. A mass-spring-damper model of a pulsating heat pipe with a non-uniform and asymmetric filling. *Appl. Therm. Eng.* **2015**, *91*, 80–90. [CrossRef]
41. Liu, J.; Xie, G.; Sundén, B. Flow pattern and heat transfer past two tandem arranged cylinders with oscillating inlet velocity. *Appl. Therm. Eng.* **2017**, *120*, 614–625. [CrossRef]
42. Tong, B.Y.; Wong, T.N.; Ooi, K.T. Closed-loop pulsating heat pipe. *Appl. Therm. Eng.* **2001**, *21*, 1845–1862. [CrossRef]

43. Jo, J.; Kim, J.; Kim, S.J. Experimental investigations of heat transfer mechanisms of a pulsating heat pipe. *Energy Convers. Manag.* **2019**, *181*, 331–341. [CrossRef]
44. Senjaya, R.; Inoue, T. Oscillating heat pipe simulation considering bubble generation Part I: Presentation of the model and effects of a bubble generation. *Int. J. Heat Mass Transf.* **2013**, *60*, 816–824. [CrossRef]
45. Yoon, A.; Kim, S.J. A deep-learning approach for predicting oscillating motion of liquid slugs in a closed-loop pulsating heat pipe. *Int. J. Heat Mass Transf.* **2021**, *181*, 121860. [CrossRef]
46. Yu, C.; Ji, Y.; Li, Y.; Liu, Z.; Chu, L.; Kuang, H.; Wang, Z. A three-dimensional oscillating heat pipe filled with liquid metal and ammonia for high-power and high-heat-flux dissipation. *Int. J. Heat Mass Transf.* **2022**, *194*, 123096. [CrossRef]
47. Thompson, S.M.; Cheng, P.; Ma, H.B. An experimental investigation of a three-dimensional flat-plate oscillating heat pipe with staggered microchannels. *Int. J. Heat Mass Transf.* **2011**, *54*, 3951–3959. [CrossRef]
48. Ji, Y.; Wu, M.; Feng, Y.; Yu, C.; Chu, L.; Chang, C.; Li, Y.; Xiao, X.; Ma, H. An experimental investigation on the heat transfer performance of a liquid metal high-temperature oscillating heat pipe. *Int. J. Heat Mass Transf.* **2020**, *149*, 119198. [CrossRef]
49. Ji, Y.; Wu, M.; Feng, Y.; Liu, H.; Yang, X.; Li, Y.; Chang, C. Experimental study on the effects of sodium and potassium proportions on the heat transfer performance of liquid metal high-temperature oscillating heat pipes. *Int. J. Heat Mass Transf.* **2022**, *194*, 123116. [CrossRef]
50. Czajkowski, C.; Nowak, A.I.; Ochman, A.; Pietrowicz, S. Flower Shaped Oscillating Heat Pipe at the thermosyphon condition: Performance at different rotational speeds, filling ratios, and heat supplies. *Appl. Therm. Eng.* **2022**, *212*, 118540. [CrossRef]
51. Qu, J.; Wu, H.; Cheng, P. Thermal performance of an oscillating heat pipe with Al_2O_3–water nanofluids. *Int. Commun. Heat Mass Transf.* **2010**, *37*, 111–115. [CrossRef]
52. Qu, J.; Wang, Q. Experimental study on the thermal performance of vertical closed-loop oscillating heat pipes and correlation modeling. *Appl. Energy* **2013**, *112*, 1154–1160. [CrossRef]
53. Tokuda, D.; Inoue, T. Heat transport characteristics of a sodium oscillating heat pipe: Thermal performance. *Int. J. Heat Mass Transf.* **2022**, *196*, 123281. [CrossRef]
54. Zhao, J.; Qu, J.; Rao, Z. Experiment investigation on thermal performance of a large-scale oscillating heat pipe with self-rewetting fluid used for thermal energy storage. *Int. J. Heat Mass Transf.* **2017**, *108*, 760–769. [CrossRef]
55. Lin, Z.; Wang, S.; Huo, J.; Hu, Y.; Chen, J.; Zhang, W.; Lee, E. Heat transfer characteristics and LED heat sink application of aluminum plate oscillating heat pipes. *Appl. Therm. Eng.* **2011**, *31*, 2221–2229. [CrossRef]
56. Qian, N.; Fu, Y.; Zhang, Y.; Chen, J.; Xu, J. Experimental investigation of thermal performance of the oscillating heat pipe for the grinding wheel. *Int. J. Heat Mass Transf.* **2019**, *136*, 911–923. [CrossRef]
57. Monroe, J.G.; Ibrahim, O.T.; Thompson, S.M.; Shamsaei, N. Energy harvesting via fluidic agitation of a magnet within an oscillating heat pipe. *Appl. Therm. Eng.* **2018**, *129*, 884–892. [CrossRef]
58. Zhao, J.; Rao, Z.; Liu, C.; Li, Y. Experimental investigation on thermal performance of phase change material coupled with closed-loop oscillating heat pipe (PCM/CLOHP) used in thermal management. *Appl. Therm. Eng.* **2016**, *93*, 90–100. [CrossRef]
59. Qu, J.; Ke, Z.; Zuo, A.; Rao, Z. Experimental investigation on thermal performance of phase change material coupled with three-dimensional oscillating heat pipe (PCM/3D-OHP) for thermal management application. *Int. J. Heat Mass Transf.* **2019**, *129*, 773–782. [CrossRef]
60. Jin, H.; Lin, G.; Zeiny, A.; Bai, L.; Cai, J.; Wen, D. Experimental study of transparent oscillating heat pipes filled with solar absorptive nanofluids. *Int. J. Heat Mass Transf.* **2019**, *139*, 789–801. [CrossRef]
61. Alqahtani, A.A.; Edwardson, S.; Marengo, M.; Bertola, V. Performance of flat-plate, flexible polymeric pulsating heat pipes at different bending angles. *Appl. Therm. Eng.* **2022**, *216*, 118948. [CrossRef]
62. Iwata, N.; Miyazaki, Y.; Yasuda, S.; Ogawa, H. Thermal performance and flexibility evaluation of metallic micro oscillating heat pipe for thermal strap. *Appl. Therm. Eng.* **2021**, *197*, 117342. [CrossRef]
63. Wei, A.; Qu, J.; Qiu, H.; Wang, C.; Cao, G. Heat transfer characteristics of plug-in oscillating heat pipe with binary-fluid mixtures for electric vehicle battery thermal management. *Int. J. Heat Mass Transf.* **2019**, *135*, 746–760. [CrossRef]
64. Tsai, C.Y.; Chien, H.T.; Ding, P.P.; Chan, B.; Luh, T.Y.; Chen, P.H. Effect of structural character of gold nanoparticles in nanofluid on heat pipe thermal performance. *Mater. Lett.* **2004**, *58*, 1461–1465. [CrossRef]
65. Ma, H.B.; Wilson, C.; Borgmeyer, B.; Park, K.; Yu, Q.; Choi, S.U.S.; Tirumala, M. Effect of nanofluid on the heat transport capability in an oscillating heat pipe. *Appl. Phys. Lett.* **2006**, *88*, 143116. [CrossRef]
66. Kang, S.W.; Wei, W.C.; Tsai, S.H.; Huang, C.C. Experimental investigation of nanofluids on sintered heat pipe thermal performance. *Appl. Therm. Eng.* **2009**, *29*, 973–979. [CrossRef]
67. Naphon, P.; Thongkum, D.; Assadamongkol, P. Heat pipe efficiency enhancement with refrigerant–nanoparticles mixtures. *Energy Convers. Manag.* **2009**, *50*, 772–776. [CrossRef]
68. Hajian, R.; Layeghi, M.; Sani, K.A. Experimental study of nanofluid effects on the thermal performance with response time of heat pipe. *Energy Convers. Manag.* **2012**, *56*, 63–68. [CrossRef]
69. Aydın, D.Y.; Aydın, E.; Guru, M. The effects of particle mass fraction and static magnetic field on the thermal performance of NiFe2O4 nanofluid in a heat pipe. *Int. J. Therm. Sci.* **2023**, *183*, 107875. [CrossRef]
70. Wang, S.; Lin, Z.; Zhang, W.; Chen, J. Experimental study on pulsating heat pipe with functional thermal fluids. *Int. J. Heat Mass Transf.* **2009**, *52*, 5276–5279. [CrossRef]

71. Qu, J.; Wu, H. Thermal performance comparison of oscillating heat pipes with SiO_2/water and Al_2O_3/water nanofluids. *Int. J. Therm. Sci.* **2011**, *50*, 1954–1962. [CrossRef]

72. Hung, Y.H.; Teng, T.P.; Lin, B.G. Evaluation of the thermal performance of a heat pipe using alumina nanofluids. *Exp. Therm. Fluid Sci.* **2013**, *44*, 504–511. [CrossRef]

73. Goshayeshi, H.R.; Goodarzi, M.; Safaei, M.R.; Dahari, M. Experimental study on the effect of inclination angle on heat transfer enhancement of a ferrofluid in a closed loop oscillating heat pipe under magnetic field. *Exp. Therm. Fluid Sci.* **2016**, *74*, 265–270. [CrossRef]

74. Davari, H.; Goshayeshi, H.R.; Öztop, H.F.; Chaer, I. Experimental investigation of oscillating heat pipe efficiency for a novel condenser by using Fe3O4 nanofluid. *J. Therm. Anal. Calorim.* **2020**, *140*, 2605–2614. [CrossRef]

75. Goshayeshi, H.R.; Safaei, M.R.; Goodarzi, M.; Dahari, M. Particle size and type effects on heat transfer enhancement of Ferro-nanofluids in a pulsating heat pipe. *Powder Technol.* **2016**, *301*, 1218–1226. [CrossRef]

76. Kang, S.W.; Wang, Y.C.; Liu, Y.C.; Lo, H.M. Visualization and thermal resistance measurements for a magnetic nanofluid pulsating heat pipe. *Appl. Therm. Eng.* **2017**, *126*, 1044–1050. [CrossRef]

77. Karthikeyan, V.K.; Ramachandran, K.; Pillai, B.C.; Solomon, A.B. Effect of nanofluids on thermal performance of closed loop pulsating heat pipe. *Exp. Therm. Fluid Sci.* **2014**, *54*, 171–178. [CrossRef]

78. Kim, H.J.; Lee, S.H.; Kim, S.B.; Jang, S.P. The effect of nanoparticle shape on the thermal resistance of a flat-plate heat pipe using acetone-based Al_2O_3 nanofluids. *Int. J. Heat Mass Transf.* **2016**, *92*, 572–577. [CrossRef]

79. Jafarmadar, S.; Azizinia, N.; Razmara, N.; Mobadersani, F. Thermal analysis and entropy generation of pulsating heat pipes using nanofluids. *Appl. Therm. Eng.* **2016**, *103*, 356–364. [CrossRef]

80. Goshayeshi, H.R.; Chaer, I. Experimental study and flow visualization of Fe_2O_3/kerosene in glass oscillating heat pipes. *Appl. Therm. Eng.* **2016**, *103*, 1213–1218. [CrossRef]

81. Goshayeshi, H.R.; Izadi, F.; Bashirnezhad, K. Comparison of heat transfer performance on closed pulsating heat pipe for Fe_3O_4 and γFe_2O_3 for achieving an empirical correlation. *Phys. E Low-Dimens. Syst. Nanostructures* **2017**, *89*, 43–49. [CrossRef]

82. Gandomkar, A.; Saidi, M.H.; Shafii, M.B.; Vandadi, M.; Kalan, K. Visualization and comparative investigations of pulsating ferro-fluid heat pipe. *Appl. Therm. Eng.* **2017**, *116*, 56–65. [CrossRef]

83. Monroe, J.G.; Kumari, S.; Fairley, J.D.; Walters, K.B.; Berg, M.J.; Thompson, S.M. On the energy harvesting and heat transfer ability of a ferro-nanofluid oscillating heat pipe. *Int. J. Heat Mass Transf.* **2019**, *132*, 162–171. [CrossRef]

84. Tanshen, M.R.; Munkhbayar, B.; Nine, M.J.; Chung, H.; Jeong, H. Effect of functionalized MWCNTs/water nanofluids on thermal resistance and pressure fluctuation characteristics in oscillating heat pipe. *Int. Commun. Heat Mass Transf.* **2013**, *48*, 93–98. [CrossRef]

85. Sadeghinezhad, E.; Mehrali, M.; Rosen, M.A.; Akhiani, A.R.; Latibari, S.T.; Mehrali, M.; Metselaar, H.S.C. Experimental investigation of the effect of graphene nanofluids on heat pipe thermal performance. *Appl. Therm. Eng.* **2016**, *100*, 775–787. [CrossRef]

86. Kim, K.M.; Bang, I.C. Effects of graphene oxide nanofluids on heat pipe performance and capillary limits. *Int. J. Therm. Sci.* **2016**, *100*, 346–356. [CrossRef]

87. Wu, Q.; Xu, R.; Wang, R.; Li, Y. Effect of C60 nanofluid on the thermal performance of a flat-plate pulsating heat pipe. *Int. J. Heat Mass Transf.* **2016**, *100*, 892–898. [CrossRef]

88. Xing, M.; Yu, J.; Wang, R. Performance of a vertical closed pulsating heat pipe with hydroxylated MWNTs nanofluid. *Int. J. Heat Mass Transf.* **2017**, *112*, 81–88. [CrossRef]

89. Zhou, Y.; Cui, X.; Weng, J.; Shi, S.; Han, H.; Chen, C. Experimental investigation of the heat transfer performance of an oscillating heat pipe with graphene nanofluids. *Powder Technol.* **2018**, *332*, 371–380. [CrossRef]

90. Nazari, M.A.; Ghasempour, R.; Ahmadi, M.H.; Heydarian, G.; Shafii, M.B. Experimental investigation of graphene oxide nanofluid on heat transfer enhancement of pulsating heat pipe. *Int. J. Heat Mass Transf.* **2018**, *91*, 90–94. [CrossRef]

91. Xu, Y.; Xue, Y.; Qi, H.; Cai, W. Experimental study on heat transfer performance of pulsating heat pipes with hybrid working fluids. *Int. J. Heat Mass Transf.* **2020**, *157*, 119727. [CrossRef]

92. Zhou, Z.; Lv, Y.; Qu, J.; Sun, Q.; Grachev, D. Performance evaluation of hybrid oscillating heat pipe with carbon nanotube nanofluids for electric vehicle battery cooling. *Appl. Therm. Eng.* **2021**, *196*, 117300. [CrossRef]

93. Zhou, Y.; Yang, H.; Liu, L.; Zhang, M.; Wang, Y.; Zhang, Y.; Zhou, B. Enhancement of start-up and thermal performance in pulsating heat pipe with GO/water nanofluid. *Powder Technol.* **2021**, *384*, 414–422. [CrossRef]

94. Zhang, D.; He, Z.; Guan, J.; Tang, S.; Shen, C. Heat transfer and flow visualization of pulsating heat pipe with silica nanofluid: An experimental study. *Int. J. Heat Mass Transf.* **2021**, *183*, 122100. [CrossRef]

95. Sadeghinezhad, E.; Akhiani, A.R.; Metselaar, H.S.C.; Latibari, S.T.; Mehrali, M.; Mehrali, M. Parametric study on the thermal performance enhancement of a thermosyphon heat pipe using covalent functionalized graphene nanofluids. *Appl. Therm. Eng.* **2020**, *175*, 115385. [CrossRef]

96. Khajehpour, E.; Noghrehabadi, A.R.; Nasab, A.E.; Nabavi, S.H. Experimental investigation of the effect of nanofluids on the thermal resistance of a thermosiphon L-shape heat pipe at different angles. *Int. Commun. Heat Mass Transf.* **2020**, *113*, 104549. [CrossRef]

97. Li, Z.; Sarafraz, M.; Mazinani, A.; Moria, H.; Tlili, I.; Alkanhal, T.A.; Goodarzi, M.; Safaei, M.R. Operation analysis, response and performance evaluation of a pulsating heat pipe for low temperature heat recovery. *Energy Convers. Manag.* **2020**, *222*, 113230. [CrossRef]

98. Choi, D.; Lee, K.Y. Experimental study on confinement effect of two-phase closed thermosyphon and heat transfer enhancement using cellulose nanofluid. *Appl. Therm. Eng.* **2021**, *183*, 116247. [CrossRef]

99. Pandey, H.; Gupta, N.K. A descriptive review of the thermal transport mechanisms in mono and hybrid nanofluid-filled heat pipes and current developments. *Therm. Sci. Eng. Prog.* **2022**, *31*, 101281. [CrossRef]

100. Zufar, M.; Gunnasegaran, P.; Ng, K.C.; Mehta, H.B. Evaluation of the thermal performance of hybrid nanofluids in pulsating heat pipe. *CFD Lett.* **2019**, *11*, 13–24.

101. Zufar, M.; Gunnasegaran, P.; Kumar, H.; Ng, K. Numerical and experimental investigations of hybrid nanofluids on pulsating heat pipe performance. *Int. J. Heat Mass Transf.* **2020**, *146*, 118887. [CrossRef]

102. Moghadasi, H.; Aminian, E.; Saffari, H.; Mahjoorghani, M.; Emamifar, A. Numerical analysis on laminar forced convection improvement of hybrid nanofluid within a U-bend pipe in porous media. *Int. J. Mech. Sci.* **2020**, *179*, 105659. [CrossRef]

103. Xu, Q.; Liu, L.; Feng, J.; Qiao, L.; Yu, C.; Shi, W.; Ding, C.; Zang, Y.; Chang, C.; Xiong, Y.; et al. A comparative investigation on the effect of different nanofluids on the thermal performance of two-phase closed thermosyphon. *Int. J. Heat Mass Transf.* **2019**, *149*, 119189. [CrossRef]

104. Mukherjee, S.; Mishra, P.C.; Aljuwayhel, N.F.; Ali, N.; Chaudhuri, P. Thermo-fluidic performance of SiO_2–ZnO/water hybrid nanofluid on enhancement of heat transport in a tube: Experimental results. *Int. J. Therm. Sci.* **2022**, *182*, 107808. [CrossRef]

105. Veeramachaneni, S.; Pisipaty, S.K.; Vedula, D.R.; Solomon, A.B.; Harsha, V.S. Effect of copper–graphene hybrid nanoplatelets in a miniature loop heat pipe. *J. Therm. Anal. Calorim.* **2022**, *147*, 5985–5999. [CrossRef]

106. Sagar, K.R.; Naik, H.B.; Mehta, H.B. Numerical study of liquid nitrogen based pulsating heat pipe for cooling superconductors. *Int. J. Refrig.* **2021**, *122*, 33–46. [CrossRef]

107. Bruce, R.; Barba, M.; Bonelli, A.; Baudouy, B. Thermal performance of a meter-scale horizontal nitrogen Pulsating Heat Pipe. *Cryogenics* **2018**, *93*, 66–74. [CrossRef]

108. Li, M.; Li, L.; Xu, D. Effect of number of turns and configurations on the heat transfer performance of helium cryogenic pulsating heat pipe. *Cryogenics* **2018**, *96*, 159–165. [CrossRef]

109. Sun, X.; Li, S.; Jiao, B.; Gan, Z.; Pfotenhauer, J. Experimental study on a hydrogen closed-loop pulsating heat pipe with two turns. *Cryogenics* **2019**, *97*, 63–69. [CrossRef]

110. Liu, Y.; Deng, H.; Pfotenhauer, J.; Gan, Z. Design of a Hydrogen Pulsating Heat Pipe. *Phys. Procedia* **2015**, *67*, 551–556. [CrossRef]

111. Liang, Q.; Li, Y.; Wang, Q. Experimental investigation on the performance of a neon cryogenic oscillating heat pipe. *Cryogenics* **2017**, *84*, 7–12. [CrossRef]

112. Liang, Q.; Li, Y.; Wang, Q. Effects of filling ratio and condenser temperature on the thermal performance of a neon cryogenic oscillating heat pipe. *Cryogenics* **2018**, *89*, 102–106. [CrossRef]

113. Barba, M.; Bruce, R.; Bouchet, F.; Bonelli, A.; Baudouy, B. Effects of filling ratio of a long cryogenic Pulsating Heat Pipe. *Appl. Therm. Eng.* **2021**, *194*, 117072. [CrossRef]

114. Sun, X.; Li, S.; Wang, B.; Jiao, B.; Pfotenhauer, J.; Miller, F.; Gan, Z. Numerical study of the thermal performance of a hydrogen pulsating heat pipe. *Int. J. Therm. Sci.* **2022**, *172*, 107302. [CrossRef]

115. Li, Y.; Wang, Q.; Chen, S.; Zhao, B.; Dai, Y. Experimental investigation of the characteristics of cryogenic oscillating heat pipe. *Int. J. Heat Mass Transf.* **2014**, *79*, 713–719. [CrossRef]

116. Xu, D.; Li, L.; Liu, H. Experimental investigation on the thermal performance of helium based cryogenic pulsating heat pipe. *Exp. Therm. Fluid Sci.* **2016**, *70*, 61–68. [CrossRef]

117. Fonseca, L.D.; Miller, F.; Pfotenhauer, J. Experimental heat transfer analysis of a cryogenic nitrogen pulsating heat Pipe at various liquid fill ratios. *Appl. Therm. Eng.* **2018**, *130*, 343–353. [CrossRef]

118. Fonseca, L.D.; Pfotenhauer, J.; Miller, F. Results of a three evaporation cryogenic helium pulsating heat pipe. *Int. J. Heat Mass Transf.* **2018**, *120*, 1275–1286. [CrossRef]

119. Senjaya, R.; Inoue, T. Effects of non-condensable gas on the performance of oscillating heat pipe, part I: Theoretical study. *Appl. Therm. Eng.* **2014**, *73*, 1387–1392. [CrossRef]

120. Senjaya, R.; Inoue, T. Effects of non-condensable gas on the performance of oscillating heat pipe, part II: Experimental study. *Appl. Therm. Eng.* **2014**, *73*, 1393–1400. [CrossRef]

121. Chen, X.; Lin, Y.; Shao, S.; Wu, W. Study on heat transfer characteristics of ethane pulsating heat pipe in middle-low temperature region. *Appl. Therm. Eng.* **2019**, *152*, 697–705. [CrossRef]

122. Sarangi, R.K.; Rane, M.V. Experimental investigations for start up and maximum heat load of closed loop pulsating heat pipe. *Procedia Eng.* **2013**, *51*, 683–687. [CrossRef]

123. Han, H.; Cui, X.; Zhu, Y.; Sun, S. A comparative study of the behavior of working fluids and their properties on the performance of pulsating heat pipes (PHP). *Int. J. Therm. Sci.* **2014**, *82*, 138–147. [CrossRef]

124. Kim, B.; Li, L.; Kim, J.; Kim, D. A study on thermal performance of parallel connected pulsating heat pipe. *Appl. Therm. Eng.* **2017**, *126*, 1063–1068. [CrossRef]

125. Miura, M.; Nagasaki, T.; Ito, Y. Experimental investigation of heat transport with oscillating liquid column in pulsating heat pipe using forced oscillation system. *Int. J. Heat Mass Transf.* **2017**, *106*, 997–1004. [CrossRef]

126. Liang, Q.; Hao, T.; Wang, K.; Ma, X.; Lan, Z.; Wang, Y. Startup and transport characteristics of oscillating heat pipe using ionic liquids. *Int. Commun. Heat Mass Transf.* **2018**, *94*, 1–13. [CrossRef]
127. Patel, V.M.; Mehta, H.B. Channel wise displacement-velocity-frequency analysis in acetone charged multi-turn Closed Loop Pulsating Heat Pipe. *Energy Convers. Manag.* **2019**, *195*, 367–383. [CrossRef]
128. Wang, P.; Cui, X.; Weng, J.; Cai, Z.; Cai, R. Experimental investigation of the heat transfer performance of an oscillating heat pipe with LiCl salt solution. *Int. J. Heat Mass Transf.* **2020**, *158*, 120033. [CrossRef]
129. Mehta, K.; Mehta, N.; Patel, V. Experimental investigation of the thermal performance of closed loop flat plate oscillating heat pipe. *Exp. Heat Transf.* **2020**, *34*, 85–103. [CrossRef]
130. Wu, L.; Chen, J.; Wang, S. Experimental study on thermal performance of a pulsating heat pipe using R1233zd(E) as working fluid. *Int. Commun. Heat Mass Transf.* **2022**, *135*, 106152. [CrossRef]
131. Takawale, A.; Abraham, S.; Sielaff, A.; Mahapatra, P.S.; Pattamatta, A.; Stephan, P. A comparative study of flow regimes and thermal performance between flat plate pulsating heat pipe and capillary tube pulsating heat pipe. *Appl. Therm. Eng.* **2018**, *149*, 613–624. [CrossRef]
132. Bastakoti, D.; Zhang, H.; Cai, W.; Li, F. An experimental investigation of thermal performance of pulsating heat pipe with alcohols and surfactant solutions. *Int. J. Heat Mass Transf.* **2018**, *117*, 1032–1040. [CrossRef]
133. Cui, X.; Zhu, Y.; Li, Z.; Shun, S. Combination study of operation characteristics and heat transfer mechanism for pulsating heat pipe. *Appl. Therm. Eng.* **2014**, *65*, 394–402. [CrossRef]
134. Bae, J.; Lee, S.Y.; Kim, S.J. Numerical investigation of effect of film dynamics on fluid motion and thermal performance in pulsating heat pipes. *Energy Convers. Manag.* **2017**, *151*, 296–310. [CrossRef]
135. Sun, Q.; Qu, J.; Li, X.; Yuan, J. Experimental investigation of thermo-hydrodynamic behavior in a closed loop oscillating heat pipe. *Exp. Therm. Fluid Sci.* **2017**, *82*, 450–458. [CrossRef]
136. Mahapatra, B.N.; Das, P.K.; Sahoo, S.S. Scaling analysis and experimental investigation of pulsating loop heat pipes. *Appl. Therm. Eng.* **2016**, *108*, 358–367. [CrossRef]
137. Xue, Z.H.; Qu, W. Experimental and theoretical research on a ammonia pulsating heat pipe: New full visualization of flow pattern and operating mechanism study. *Int. J. Heat Mass Transf.* **2017**, *106*, 149–166. [CrossRef]
138. Liu, X.; Chen, Y.; Shi, M. Dynamic performance analysis on start-up of closed-loop pulsating heat pipes (CLPHPs). *Int. J. Therm. Sci.* **2013**, *65*, 224–233. [CrossRef]
139. Hao, T.; Ma, H.; Ma, X. Heat transfer performance of polytetrafluoroethylene oscillating heat pipe with water, ethanol, and acetone as working fluids. *Int. J. Heat Mass Transf.* **2018**, *131*, 109–120. [CrossRef]
140. Pachghare, P.R.; Mahalle, A.M. Effect of pure and binary fluids on closed loop pulsating heat pipe thermal performance. *Procedia Eng.* **2013**, *51*, 624–629. [CrossRef]
141. Su, Q.; Chang, S.; Song, M.; Zhao, Y.; Dang, C. An experimental study on the heat transfer performance of a loop heat pipe system with ethanol-water mixture as working fluid for aircraft anti-icing. *Int. J. Heat Mass Transf.* **2019**, *139*, 280–292. [CrossRef]
142. Zhu, Y.; Cui, X.; Han, H.; Sun, S. The study on the difference of the start-up and heat-transfer performance of the pulsating heat pipe with water−acetone mixtures. *Int. J. Heat Mass Transf.* **2014**, *77*, 834–842. [CrossRef]
143. Han, H.; Cui, X.; Zhu, Y.; Xu, T.; Sui, Y.; Sun, S. Experimental study on a closed-loop pulsating heat pipe (CLPHP) charged with water-based binary zeotropes and the corresponding pure fluids. *Energy* **2016**, *109*, 724–736. [CrossRef]
144. Shi, S.; Cui, X.; Han, H.; Weng, J.; Li, Z. A study of the heat transfer performance of a pulsating heat pipe with ethanol-based mixtures. *Appl. Therm. Eng.* **2016**, *102*, 1219–1227. [CrossRef]
145. Cui, X.; Qiu, Z.; Weng, J.; Li, Z. Heat transfer performance of closed loop pulsating heat pipes with methanol-based binary mixtures. *Exp. Therm. Fluid Sci.* **2016**, *76*, 253–263. [CrossRef]
146. Xu, R.; Zhang, C.; Chen, H.; Wu, Q.; Wang, R. Heat transfer performance of pulsating heat pipe with zeotropic immiscible binary mixtures. *Int. J. Heat Mass Transf.* **2019**, *137*, 31–41. [CrossRef]
147. Chang, G.; Li, Y.; Zhao, W.; Xu, Y. Performance investigation of flat-plate CLPHP with pure and binary working fluids for PEMFC cooling. *Int. J. Hydrogen Energy* **2021**, *46*, 30433–30441. [CrossRef]
148. Markal, B.; Varol, R. Thermal investigation and flow pattern analysis of a closed-loop pulsating heat pipe with binary mixtures. *J. Braz. Soc. Mech. Sci. Eng.* **2020**, *42*, 549. [CrossRef]
149. Markal, B.; Varol, R. Experimental investigation and force analysis of flat-plate type pulsating heat pipes having ternary mixtures. *Int. Commun. Heat Mass Transf.* **2020**, *121*, 105084. [CrossRef]
150. Markal, B.; Varol, R. Investigation of the effects of miscible and immiscible binary fluids on thermal performance of pulsating heat pipes. *Heat Mass Transf.* **2021**, *57*, 1527–1542. [CrossRef]
151. Wang, X.H.; Zheng, H.C.; Si, M.Q.; Han, X.H.; Chen, G.M. Experimental investigation of the influence of surfactant on the heat transfer performance of pulsating heat pipe. *Int. J. Heat Mass Transf.* **2015**, *83*, 586–590. [CrossRef]
152. Patel, V.M.; Mehta, H.B. Influence of working fluids on startup mechanism and thermal performance of a closed loop pulsating heat pipe. *Appl. Therm. Eng.* **2017**, *110*, 1568–1577. [CrossRef]
153. Hao, T.; Ma, X.; Lan, Z.; Li, N.; Zhao, Y.; Ma, H. Effects of hydrophilic surface on heat transfer performance and oscillating motion for an oscillating heat pipe. *Int. J. Heat Mass Transf.* **2014**, *72*, 50–65. [CrossRef]
154. Xing, M.; Wang, R.; Xu, R. Experimental study on thermal performance of a pulsating heat pipe with surfactant aqueous solution. *Int. J. Heat Mass Transf.* **2018**, *127*, 903–909. [CrossRef]

155. Wang, J.; Li, F.C. Experimental study on the characteristics of CHF and pressure fluctuations of surfactant solution flow boiling. *Int. J. Heat Mass Transf.* **2017**, *115*, 1004–1010. [CrossRef]
156. Wang, J.; Xie, J.; Liu, X. Investigation on the performance of closed-loop pulsating heat pipe with surfactant. *Appl. Therm. Eng.* **2019**, *160*, 113998. [CrossRef]
157. Bao, K.; Wang, X.; Fang, Y.; Ji, X.; Han, X.; Chen, G. Effects of the surfactant solution on the performance of the pulsating heat pipe. *Appl. Therm. Eng.* **2020**, *178*, 115678. [CrossRef]
158. Abe, Y.; Iwasaki, A.; Tanaka, K. Microgravity Experiments on Phase Change of Self-Rewetting Fluids. *Ann. N. Y. Acad. Sci.* **2004**, *1027*, 269–285. [CrossRef]
159. Singh, B.; Kumar, P. Heat transfer enhancement in pulsating heat pipe by alcohol-water based self-rewetting fluid. *Therm. Sci. Eng. Prog.* **2020**, *22*, 100809. [CrossRef]
160. Hu, Y.; Liu, T.; Li, X.; Wang, S. Heat transfer enhancement of micro oscillating heat pipes with self-rewetting fluid. *Int. J. Heat Mass Transf.* **2014**, *70*, 496–503. [CrossRef]
161. Wu, S.C. Study of self-rewetting fluid applied to loop heat pipe. *Int. J. Therm. Sci.* **2015**, *98*, 374–380. [CrossRef]
162. Su, X.; Zhang, M.; Han, W.; Guo, X. Enhancement of heat transport in oscillating heat pipe with ternary fluid. *Int. J. Heat Mass Transf.* **2015**, *87*, 258–264. [CrossRef]
163. Su, X.; Zhang, M.; Han, W.; Guo, X. Experimental study on the heat transfer performance of an oscillating heat pipe with self-rewetting nanofluid. *Int. J. Heat Mass Transf.* **2016**, *100*, 378–385. [CrossRef]
164. Savino, R.; Di Paola, R.; Cecere, A.; Fortezza, R. Self-rewetting heat transfer fluids and nanobrines for space heat pipes. *Acta Astronaut.* **2010**, *67*, 1030–1037. [CrossRef]
165. Wang, J.; Xie, J.; Liu, X. Investigation of wettability on performance of pulsating heat pipe. *Int. J. Heat Mass Transf.* **2020**, *150*, 119354. [CrossRef]
166. Lin, Z.; Wang, S.; Chen, J.; Huo, J.; Hu, Y.; Zhang, W. Experimental study on effective range of miniature oscillating heat pipes. *Appl. Therm. Eng.* **2011**, *31*, 880–886. [CrossRef]
167. Xian, H.; Xu, W.; Zhang, Y.; Du, X.; Yang, Y. Thermal characteristics and flow patterns of oscillating heat pipe with pulse heating. *Int. J. Heat Mass Transf.* **2014**, *79*, 332–341. [CrossRef]
168. Zhao, J.; Jiang, W.; Liu, C.; Rao, Z. Thermal performance enhancement of an oscillating heat pipe with external expansion structure for thermal energy recovery and storage. *Appl. Therm. Eng.* **2019**, *155*, 667–675. [CrossRef]
169. Taft, B.S.; Rhodes, M. Experimental investigation of oscillating heat pipes under direct current and pulse width modulation heating input conditions. *Appl. Therm. Eng.* **2017**, *126*, 1018–1022. [CrossRef]
170. Chu, L.; Ji, Y.; Liu, Z.; Yu, C.; Wu, Z.; Wang, Z.; Yang, Y.; Yang, X. Structure optimization of a three-dimensional coil oscillating heat pipe. *Int. J. Heat Mass Transf.* **2022**, *183*, 122229. [CrossRef]
171. Chen, X.; Chen, S.; Zhang, Z.; Sun, D.; Liu, X. Heat transfer investigation of a flat-plate oscillating heat pipe with tandem dual channels under nonuniform heating. *Int. J. Heat Mass Transf.* **2021**, *180*, 121830. [CrossRef]
172. Mangini, D.; Mameli, M.; Fioriti, D.; Filippeschi, S.; Araneo, L.; Marengo, M. Hybrid pulsating heat pipe for space applications with non-uniform heating patterns: Ground and microgravity experiments. *Appl. Therm. Eng.* **2017**, *126*, 1029–1043. [CrossRef]
173. Peng, H.; Pai, P.F.; Ma, H. Nonlinear thermomechanical finite-element modeling, analysis and characterization of multi-turn oscillating heat pipes. *Int. J. Heat Mass Transf.* **2014**, *69*, 424–437. [CrossRef]
174. Qu, J.; Zhao, J.; Rao, Z. Experimental investigation on thermal performance of multi-layers three-dimensional oscillating heat pipes. *Int. J. Heat Mass Transf.* **2017**, *115*, 810–819. [CrossRef]
175. Yasuda, Y.; Nabeshima, F.; Horiuchi, K.; Nagai, H. Visualization of the working fluid in a flat-plate pulsating heat pipe by neutron radiography. *Int. J. Heat Mass Transf.* **2022**, *185*, 122336. [CrossRef]
176. Lim, J.; Kim, S.J. A channel layout of a micro pulsating heat pipe for an excessively localized heating condition. *Appl. Therm. Eng.* **2021**, *196*, 117266. [CrossRef]
177. Jiaqiang, E.; Zhao, X.; Deng, Y.; Zhu, H. Pressure distribution and flow characteristics of closed oscillating heat pipe during the starting process at different vacuum degrees. *Appl. Therm. Eng.* **2016**, *93*, 166–173.
178. Kharangate, C.R.; Mudawar, I. Review of computational studies on boiling and condensation. *Int. J. Heat Mass Transf.* **2017**, *108*, 1164–1196. [CrossRef]
179. Tanguy, S.; Sagan, M.; Lalanne, B.; Couderc, F.; Colin, C. Benchmarks and numerical methods for the simulation of boiling flows. *J. Comput. Phys.* **2014**, *264*, 1–22. [CrossRef]
180. Magnini, M.; Pulvirenti, B.; Thome, J. Numerical investigation of hydrodynamics and heat transfer of elongated bubbles during flow boiling in a microchannel. *Int. J. Heat Mass Transf.* **2013**, *59*, 451–471. [CrossRef]
181. Barba, M.; Bruce, R.; Baudouy, B. Numerical simulation of the thermal and fluid-dynamic behavior of a cryogenic capillary tube. *Cryogenics* **2020**, *106*, 103044. [CrossRef]
182. Nuntaphan, A.; Vithayasai, S.; Vorayos, N.; Vorayos, N.; Kiatsiriroat, T. Use of oscillating heat pipe technique as extended surface in wire-on-tube heat exchanger for heat transfer enhancement. *Int. Commun. Heat Mass Transf.* **2010**, *37*, 287–292. [CrossRef]
183. Qian, N.; Wang, X.; Fu, Y.; Zhao, Z.; Xu, J.; Chen, J. Predicting heat transfer of oscillating heat pipes for machining processes based on extreme gradient boosting algorithm. *Appl. Therm. Eng.* **2020**, *164*, 114521. [CrossRef]
184. Sun, Q.; Qu, J.; Wang, Q.; Yuan, J. Operational characteristics of oscillating heat pipes under micro-gravity condition. *Int. Commun. Heat Mass Transf.* **2017**, *88*, 28–36. [CrossRef]

185. Nemati, R.; Shafii, M.B. Advanced heat transfer analysis of a U-shaped pulsating heat pipe considering evaporative liquid film trailing from its liquid slug. *Appl. Therm. Eng.* **2018**, *138*, 475–489. [CrossRef]
186. Daimaru, T.; Nagai, H.; Ando, M.; Tanaka, K.; Okamoto, A.; Sugita, H. Comparison between numerical simulation and on-orbit experiment of oscillating heat pipes. *Int. J. Heat Mass Transf.* **2017**, *109*, 791–806.
187. Adachi, T.; Fujita, K.; Nagai, H. Numerical study of temperature oscillation in loop heat pipe. *Appl. Therm. Eng.* **2019**, *163*, 114281. [CrossRef]
188. Odagiri, K.; Wolk, K.; Cappucci, S.; Morellina, S.; Roberts, S.; Pate, A.; Furst, B.; Sunada, E.; Daimaru, T. Three-dimensional heat transfer analysis of flat-plate oscillating heat pipes. *Appl. Therm. Eng.* **2021**, *195*, 117189. [CrossRef]
189. Jahan, S.A.; Ali, M.; Islam, M.Q. Effect of inclination angles on heat transfer characteristics of a closed loop pulsating heat pipe. *Procedia Eng.* **2013**, *56*, 82–87. [CrossRef]
190. Mameli, M.; Araneo, L.; Filippeschi, S.; Marelli, L.; Testa, R.; Marengo, M. Thermal response of a closed loop pulsating heat pipe under a varying gravity force. *Int. J. Therm. Sci.* **2014**, *80*, 11–22. [CrossRef]
191. Mameli, M.; Manno, V.; Filippeschi, S.; Marengo, M. Thermal instability of a Closed Loop Pulsating Heat Pipe: Combined effect of orientation and filling ratio. *Exp. Therm. Fluid Sci.* **2014**, *59*, 222–229. [CrossRef]
192. Ayel, V.; Araneo, L.; Scalambra, A.; Mameli, M.; Romestant, C.; Piteau, A.; Marengo, M.; Filippeschi, S.; Bertin, Y. Experimental study of a closed loop flat plate pulsating heat pipe under a varying gravity force. *Int. J. Therm. Sci.* **2015**, *96*, 23–34. [CrossRef]
193. Mangini, D.; Mameli, M.; Georgoulas, A.; Araneo, L.; Filippeschi, S.; Marengo, M. A pulsating heat pipe for space applications: Ground and microgravity experiments. *Int. J. Therm. Sci.* **2015**, *95*, 53–63. [CrossRef]
194. Cecere, A.; De Cristofaro, D.; Savino, R.; Ayel, V.; Sole-Agostinelli, T.; Marengo, M.; Romestant, C.; Bertin, Y. Experimental analysis of a flat plate pulsating heat pipe with self-rewetting fluids during a parabolic flight campaign. *Acta Astronaut.* **2018**, *147*, 454–461. [CrossRef]
195. Xing, M.; Wang, R.; Yu, J. The impact of gravity on the performance of pulsating heat pipe using surfactant solution. *Int. J. Heat Mass Transf.* **2020**, *151*, 119466. [CrossRef]
196. Pagliarini, L.; Cattani, L.; Bozzoli, F.; Mameli, M.; Filippeschi, S.; Rainieri, S.; Marengo, M. Thermal characterization of a multi-turn pulsating heat pipe in microgravity conditions: Statistical approach to the local wall-to-fluid heat flux. *Int. J. Heat Mass Transf.* **2021**, *169*, 120930. [CrossRef]
197. Lv, Y.; Xia, G.; Cheng, L.; Ma, D. Experimental investigation into unstable two phase flow phenomena during flow boiling in multi-microchannels. *Int. J. Therm. Sci.* **2021**, *166*, 106985. [CrossRef]
198. Yuan, D.; Qu, W.; Ma, T. Flow and heat transfer of liquid plug and neighboring vapor slugs in a pulsating heat pipe. *Int. J. Heat Mass Transf.* **2010**, *53*, 1260–1268. [CrossRef]
199. Karthikeyan, V.K.; Khandekar, S.; Pillai, B.C.; Sharma, P.K. Infrared thermography of a pulsating heat pipe: Flow regimes and multiple steady states. *Appl. Therm. Eng.* **2014**, *62*, 470–480. [CrossRef]
200. Spinato, G.; Borhani, N.; D'Entremont, B.P.; Thome, J.R. Time-strip visualization and thermo-hydrodynamics in a Closed Loop Pulsating Heat Pipe. *Appl. Therm. Eng.* **2015**, *78*, 364–372. [CrossRef]
201. Xian, H.; Xu, W.; Zhang, Y.; Du, X.; Yang, Y. Experimental investigations of dynamic fluid flow in oscillating heat pipe under pulse heating. *Appl. Therm. Eng.* **2015**, *88*, 376–383. [CrossRef]
202. Pouryoussefi, S.M.; Zhang, Y. Numerical investigation of chaotic flow in a 2D closed-loop pulsating heat pipe. *Appl. Therm. Eng.* **2016**, *98*, 617–627. [CrossRef]
203. Feldmann, D.; Wagner, C. On the influence of computational domain length on turbulence in oscillatory pipe flow. *Int. J. Heat Fluid Flow* **2016**, *61*, 229–244. [CrossRef]
204. Pouryoussefi, S.M.; Zhang, Y. Analysis of chaotic flow in a 2D multi-turn closed-loop pulsating heat pipe. *Appl. Therm. Eng.* **2017**, *126*, 1069–1076. [CrossRef]
205. Mangini, D.; Marengo, M.; Araneo, L.; Mameli, M.; Fioriti, D.; Filippeschi, S. Infrared analysis of the two phase flow in a single closed loop pulsating heat pipe. *Exp. Therm. Fluid Sci.* **2018**, *97*, 304–312. [CrossRef]
206. Xia, G.; Lv, Y.; Cheng, L.; Ma, D.; Jia, Y. Experimental study and dynamic simulation of the continuous two-phase instable boiling in multiple parallel microchannels. *Int. J. Heat Mass Transf.* **2019**, *138*, 961–984. [CrossRef]
207. Yoon, A.; Kim, S.J. Experimental and theoretical studies on oscillation frequencies of liquid slugs in micro pulsating heat pipes. *Energy Convers. Manag.* **2019**, *181*, 48–58. [CrossRef]
208. Ling, Y.Z.; Zhang, X.S.; Wang, X. Study of flow characteristics of an oscillating heat pipe. *Appl. Therm. Eng.* **2019**, *160*, 113995. [CrossRef]
209. Noh, H.Y.; Kim, S.J. Numerical simulation of pulsating heat pipes: Parametric investigation and thermal optimization. *Energy Convers. Manag.* **2020**, *203*, 112237. [CrossRef]
210. Ahmad, H.; Kim, S.K.; Jung, S.Y. Analysis of thermally driven flow behaviors for two-turn closed-loop pulsating heat pipe in ambient conditions: An experimental approach. *Int. J. Heat Mass Transf.* **2020**, *150*, 119245. [CrossRef]
211. Vo, D.T.; Kim, H.T.; Ko, J.; Bang, K.H. An experiment and three-dimensional numerical simulation of pulsating heat pipes. *Int. J. Heat Mass Transf.* **2020**, *150*, 119317. [CrossRef]
212. Schwarz, F.; Uddehal, S.R.; Lodermeyer, A.; Bagheri, E.M.; Forster-Heinlein, B.; Becker, S. Interaction of flow pattern and heat transfer in oscillating heat pipes for hot spot applications. *Appl. Therm. Eng.* **2021**, *196*, 117334. [CrossRef]

213. Rao, M.; Lefèvre, F.; Czujko, P.-C.; Khandekar, S.; Bonjour, J. Numerical and experimental investigations of thermally induced oscillating flow inside a capillary tube. *Int. J. Therm. Sci.* **2017**, *115*, 29–42. [CrossRef]
214. Yoon, A.; Kim, S.J. Characteristics of oscillating flow in a micro pulsating heat pipe: Fundamental-mode oscillation. *Int. J. Heat Mass Transf.* **2017**, *109*, 242–253. [CrossRef]
215. Spinato, G.; Borhani, N.; Thome, J.R. Understanding the self-sustained oscillating two-phase flow motion in a closed loop pulsating heat pipe. *Energy* **2015**, *90*, 889–899. [CrossRef]
216. Dilawar, M.; Pattamatta, A. A parametric study of oscillatory two-phase flows in a single turn Pulsating Heat Pipe using a non-isothermal vapor model. *Appl. Therm. Eng.* **2013**, *51*, 1328–1338. [CrossRef]
217. Kato, S.; Okuyama, K.; Ichikawa, T.; Mori, S. A single, straight-tube pulsating heat pipe (examination of a mechanism for the enhancement of heat transport). *Int. J. Heat Mass Transf.* **2013**, *64*, 254–262. [CrossRef]
218. Miura, M.; Nagasaki, T.; Ito, Y. Experimental study on heat transport induced by phase changes associated with liquid column oscillation in pulsating heat pipes. *Int. J. Heat Mass Transf.* **2019**, *133*, 652–661. [CrossRef]
219. Daimaru, T.; Yoshida, S.; Nagai, H. Study on thermal cycle in oscillating heat pipes by numerical analysis. *Appl. Therm. Eng.* **2017**, *113*, 1219–1227. [CrossRef]
220. Das, S.P.; Nikolayev, V.S.; Lefèvre, F.; Pottier, B.; Khandekar, S.; Bonjour, J. Thermally induced two-phase oscillating flow inside a capillary tube. *Int. J. Heat Mass Transf.* **2010**, *53*, 3905–3913. [CrossRef]
221. Jung, C.; Kim, S.J. Effects of oscillation amplitudes on heat transfer mechanisms of pulsating heat pipes. *Int. J. Heat Mass Transf.* **2021**, *165*, 120642. [CrossRef]
222. Sarangi, R.K.; Swain, A.; Kar, S.P.; Sekhar, P.C. Modeling for liquid plug oscillation frequency and amplitude of Pulsating heat pipe. *Mater. Today Proc.* **2022**, *49*, 372–377. [CrossRef]
223. Pai, P.F.; Peng, H.; Ma, H. Thermomechanical finite-element analysis and dynamics characterization of three-plug oscillating heat pipes. *Int. J. Heat Mass Transf.* **2013**, *64*, 623–635. [CrossRef]
224. Perna, R.; Abela, M.; Mameli, M.; Mariotti, A.; Pietrasanta, L.; Marengo, M.; Filippeschi, S. Flow characterization of a pulsating heat pipe through the wavelet analysis of pressure signals. *Appl. Therm. Eng.* **2020**, *171*, 115128. [CrossRef]
225. Recklin, V.; Pattamatta, A.; Stephan, P. Experimental investigation on the thermo-hydrodynamics of oscillatory meniscus in a capillary tube using FC-72 as working fluid. *Int. J. Multiph. Flow* **2015**, *75*, 82–87. [CrossRef]
226. Rao, M.; Lefèvre, F.; Khandekar, S.; Bonjour, J. Understanding transport mechanism of a self-sustained thermally driven oscillating two-phase system in a capillary tube. *Int. J. Heat Mass Transf.* **2013**, *65*, 451–459. [CrossRef]
227. Rao, M.; Lefèvre, F.; Khandekar, S.; Bonjour, J. Heat and mass transfer mechanisms of a self-sustained thermally driven oscillating liquid–vapour meniscus. *Int. J. Heat Mass Transf.* **2015**, *86*, 519–530. [CrossRef]
228. Fourgeaud, L.; Nikolayev, V.S.; Ercolani, E.; Duplat, J.; Gully, P. In situ investigation of liquid films in pulsating heat pipe. *Appl. Therm. Eng.* **2017**, *126*, 1023–1028. [CrossRef]
229. Jiaqiang, E.; Zhao, X.; Liu, H.; Chen, J.; Zuo, W.; Peng, Q. Field synergy analysis for enhancing heat transfer capability of a novel narrow-tube closed oscillating heat pipe. *Appl. Energy* **2016**, *175*, 218–228. [CrossRef]
230. Ando, M.; Okamoto, A.; Tanaka, K.; Maeda, M.; Sugita, H.; Daimaru, T.; Nagai, H. On-orbit demonstration of oscillating heat pipe with check valves for space application. *Appl. Therm. Eng.* **2018**, *130*, 552–560. [CrossRef]
231. Mito, T.; Natsume, K.; Yanagi, N.; Tamura, H.; Tamada, T.; Shikimachi, K.; Hirano, N.; Nagaya, S. Achievement of high heat removal characteristics of superconducting magnets with imbedded oscillating heat pipes. *IEEE Trans. Appl. Supercond.* **2011**, *21*, 2470–2473. [CrossRef]
232. Shi, W.; Li, W.; Pan, L.; Tan, X. Heat transfer properties and chaotic analysis of parallel type pulsating heat pipe. *Trans. Tianjin Univ.* **2011**, *17*, 435–439. [CrossRef]
233. Zhao, J.; Rao, Z.; Liu, C.; Li, Y. Experiment study of oscillating heat pipe and phase change materials coupled for thermal energy storage and thermal management. *Int. J. Heat Mass Transf.* **2016**, *99*, 252–260. [CrossRef]
234. Qu, J.; Wang, C.; Li, X.; Wang, H. Heat transfer performance of flexible oscillating heat pipes for electric/hybrid-electric vehicle battery thermal management. *Appl. Therm. Eng.* **2018**, *135*, 1–9. [CrossRef]
235. Zhao, J.; Jiang, W.; Rao, Z. Thermal performance investigation of an oscillating heat pipe with external expansion structure used for thermal energy recovery and storage. *Int. J. Heat Mass Transf.* **2019**, *132*, 920–928. [CrossRef]
236. Saha, N.; Das, P.K.; Sharma, P.K. Influence of process variables on the hydrodynamics and performance of a single loop pulsating heat pipe. *Int. J. Heat Mass Transf.* **2014**, *74*, 238–250. [CrossRef]
237. Senjaya, R.; Inoue, T. Oscillating heat pipe simulation considering bubble generation Part II: Effects of fitting and design parameters. *Int. J. Heat Mass Transf.* **2013**, *60*, 825–835. [CrossRef]
238. Lips, S.; Bensalem, A.; Bertin, Y.; Ayel, V.; Romestant, C.; Bonjour, J. Experimental evidences of distinct heat transfer regimes in pulsating heat pipes (PHP). *Appl. Therm. Eng.* **2010**, *30*, 900–907. [CrossRef]
239. Sun, Q.; Qu, J.; Yuan, J.; Wang, Q. Operational characteristics of an MEMS-based micro oscillating heat pipe. *Appl. Therm. Eng.* **2017**, *124*, 1269–1278. [CrossRef]
240. Nikolayev, V.S. Effect of tube heat conduction on the single branch pulsating heat pipe start-up. *Int. J. Heat Mass Transf.* **2016**, *95*, 477–487. [CrossRef]
241. Pandey, H.; Agarwal, S.; Gupta, N.K. Temporal performance evaluation of CuO + GO hybrid nanofluids in heat pipe. *Heat Transf. Res.* **2022**, *53*, 75–96. [CrossRef]

242. Abbasi, A.; Al-Khaled, K.; Khan, M.I.; Khan, S.U.; El-Refaey, A.M.; Farooq, W.; Jameel, M.; Qayyum, S. Optimized analysis and enhanced thermal efficiency of modified hybrid nanofluid (Al_2O_3, CuO, Cu) with nonlinear thermal radiation and shape features. *Case Stud. Therm. Eng.* **2021**, *28*, 101425. [CrossRef]

243. Shanmugapriya, M.; Sundareswaran, R.; Kumar, P.S. Heat and Mass Transfer Enhancement of MHD Hybrid Nanofluid Flow in the Presence of Activation Energy. *Int. J. Chem. Eng.* **2021**, *2021*, 9473226. [CrossRef]

244. Rao, Z.; Wang, Q.; Zhao, J.; Huang, C. Experimental investigation on the thermal performance of a closed oscillating heat pipe in thermal management. *Heat Mass Transf.* **2017**, *53*, 3059–3071. [CrossRef]

245. Zamani, R.; Kalan, K.; Shafii, M.B. Experimental investigation on thermal performance of closed loop pulsating heat pipes with soluble and insoluble binary working fluids and a proposed correlation. *Heat Mass Transf.* **2018**, *55*, 375–384. [CrossRef]

246. Ayel, V.; Slobodeniuk, M.; Bertossi, R.; Karmakar, A.; Martineau, F.; Romestant, C.; Bertin, Y.; Khandekar, S. Thermal performances of a flat-plate pulsating heat pipe tested with water, aqueous mixtures and surfactants. *Int. J. Therm. Sci.* **2022**, *178*, 107599. [CrossRef]

247. Rho, H.; Lee, S.; Bae, S.; Kim, T.W.; Lee, D.S.; Lee, H.J.; Hwang, J.Y.; Jeong, T.; Kim, S.; Ha, J.S.; et al. Three-Dimensional Porous Copper-Graphene Heterostructures with Durability and High Heat Dissipation Performance. *Sci. Rep.* **2015**, *5*, 12710. [CrossRef]

248. Malla, L.K.; Dhanalakota, P.; Mahapatra, P.S.; Pattamatta, A. Thermal and flow characteristics in a flat plate pulsating heat pipe with ethanol-water mixtures: From slug-plug to droplet oscillations. *Int. J. Heat Mass Transf.* **2022**, *194*, 123066. [CrossRef]

249. Li, Q.; Wang, C.; Wang, Y.; Wang, Z.; Li, H.; Lian, C. Study on the effect of the adiabatic section parameters on the performance of pulsating heat pipes. *Appl. Therm. Eng.* **2020**, *180*, 115813. [CrossRef]

250. Błasiak, P.; Opalski, M.; Parmar, P.; Czajkowski, C.; Pietrowicz, S. The Thermal—Flow Processes and Flow Pattern in a Pulsating Heat Pipe—Numerical Modelling and Experimental Validation. *Energies* **2021**, *14*, 5952. [CrossRef]

251. Zhao, J.; Wu, C.; Rao, Z. Numerical study on heat transfer enhancement of closed loop oscillating heat pipe through active incentive method. *Int. Commun. Heat Mass Transf.* **2020**, *115*, 104612. [CrossRef]

252. Dreiling, R.; Dubois, V.; Zimmermann, S.; Nguyen-Xuan, T.; Schreivogel, P.; di Mare, F. Numerical investigation of slug flow in pulsating heat pipes using an interface capturing approach. *Int. J. Heat Mass Transf.* **2022**, *199*, 123459. [CrossRef]

253. Wang, X.; Yan, Y.; Meng, X.; Chen, G. A general method to predict the performance of closed pulsating heat pipe by artificial neural network. *Appl. Therm. Eng.* **2019**, *157*, 113761. [CrossRef]

254. Jalilian, M.; Kargarsharifabad, H.; Abbasi Godarzi, A.; Ghofrani, A.; Shafii, M.B. Simulation and optimization of pulsating heat pipe flat-plate solar collectors using neural networks and genetic algorithm: A semi-experimental investigation. *Clean Technol. Environ. Policy* **2016**, *18*, 2251–2264. [CrossRef]

255. Wen, J. Thermal resistance modeling of oscillating heat pipes filled with acetone by using artificial neural network. *J. Therm. Anal. Calorim.* **2021**, *144*, 1873–1881. [CrossRef]

256. Jokar, A.; Godarzi, A.A.; Saber, M.; Shafii, M.B. Simulation and optimization of a pulsating heat pipe using artificial neural network and genetic algorithm. *Heat Mass Transf.* **2016**, *52*, 2437–2445. [CrossRef]

Article

Numerical Investigation on the Effect of Blockage on the Icing of Airfoils

Daixiao Lu [1,2], Zhiliang Lu [1,*], Zhirong Han [2], Xian Xu [2] and Ying Huang [2]

1   Key Laboratory of Unsteady Aerodynamics and Flow Control, Ministry of Industry and Information Technology, Nanjing University of Aeronautics and Astronautics, Nanjing 210016, China
2   Shanghai Aircraft Design and Research Institute, Shanghai 201210, China
*   Correspondence: luzl@nuaa.edu.cn

Abstract: The blockage is one of the important factors affecting the icing of airfoils in wind tunnel tests. In this paper, numerical simulations are conducted to study the effect of blockage on the icing of different airfoils. By reducing the height of testing wind tunnels, the blockage is increased, and the changes in the height and angle of the ice horn are numerically investigated. The simulation results indicate that as the blockage increases, the flow velocity above the stagnation point of the airfoil increases, leading to larger pressure coefficients distribution and stronger heat transfer capacity. As a result, the position of icing moves forward, and the angle of the upper ice horn becomes smaller. In addition, the increased flow velocity facilitates the collection of water droplets in the area, which improves the icing and increases the height of the upper ice horn. It is also found that the blockage increases the angle of attack of the airfoil, moving the stagnation point backward and decreasing the angle of the upper ice horn. When the blockage is above 15%, the joint influence of the opening angle and height of the upper ice horn significantly reduces the projection height of the upper ice horn in the direction of the incoming flow, leading to unacceptable criticality of the ice shape.

Keywords: blockage; icing of airfoils; stagnation point; angle of attack; criticality of the ice shape

Citation: Lu, D.; Lu, Z.; Han, Z.; Xu, X.; Huang, Y. Numerical Investigation on the Effect of Blockage on the Icing of Airfoils. *Aerospace* **2022**, *9*, 587. https://doi.org/10.3390/aerospace9100587

Academic Editor: Sergey Leonov

Received: 6 September 2022
Accepted: 7 October 2022
Published: 10 October 2022

Publisher's Note: MDPI stays neutral with regard to jurisdictional claims in published maps and institutional affiliations.

1. Introduction

Ice accretion occurs immediately on wing, tail, propeller, rotor and even antenna when aircraft fly through the cloud full of supercooled water droplets or encountering precipitations such as freezing rain or drizzle. Ice accumulating on the aerodynamic sensitive surfaces leads safety issue because of its detrimental effect on the performance such as maximum lift penalty, stall angle reduction, and parasite drag increase. Aviation accident data showed that 583 icing accidents occurred and caused more than 800 fatalities from 1982 to 2000 in the United States [1]. For this reason, aircraft icing was even recognized as the "most wanted aviation transportation safety improvement" by the National Transportation Safety Board (NTSB) [2]. The Federal Aviation Regulations (FAR) Part 25 Appendix C define icing envelopes (in terms of air temperature, liquid water content and droplets' size) for aircraft certification, corresponding to 99.9% of icing conditions found in stratiform clouds [3]. Most light aircraft are not required to pass this certification and do not usually have the required ice-protection systems for flying into icing conditions. Larger aircraft are equipped with a variety of anti-icing or de-icing systems (heaters, pneumatic boots and liquid flows) to help them prevent ice formation [4].

In order to guarantee safe and on schedule operations of transportation aircraft, accurate prediction of ice accretion, related performance degradation and anti/de-icing systems development play an important role for the design of aircraft. Over the years, test approaches including flight and wind-tunnel tests, and numerical simulations have been adopted in the prediction of ice accretion. Following the development and progress of computational science as well as the great advantages of numerical simulation in the economic cost, many ice-prediction codes have been successfully developed and serve as valuable design and certification tools for the aircraft manufacturers such as LEWICE developed by NASA Lewis Research Center [5], ONERA 2-D/3-D developed by ONERA [6], IMPIN3D developed by Italy [7], and the second-generation 3-D icing simulation system FENSAP-ICE developed by Numerical Technologies International [8], etc. Recently, a large number of articles have been published that are devoted to the application of machine learning methods, deep neural networks in the field of fluid dynamics and even the procedure and method for the ice accretion prediction for different airfoils using artificial neural networks (ANNs) are discussed [9]. It is worth expecting that the combination of two approaches: machine learning and numerical simulation, will speed up the prediction of the icing shape and significantly reduce computational costs.

Considering that the natural icing conditions are very difficult and costly to obtain, the icing test in wind tunnels is not only an important means to study aircraft icing (an essential step to obtaining civil aircraft airworthiness certification) [10,11] but also an important method to develop and verify numerical tools used for icing [11–14]. The civil aircraft authorities have set clear approval requirements for the wind tunnels and numerical tools used for icing in the design of aircraft [15,16]. Previous wind tunnel tests show that wall interference is an important factor affecting the accuracy of testing results [17–19]. In a closed wind tunnel, the existence of walls limits the bending and diffusion of streamlines. Therefore, the area through which the airflow between adjacent streamlines is smaller than that in the free space, and the average flow velocity between the airfoil and tunnel wall is larger. This phenomenon is called the blocking effect. In reality, the limitation of the bending of streamlines is equivalent to inducing an upward angle to the airfoil, thereby increasing the angle of attack as compared with that defined using the axis of the wind tunnel [18]. The increased angle of attack will move the icing from the upper surface of the airfoil to the lower surface, leading to overestimated aerodynamic performances of airfoils. Consequently, the criticality of the ice shape is not guaranteed, and the test results will not be recognized.

The blockage is defined as the ratio of the maximum incoming airflow area to the cross-sectional area of the test tunnel. The US Federal Aviation Administration [20] reported that the blockage should not exceed 10% in the icing test of airfoils in wind tunnels. However, in practice, the chord length of the wings of large passenger aircraft is generally long (may exceed 5 m). Even the world's largest wind tunnel (the FL-16 ice wind tunnel [21] of the China Aerodynamic Research and Development Center, with a size of 4.8 m × 3.2 m) is not suitable for such experiments. On the other hand, the aircraft airworthiness examiners generally do not allow the use of scaled test models in icing wind tunnel tests. Even though the hybrid wing design technology is adopted [22,23], the blockage will still exceed 10%, which makes the reliability of icing test results questionable under certain circumstances (e.g., with a large angle of attack).

Guo Qiling et al. [24,25] of the Key Laboratory of Icing and Deicing of CARDC studied the influence of walls on the impinging characteristics of water droplets and icing of airfoils. Their research shows that the blockage will affect the impinging characteristics of water droplets, and the increased blockage will improve icing on the surface of airfoils. Zocca et al. [26] found that the blockage has a significant impact on the icing of airfoils and brings a large deviation to the ice shape. Qin et al. [27] numerically examined the influence of cavity walls on the impinging characteristics of water droplets. They concluded that the blockage would shift the stagnation point of airflows on the airfoil, which significantly influences the collection of water droplets on the airfoil surface. The NRC wind tunnel in Canada also noticed that the increased blockage is beneficial for the collection of water droplets. They revealed the relationships between the water droplet collection rate and the Reynolds number of water droplets at different blockages [28]. It can be seen from the studies mentioned above that the impinging characteristics of water droplets and resultant ice shape are both affected by the blockage. Nonetheless, there is still no recommended value of the maximum blockage acceptable in wind tunnel tests. In this paper, the NACA0012 and GLC305 airfoils in the literature are used as examples and numerically investigated.

By adjusting the height of the testing wind tunnel, the icing characteristics of airfoils at different blockages are simulated. Based on the ice shapes obtained from the tests, the influence of the blockage on the ice shape is studied, and the maximum blockage allowed in the icing tests of airfoils is explored. The outline of this article is organized as follows. Section 2 introduces the numerical methodology employed in this research. Section 3 validates the numerical model. Section 4 presents and discusses the simulation results. Finally, concluding remarks are drawn in Section 5.

2. Numerical Methodology

2.1. Model Description and Mesh Configuration

Figure 1a presents the sketch of the airfoil in a wind tunnel under investigation in this study. In the numerical simulations, for simplicity, the airfoil is installed without fixed boundary conditions, and its left and right supporting structures are omitted. The blockage is changed by adjusting the height between the upper and lower tunnel walls, ignoring the interference of the side tunnel walls. For instance, If the height of the teste section is decrease and the installation angle of the wing model remains the same, the blockage is increased. As shown in Figure 1a, two-dimensional unstructured grids are adopted to mesh the airfoil and the surrounding flow field. To save computational cost, denser grids are constructed in the airfoil region as compared with the surrounding flow field that employs relatively coarse grids. As such, triangular prism boundary layer grids are used on the surface of the airfoil, as displayed in Figure 1b. Likewise, triangular pyramid grids are employed near the leading edge of the airfoil, as shown in Figure 1c. The prism layers are 40, the first layer is about 1×10^{-6} m high, and the expansion ratio is 1.2, the number of grid cells along the airfoil is 120.

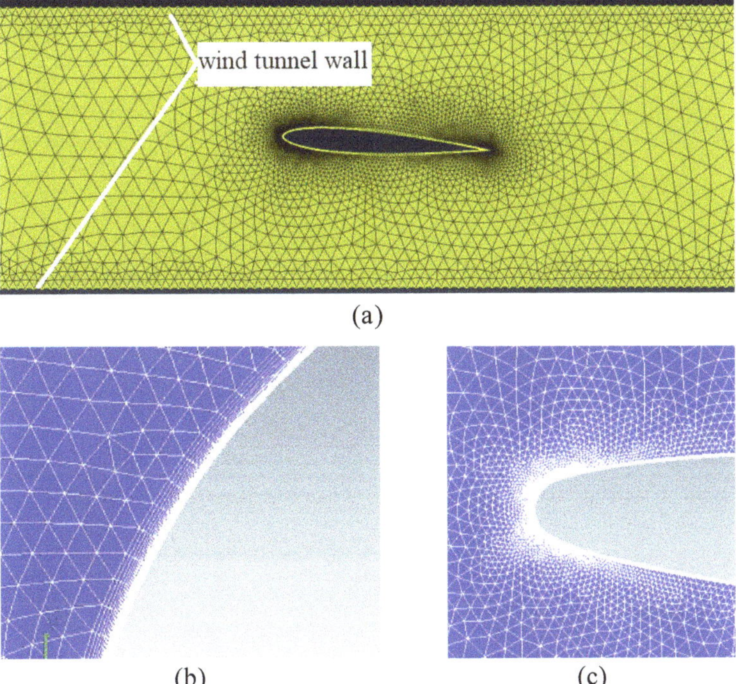

Figure 1. Model description and mesh configuration. (**a**) Sketch of the airfoil in a wind tunnel. (**b**) Boundary layer grids. (**c**) Grids near the edge of the airfoil.

2.2. Governing Equations

Following the construction of mesh for the airfoil in the wind tunnel, numerical calculations are conducted to simulate the ice accretion. The working fluid is air and follows the Navier-Stokes equations and the equation of state, which are,

$$\frac{\partial \overline{\rho}}{\partial t} + \frac{\partial}{\partial x_i}(\overline{\rho}\widetilde{u}_i) = 0 \tag{1}$$

$$\frac{\partial}{\partial t}(\overline{\rho}\widetilde{u}_i) + \frac{\partial}{\partial x_j}\left(\overline{\rho}\widetilde{u}_i\widetilde{u}_j + \overline{p}\delta_{ij} - \widetilde{\tau_{ij}^{tot}}\right) = 0 \tag{2}$$

$$\frac{\partial}{\partial t}(\overline{\rho}\widetilde{e}_{tot}) + \frac{\partial}{\partial x_j}\left(\overline{\rho}\widetilde{u}_j\widetilde{e}_{tot} + \widetilde{u}_j\overline{p} + \widetilde{q_j^{tot}} - \widetilde{u}_i\widetilde{\tau_{ij}^{tot}}\right) = 0 \tag{3}$$

$$p = \rho R_g T \tag{4}$$

where the overbars "~" and "−" denote the density-weighted time averaging (Favre averaging) and classical time averaging (Reynolds averaging), respectively. In this case, p, ρ and T are pressure, density and temperature, u_i or u_j stands for the velocity component in corresponding direction, $e_{tot} = e + u_i u_i / 2$ is the total energy with e being the specific internal energy, τ_{ij}^{tot} is the total stress tensor, q_j^{tot} is the heat flux, δ_{ij} is the Kronecker delta, and R_g is the specific gas constant. The Spalart–Allmaras turbulence model is used in the flow field calculation. In the calculation of droplet trajectory, a single average diameter called Median Volumetric Diameter (MVD) is used, and the coupled heat transfer calculation related to the anti-ice system is not involved.

The governing equations stated above are resolved numerically using the commercial CFD (computational fluid dynamics) package ANSYS Fensap-ice R19.0. Fensap-ice R19.0 includes many modules, such as flow field calculation, droplet collision calculation, ice accretion, and heat transfer calculation, which can be used for icing analysis.

2.3. Boundary Conditions

The boundary conditions in the numerical simulations are defined referring to the NACA0012 and GLC305 airfoils. Table 1 lists the parameters of NACA0012 and GLC305 airfoils used in the icing test in NASA Lewis icing Research Tunnel (IRT) [13].

Table 1. Parameters of airfoils used in the numerical simulations.

	NACA0012 Airfoil	GLC305 Airfoil
Chord length	21 in	36 in
Velocity	102.8 m/s	90 m/s
Angle of attack	4°	4.5°
Time	7 min	16.7 min
LWC	0.55 g/m^3	0.405 g/m^3
MVD	20 μm	20 μm
Temperature	265.37 K	267.40 K

2.4. Sensitivity Studies

Sensitivity studies on the influence of grid number and icing step number on the simulation results are performed. In the multistep icing calculation, the multi-step method based on artificial mesh reconstruction is adopted to ensure that the reconstructed mesh can accurately describe the ice shape and maintain sufficient mesh quality. The time interval of each step of mesh reconstruction is equal: the icing time of each step is defined as the total icing time divided by mesh reconstruction steps.

Figure 2 shows the ice shapes calculated at different grid numbers and mesh reconstruction steps. The "160k 6steps" in the figure represents the case in which the total grid number is 160,000 and the icing step number is 6. First, by comparing the ice shapes at grid numbers of 160,000, 300,000 and 476,000, It has been found that the grid number of 476,000 is sufficient enough to accurately describe the ice shape while minimizing the computational cost simultaneously. Therefore, the optimal grid number is chosen at 476,000. Subsequently, by comparing the ice shapes at icing step numbers of 3, 6 and 7 when the grid number is 476,000, It has been noticed that the mesh reconstruction time has a great impact on the ice horn height. When the icing step number reaches 6, the ice shape becomes less sensitive. Hence, 6 steps of icing are adopted in the subsequent simulations in this paper.

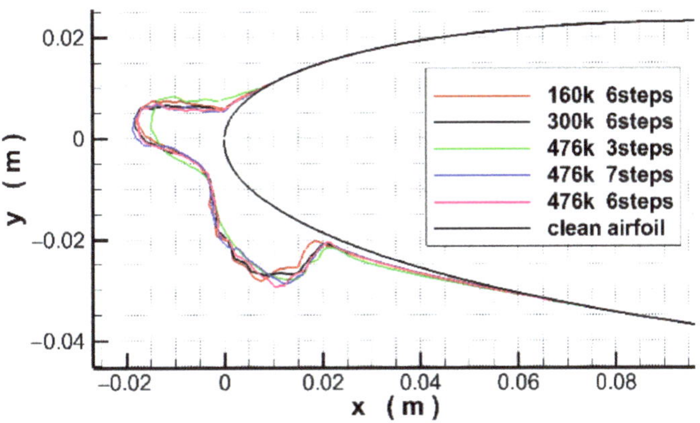

Figure 2. Calculated ice shapes at different grid numbers and icing steps.

3. Verification of Numerical Results

Comparisons are made between the experimental data of the NACA0012 airfoil in the literature [13] and the ice shape calculated by ANSYS Fensap-ice R19.0 using the same parameters. The results in Figure 3 indicate that the criticality of ice shape from numerical simulations in terms of ice horn height and angle is consistent with that of the experiment, which demonstrates the validity of the numerical methodologies proposed in this research.

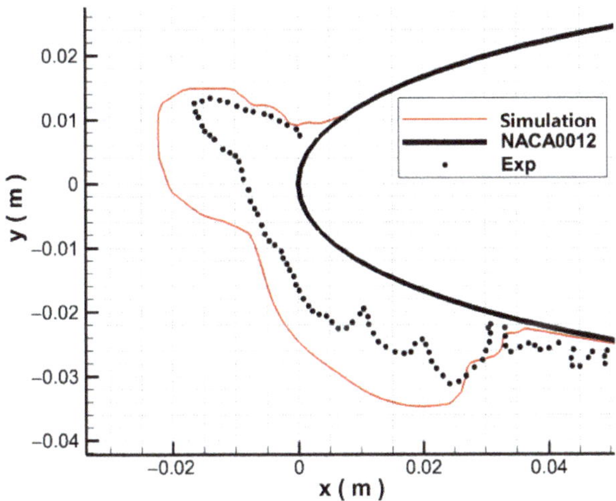

Figure 3. Comparison of ice shapes from the experiment and numerical simulations.

4. Results and Discussion

4.1. Criticality of Ice Shape

Gray [29] demonstrated that the height, position, and angle of the upper ice horn are the critical parameters that affect the formation of ice shape which in return influences the aerodynamic performance of the airfoil. The angle of the upper ice horn has a direct impact on the size and position of bubbles separating on the surface of the airfoil while the parameters of the lower ice horn generally affect the range of negative angle of attack, having little influence on the aerodynamic characteristics of the airfoil. Figure 4a displays the height of the upper ice horn h, the chord length of the upper ice horn c, and the opening angle of the upper ice horn θ. The characteristic position x of the ice shape is determined by the intersection of the trailing edge of the ice horn and the surface of the airfoil. The characteristic height h is the distance between the intersection described above and the highest point of the ice horn. In practical work, the projection height H of the upper ice horn (see Figure 4b) in the incoming flow direction is generally used to judge the criticality of the ice shape. Specifically, the larger H is, the more critical the ice shape is [30]. The parameter H reflects the joint influence of h, x, and θ.

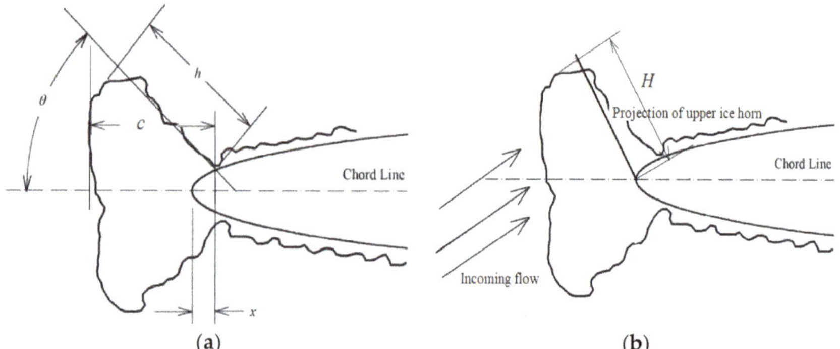

Figure 4. Characteristic parameters of the ice horn. (**a**) Ice horn dimensions. (**b**) Projection of upper ice horn in the direction of the incoming flow.

4.2. Effect of Blockage on the Criticality of Ice Shape

To simulate the installation of supports (left and right) of the airfoil in practical work, the upper and lower tunnel walls are moved toward the airfoil, thus increasing the blockage. Three blockages (10%, 15.2%, and 20%) are investigated in the simulations and the pressure coefficient distribution of each blockage is calculated. The results show that when the angle of attack is unchanged, as the blockage increases, the peak of the pressure coefficient distribution increases (see Figure 5a), and the stagnation point moves backward (see Figure 5b). This phenomenon is also reported in reference [17] which concludes that the average flow velocity between the airfoil and the tunnel wall increases after the blockage increases, leading to more restrictions on the streamline bending by the tunnel wall. Thus, the angle of attack increases because of the increase of the angle (upward) of the incoming flow.

In the simulations, the ice shapes at different blockages are calculated and presented in Figure 6. By comparing the calculated ice shape without a fixed boundary with that measured in the experiment, it has been found that as the blockage increases, the height of the upper ice horn increases, and the opening angle of the upper ice horn decreases. The increase in the height of the upper ice corner is due to the increase of local velocity around the head of the airfoil, thereby facilitating the collection of water droplets.

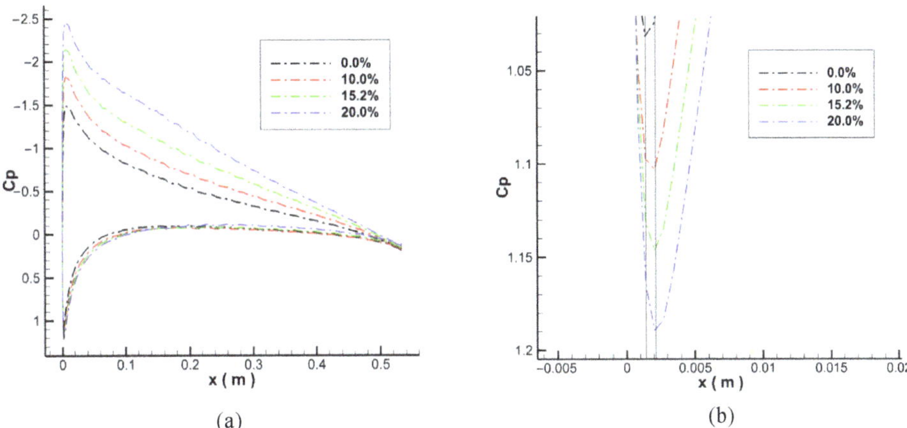

Figure 5. (**a**) Pressure coefficients distribution at different blockages. (**b**) Positions of stagnation point at different blockages.

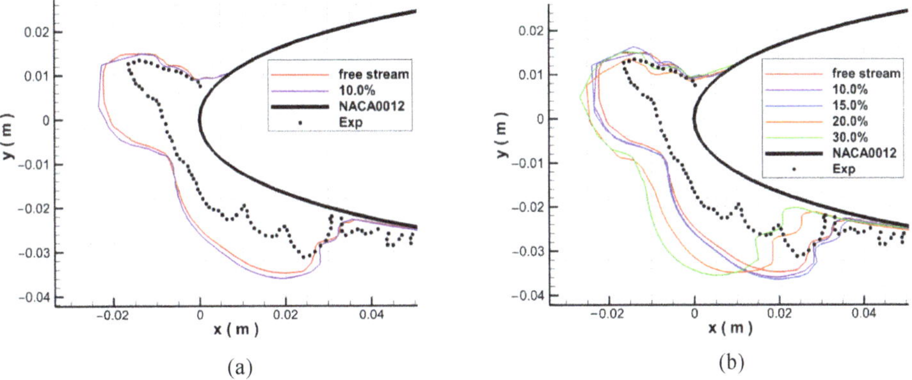

Figure 6. Ice shapes at different blockages. (**a**) Ice shape with 10% blockage. (**b**) Ice shape of NACA0012 airfoil at different blockages.

The local droplet collection efficiency β is defined as the ratio, for a given mass of water, of the area of impingement to the area through which the water passes at some distance upstream of the airfoil. Taking a unit width as one dimension of both area terms, the local droplet collection efficiency can then be defined as,

$$\beta = \frac{dy}{ds} = \frac{\Delta y_0}{\Delta s} \tag{5}$$

where Δy_0 is the spacing between water droplets at the release plane and Δs is the distance along the airfoil surface between the impact locations of the same two droplets.

As can be seen from Figure 7, the local droplet collection coefficient β increases with the increase of blockage, and the peak of β gradually moves towards the leading edge. Meanwhile, due to the increase of flow velocity in this area, the heat transfer rate is enhanced which results in faster icing of water film, smaller flow distance, and reduced angle of the upper ice horn. Figure 6b shows that when the blockage reaches 20% or even 30%, the change in the opening angle of the upper ice horn becomes more obvious.

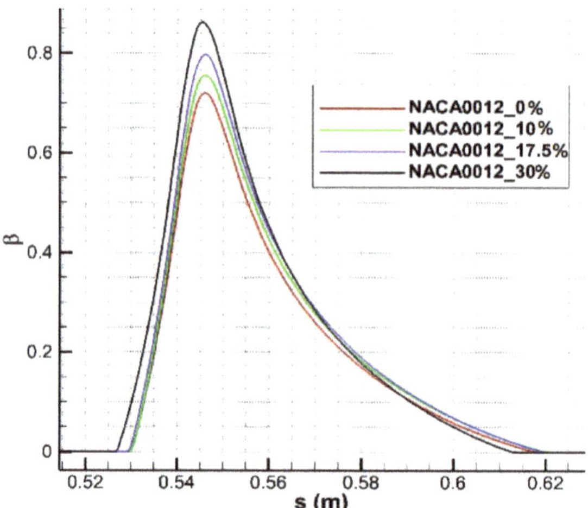

Figure 7. Droplet collection coefficients at different blockages of NACA0012 airfoil.

Numerical simulations are also performed on the airfoil of GLC305 reported in the literature [13]. As can be seen in Figure 8, as the blockage increases, the opening angle of the upper ice horn decreases. In particular, when the blockage increases to 17.6%, there is an obvious decline in the opening angle of the upper ice horn. Figure 9 shows the relationship between the blockage and the dimensionless projection height H/c of the upper ice horn in the direction of incoming flow, where c represents the chord length of the airfoil. It is found that H/c is small when the blockage falls between 10% and 15%. When the blockage is greater than 15%, H/c decreases remarkably, which significantly affects the criticality of the ice shape.

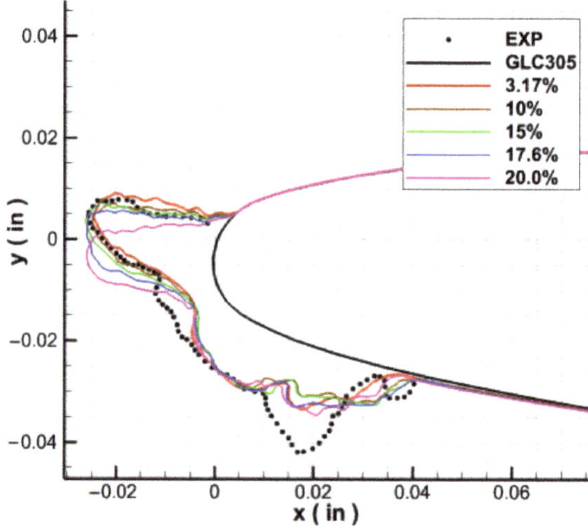

Figure 8. Ice shape of GLC305 airfoil at different blockages.

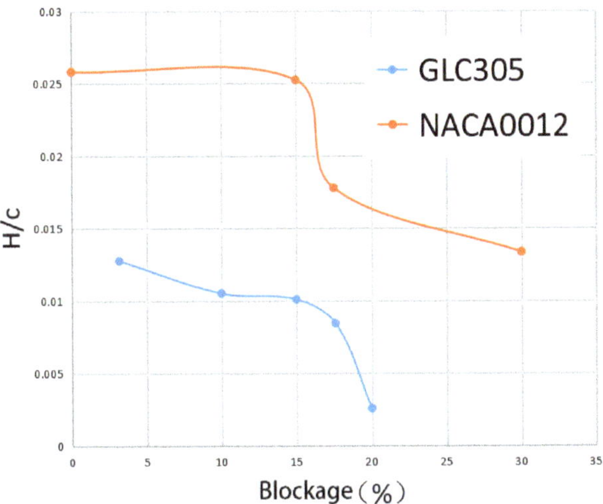

Figure 9. Relationship between the blockage and the dimensionless projection height.

5. Conclusions

This study investigates the effect of blockage on the icing of airfoils through computational fluid dynamics. Simulations are performed on the airfoils of NACA0012 and GLC305 tested in the wind tunnels, and comparisons are made between the calculated results and experimental data in the literature. The criticality of ice shape from numerical simulations in terms of ice horn height and angle is consistent with that of the experiment, which demonstrates the validity of the numerical methodologies proposed in this study. The key conclusions of this research are summarized as follows.

(1) The increase of ice horn height will increase the criticality of ice shape, while the decrease of ice horn angle will reduce the criticality of ice shape

(2) As the blockage increases, the peak of the pressure coefficient distribution increases, and the stagnation point moves backward. This is because the average flow velocity between the airfoil and the tunnel wall increases after the blockage increases, leading to more restrictions of the streamline bending by the tunnel wall.

(3) With the increase of the blockage of the ice tunnel, the joint influence of the opening angle and height of the upper ice horn significantly reduces the projection height of the upper ice horn in the direction of the incoming flow. As a result, the criticality of the ice shape is reduced. To ensure the criticality of the ice shape, the blockage should be below 15%.

Author Contributions: Conceptualization and methodology, Z.L. and D.L.; methodology and software Z.H.; validation, X.X. and Y.H.; formal analysis and investigation, D.L.; resources and data curation, Z.H.; writing—original draft preparation, D.L.; writing—review and editing, Z.L.; visualization, Z.H.; supervision, Z.L.; project administration, D.L.; funding acquisition, Z.L. All authors have read and agreed to the published version of the manuscript.

Funding: This research was funded by the National Natural Science Fundation of China (grant nos. 12102187 and 11872212), the National Key R&D Program of China (Project No. 2020YFA0712000), and a project funded by the Priority Academic Program Development of Jiangsu Higher Education Institutions.

Conflicts of Interest: The authors declare that they have no known competing financial interest or personal relationships that could have appeared to influence the work reported in this paper.

References

1. Petty, K.R.; Floyd, C.D.J. A Statistical Review of Aviation Airframe Icing Accidents in the US. In Proceedings of the 11th Conference on Aviation, Range, and Aerospace, National Trans-portation Safety Board, Hyannis, MA, USA, 7 October 2004. Available online: https://ams.confex.com/ams/11aram22sls/techprogram/paper_81425.htm (accessed on 8 October 2022).
2. Ratvasky, T.P.; Barnhart, B.P.; Lee, S. Current Methods Modeling and Simulating Icing Effects on Aircraft Performance, Stability, Control. *J. Aircr.* **2010**, *47*, 201–211. [CrossRef]
3. Heinrich, A.; Ross, R.; Zumwalt, G. *Aircraft Icing Handbook*; Civil Aviation Authority: Lower Hutt, New Zealand, 2000; pp. 13–14.
4. Shinkafi, A.; Lawson, C.; Seresinhe, R.; Quaglia, D.; Madani, I. An Intelligent Ice Protection System for Next Generation Aircraft Trajectory Optimisation. In Proceedings of the 29th Congress of the International Council of Aeronautical Sciences, ICAS 2014, St. Petersburg, Russia, 7–12 September 2014. Available online: https://www.researchgate.net/publication/290354048_An_intelligent_ice_protection_system_for_next_generation_aircraft_trajectory_optimisation (accessed on 10 March 2019).
5. Wright, W.B. *User's Manual for LEWICE 3.2*; NASA/CR-2008-214255; NASA: Washington, DC, USA, 2008.
6. Hedde, T.; Guffond, D. Improvement of the ONERA 3D icing code, comparison with 3D experiment shapes. In Proceedings of the 31st Aerospace Sciences Meeting, Reno, NV, USA, 11–14 January 1993; AIAA 93-0169.
7. Mingione, G.; Brandi, V.; Saporiti, A. A 3D ice accretion simulation code. In Proceedings of the 37th Aerospace Sciences Meeting and Exhibit, Reno, NV, USA, 11–14 January 1999; AIAA 99-0247.
8. Habashi, W.G.; Morency, F.; Beaugendre, H. *FENASP-ICE: A Second Generation 3D CFD-Based In-Flight Icing Simulation System*; SAE Technical Paper 2003-01-2157, 2003. Available online: https://saemobilus.sae.org/content/2003-01-2157/ (accessed on 5 September 2022). [CrossRef]
9. Strijhak, S.; Ryazanov, D.; Koshelev, K.; Ivanov, A. Neural Network Prediction for Ice Shapes on Airfoils Using iceFoam Simulations. *Aerospace* **2022**, *9*, 96. [CrossRef]
10. Zhou, F.; Feng, L.; Xu, C.; Zhao, K.; Han, Z. Determination and verification of critical ice shape for the certification of civil aircraft. *J. Exp. Fluid Mech.* **2016**, *30*, 8–13.
11. Zhan, P. Review on the system of icing facilities in NASA. *Aeronaut. Sci. Technol.* **2021**, *32*, 1–6.
12. Hedde, T.; Guffond, D. Development of a three-dimensional icing code—Comparison with experimental shapes. In Proceedings of the 30th Aerospace Sciences Meeting and Exhibit, Reno, NV, USA, 6–9 January 1992.
13. Weight, W.B.; Rutkowski, A. *Validation Results for LEWICE2.0*; NASA/CR-1999-208690; NASA: Washington, DC, USA, 1999.
14. Weight, W.B. Validation Results for LEWICE 3.0. In Proceedings of the 43rd AIAA Aerospace Sciences Meeting and Exhibit, Reno, NV, USA, 10–13 January 2005.
15. *SAE ARP5903-2003*; Droplet Impingement and Ice Accretion Computer Codes. SAE International: Warrendale, PA, USA, 2003. [CrossRef]
16. *SAE ARP5905-2003*; Calibration and Acceptance of Icing Wind Tunnels. SAE International: Warrendale, PA, USA, 2005. [CrossRef]
17. Li, Z. *Wind Tunnel Test Manual*; Aviation Industry Press: Beijing, China, 2015; pp. 853–854.
18. Editorial Committee of military training materials of PLA General equipment department. *Low Speed Wind Tunnel Test*; National Defense Industry Press: Beijing, China, 2002; pp. 332–333.
19. Barlow, J.B., Jr.; Rae, W.H.; Pope, A. *Low-Speed Wind Tunnel Testing*, 3rd ed.; Wiley: New York, NY, USA, 1999; pp. 234–366.
20. *Aircraft Ice Protection, AC20-73A*; Federal Aviation Administration: Washington, DC, USA, 2006.
21. Guo, X.; Zhang, P.; Zhao, X.; Yang, S.; Lin, W. Validation of thermal flow field in large icing wind tunnel. *Exp. Fluid Dyn.* **2020**, *34*, 82–91.
22. Zhao, K.L.; Lu, Z.L.; Ding, L.; Tan, G.; Feng, L.; Guo, T. A design method of hybrid airfoil applied in icing wind tunnel test. *Acta Aerodyn. Sin.* **2013**, *31*, 718–722.
23. Han, Z.R.; Zhao, K.L.; Yan, W. Application of gurney flap in hybrid wing design for icing wind tunnel. *Acta Aeronaut. Et Astronaut. Sin.* **2019**, *40*, 522387, In Chinese. [CrossRef]
24. Guo, Q.; Liu, S.; Guo, X.; Liu, Y.; Yang, S. Influence of sidewall interference on the impact characteristics of water droplets on wing surface in icing wind tunnel. In Proceedings of the 11th National Conference on Hydrodynamics, Shenzhen, China, 3 December 2020.
25. Guo, Q.; Liu, S.; Yi, X.; GUO, X.; Liu, Y. Influence of wind tunnel sidewall interference on NACA0012 airfoil icing growth. Key Laboratory of icing and anti icing, China Aerodynamics Research and development center. In Proceedings of the Fourth National Symposium on Icing and Anti Icing, Chongqing, China, 12–14 November 2020.
26. Zocca, M.; Gori, G.; Guardone, A. Blockage and three-dimensional effects in wind-tunnel testing of ice accretion over wings. *J. Aircr.* **2016**, *54*, 1–9. [CrossRef]
27. Qin, C.; Loth, E. Numerical Study of Droplet Trajectory and Collection Efficiency in IRT with Large Blockage Effects. In Proceedings of the 9th AIAA Atmospheric and Space Environments Conference, Denver, CO, USA, 5–9 June 2017.
28. Catherine, C. Effects of Aerodynamic Blockage on Stagnation Collection Efficiency in a Wind Tunnel Icing Environment. In Proceedings of the 2018 Atmospheric and Space Environments Conference, Atlanta, GA, USA, 25–29 June 2018.
29. Gray, V.H. *Correlations among Ice Measurements, Impingement Rates, Icing Conditions and Drag of a 65A004 Airfoil*; NACA TN 4151; NASA: Washington, DC, USA, 1958.
30. *Report of the 12A Working Group on Determination of Critical Ice Shapes for the Certification of Aircraft*; DOT/FAA/AR-00/37; Federal Aviation Administration: Washington, DC, USA, 2000.

Article

Contrast Icing Wind Tunnel Tests between Normal Droplets and Supercooled Large Droplets

Zhirong Han *, Jiangtao Si and Dawei Wu

Shanghai Aircraft Design and Research Institute, Shanghai 201210, China
* Correspondence: hanzhirong@comac.cc

Abstract: In order to compare and analyze the similarities and differences between normal droplet icing shapes and supercooled large droplet icing shapes, SADRI carried out normal droplet and supercooled large droplet icing wind tunnel tests in the NRC−AIWT icing wind tunnel. Taking the typical glaze ice in normal droplet icing conditions as the reference, the freezing drizzle and freezing rain icing tests under the supercooled large droplet conditions were carried out. The test results show that compared with normal droplets, the ice horn height of supercooled large droplets decreases with the increase in droplet particle size, and even the ice horn characteristics are not obvious when the icing condition is freezing rain. At the same time, the range and height of rough element ice shape after the main ice horn of supercooled large droplets are significantly larger and higher than those of the normal droplets, while the difference in the rough element in different supercooled large droplet icing conditions is small.

Keywords: supercooled large droplets; icing wind tunnel tests; glaze ice; freezing rain; rough element

Citation: Han, Z.; Si, J.; Wu, D. Contrast Icing Wind Tunnel Tests between Normal Droplets and Supercooled Large Droplets. *Aerospace* **2022**, *9*, 844. https://doi.org/10.3390/aerospace9120844

Academic Editors: Dan Zhao, Chenzhen Ji and Hexia Huang

Received: 25 November 2022
Accepted: 16 December 2022
Published: 18 December 2022

Publisher's Note: MDPI stays neutral with regard to jurisdictional claims in published maps and institutional affiliations.

1. Introduction

When a plane crosses clouds containing supercooled droplets which hit the surface, the droplets will ice and accrete on the surface when the temperature is below freezing. Ice not only changes the aerodynamic configuration of the aircraft (especially on the wings), it has a bad impact on the handling and stability of the aircraft, but also may fall off into the engine or hit the body. So, aircraft icing is one of the main threats to flight safety [1]. Johnson stated [2], "The icing problem is one of the most important ones facing the aviation industry today". In 1938, Gulick [3] tested an aspect ratio 6 wing in the Langley Full-Scale Tunnel with roughness intended to simulate an ice accretion. He found a 25% reduction in maximum lift and a 90% increase in drag for the conditions tested. Therefore, icing seriously affects the flight safety of aircraft. As for supercooled large droplet (SLD) ice, Ashenden et al. [4] found a similar result in wind tunnel tests with simulated ice accretions. The results showed more severe aerodynamic penalties due to the freezing drizzle case when operation of the deicing boot was simulated. Lee and Bragg [5] found that when the simulated ice shape was located at critical chordwise locations, a long separation bubble formed downstream of the shape and effectively eliminated the formation of a large leading-edge suction peak that was observed on the clean NACA 23012 airfoil. This resulted in a significant reduction in the maximum lift coefficient. For more information on the impact of icing on aircraft aerodynamics, refer to Reference [6].

Currently, civil aviation bureaus attach great importance to the safety of aircraft icing. In the airworthiness regulations of transport aircraft, there are a large number of items concerning the flight safety under icing conditions [7,8].

On 31 October 1994, an ATR72 model transport aircraft was wrecked in icing weather conditions [9]. The NTSB investigated the accident and confirmed that the cause of the accident was local freezing rain weather conditions that exceeded the normal drop icing environment in the FAA airworthiness regulations (Appendix C icing environment). Later, after 20 years of meteorological research, the FAA officially issued a new airworthiness icing requirement (the 140 amendment) in 2014, adding Appendix O SLD icing conditions.

With the release of new airworthiness regulations and the increasing importance of the icing status and anti/de-icing devices under SLD icing conditions, the analysis and test verification methods of icing conditions in SLD also face new challenges [10–16]. Numerical simulation and icing wind tunnel experiments are the common methods to verify the aerodynamic performance and the effect of anti/de-icing devices under icing conditions. When the aircraft manufacturer indicates the aircraft's compliance with airworthiness to a civil aviation bureau, it generally needs to use more than one method to verify safety. For flight safety under icing conditions, an icing wind tunnel test is one of the most important methods. When additional SLD icing conditions are added to the airworthiness regulations, the icing wind tunnel test becomes one of the most important and feasible means to understand the similarities and differences between the normal droplet ice shape and SLD ice shape. The icing wind tunnel can be used in any season with stable test conditions and good repeatability. Compared with the natural icing test and icing tank test, the cost is low and the test period is short, and is very safe. However, the test capability envelope is limited, subject to the capability of the test equipment. At present, the icing wind tunnel test plays an irreplaceable role in solving the three-dimensional icing ice type prediction, verifying the effect of the anti/de-icing device, and verifying the accuracy of the numerical simulation results [17–20].In order to obtain the similarities and differences between frozen ice shapes and SLD ice freezing conditions, the Shanghai Aircraft Design and Research Institute has carried out a phase of normal droplets and SLD at the high-Altitude Icy Wind Tunnel (AIWT) of the National Research Council of Canada (NRC) [21,22] to compare the icy air tunnel tests.

2. Icing Wind Tunnel

The AIWT belonging to the NRC is a refrigerated closed-loop low-speed wind tunnel oriented in a vertical plane as shown in Figure 1. The wind tunnel has two available test section sizes of 0.57 m × 0.57 m, with a demonstrated top simulated wind speed of over 100 m/s, and 0.52 m × 0.33 m, that increases the top wind speed to 180 m/s. The wind tunnel has a height simulation capacity of 0~9100 m [10]. The air temperature in the AIWT is controlled by varying the amount of refrigerant flowing through the heat exchanger in the channel loop to achieve a static air temperature at a Mach number of 0.3, ranging from −40 °C to + 30 °C or higher. The pressure in the wind tunnel can be controlled between about 101 and 30 kPa, which allows the simulation to rise from the ground to at least 40,000 ft. The wind tunnel uses 2 nozzle systems to simulate the icing conditions of SLD [23,24]. The small nozzle and large nozzle each have a set of independent water supply and gas supply systems. In the development of SLD icing condition simulation ability, the wind tunnel researchers first used computational methods to simulate the granularity distribution characteristics obtained by two nozzle joint injections with Langmuir-D distribution, and compared them with the curve of Appendix O to obtain the calculation results very close to the Appendix O curve in FAA Amendment 140. Based on the calculation results, the verification work of the combined jet in the wind tunnel was conducted. By continuously adjusting the water supply and gas supply pressure of the nozzle, the bimodal distribution characteristics consistent with the SLD conditions were well realized, as shown in Figure 2. The "App. O ZLE < 40" represents mean volume diameter (MVD) of freezing drizzle (FZDZ) less than 40 μm. The "App. O ZRE < 40" represents mean volume diameter (MVD) of freezing rain (FZRA) less than 40 μm.

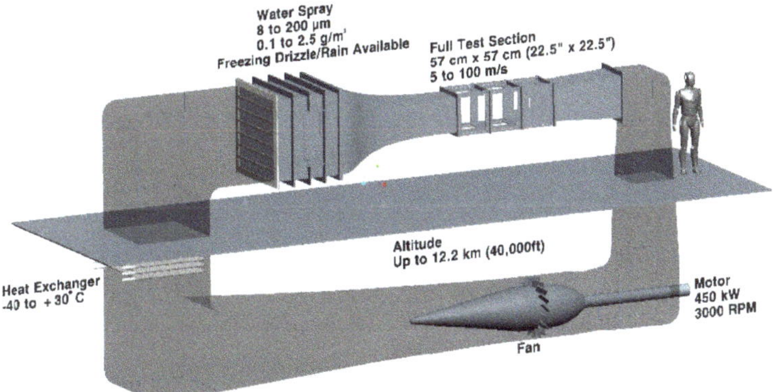

Figure 1. NRC−AIWT schematic diagram.

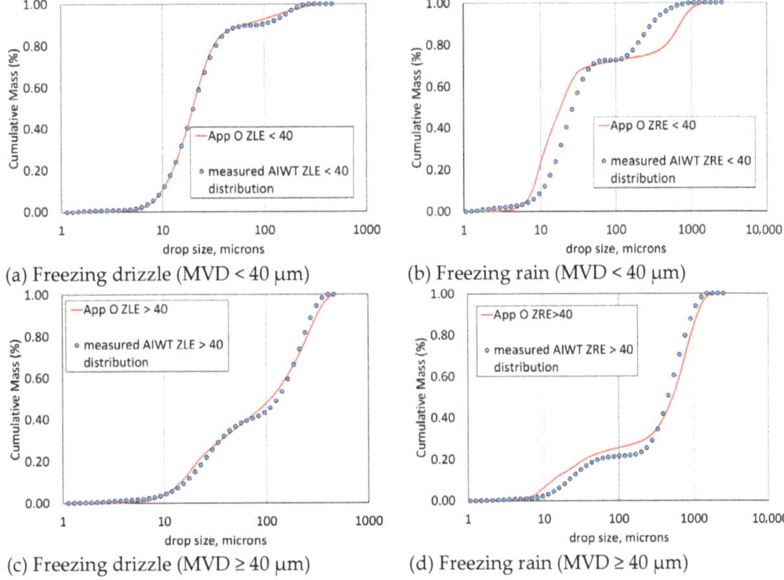

(a) Freezing drizzle (MVD < 40 µm) (b) Freezing rain (MVD < 40 µm)

(c) Freezing drizzle (MVD ≥ 40 µm) (d) Freezing rain (MVD ≥ 40 µm)

Figure 2. NRC−AIWT drop cumulative mass distribution for freezing drizzle and freezing rain.

3. Model

Considering the blockage in the icing wind tunnel in the 0.57 m × 0.57 m test section, a NACA0012 airfoil with 0.533 m chord length and 0.565 m span length was selected as the test model as shown in Figure 3 with details. The model was designed and manufactured by the NRC. The model has two leading edges, one is equipped with pressure taps as shown in Figure 4 to match the angle of attack by measuring the upper and lower surface pressure distribution. The other leading edge is for all icing tests. As shown in Figure 5, to reduce the weight, the first 1/4 of the airfoil is made of aluminum alloy (hollowed out), and the aft 3/4 of the airfoil is made of acrylic.

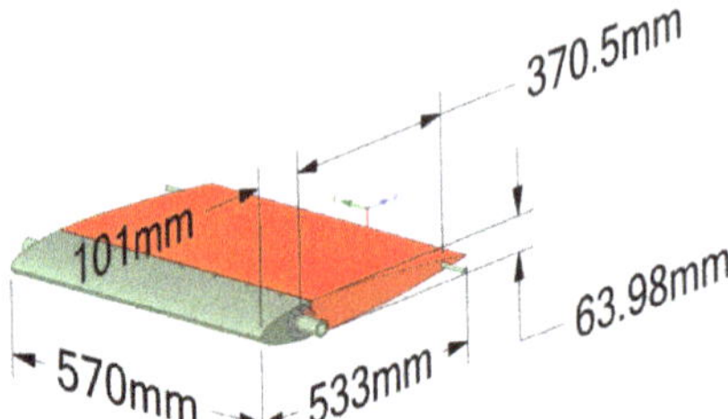

Figure 3. Test model and size.

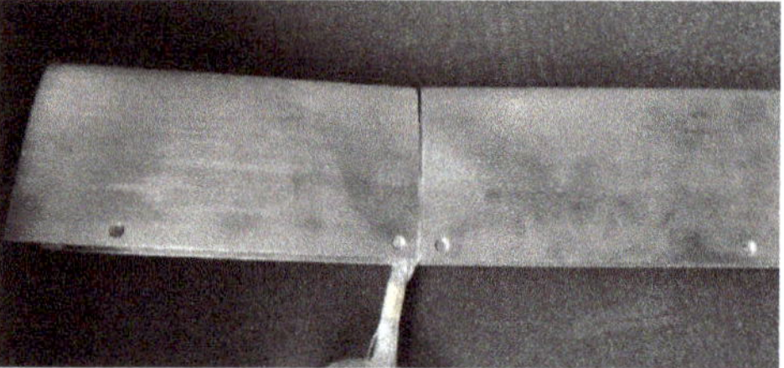

Figure 4. Pressure taps at one leading edge.

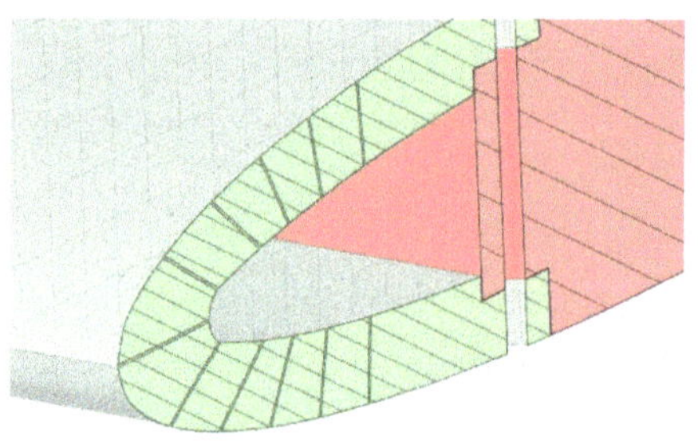
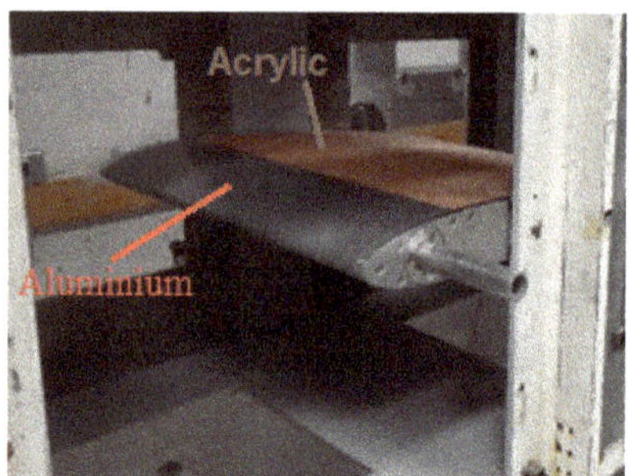

Figure 5. Two different materials of the model: aluminium alloy (hollowed out) and acrylic.

4. Test Conditions

The tests were performed in accordance with Appendixes C and O of CFR 14 Part 25. It should be noted that the AIWT does not fully guarantee that the in-flight icing simulation meets the Appendix O conditions due to the difficult decreasing temperature and velocity balance with the surrounding airflow.

As shown in Table 1, the test conditions include 10 test points (excluding repetitive runs). Target test points include the following conditions: (1) Appendix C [25,26] (double ice horns and single ice horn); (2) Appendix O glaze ice (double ice horns and single ice horn). When the droplets collide on the surface of the component, they do not immediately freeze and the water film (formed by droplets) gradually freezes while flowing along the material surface, forming glaze ice. Water film squeezes small air bubbles during the flow, so the ice is transparent and dense. Moreover, due to the aero-heating in the leading-edge area, especially the stagnation point, the thickness of the ice is small, and the large-scale icing occurs behind the two sides of the stagnation point, and it generally forms an ice horn. When near the 0° angle of attack, the ice forms a significant double ice angle; when the angle of attack is near the non-0° angle of attack, there is generally only one obvious angle of ice (main ice angle). In Table 1, the App. C-CM represents the continuous maximum icing conditions in Appendix C of Part 25 of CFR 14. App. O represents CFR 14 Part 25 Appendix

O icing condition, while FZDZ/L is the freezing drizzle (MVD < 40 µm), and FZDZ/G is the continuous maximum icing condition of the freezing drizzle (MVD ≥ 40 µm). FZRA is the icing condition of Appendix O freezing rain. As the 14,000 ft altitude exceeds the envelope of App. O when the static temperature T_s equals to −6.6 °C, altitude was set as H = 5000 ft in the uniform. The value of the median volume diameter (MVD) is taken when the cumulative mass of the droplets reaches 50%. The liquid water content (LWC) takes the maximum value corresponding to Appendix O.

Table 1. Test conditions.

Run No.	H (ft)	V (m/s)	AOA (°)	T_s (°C)	MVD (µm)	LWC (g/m³)	Time (min)	Cloud
1	14,000	95	4.9	−6.6	20	0.50	11 min	App.C-CM
2	14,000	95	0.0	−6.6	20	0.50	11 min	App.C-CM
3	14,000	95	4.9	−6.6	20	0.40	11 min	App.O-FZDZ/L
4	14,000	95	0.0	−6.6	20	0.40	11 min	App.O-FZDZ/L
5	14,000	95	4.9	−6.6	110	0.25	11 min	App.O-FZDZ/G
6	14,000	95	0.0	−6.6	110	0.25	11 min	App.O-FZDZ/G
7	5900	95	4.9	−6.6	19	0.28	11 min	App.O-FZRA/L
8	5900	95	0.0	−6.6	19	0.28	11 min	App.O-FZRA/L
9	5900	95	4.9	−6.6	526	0.24	11 min	App.O-FZRA/G
10	5900	95	0.0	−6.6	526	0.24	11 min	App.O-FZRA/G

The test conditions in Table 1 are very typical glaze ice icing conditions. In particular, the angle of attack of 4.9 degrees and H = 14,000 ft are the typical angle of attack and altitude in the holding stage of a large aircraft, which will form an obvious upper ice horn (usually on the wing). The angle of attack of 0 degrees will form obvious double ice horns (usually on the horizontal tail). The MVD = 20 µm is the most critical MVD condition after sensitivity analysis in Appendix C in FAR25. Ts = −6.6 °C is very close to the most critical static temperature (−4.1 °C). Considering the problem of ice shedding, Ts is designed to be 2.5 °C lower. The LWC is determined jointly by Ts and MVD in Appendix C in FAR25. As the model is relatively small, the ice accretion is relatively fast, so the icing time cannot reach the holding icing time (in Appendix C in FAR25 holding icing time is 45 min). Considering that the ice horn is easy to break when it is too high, after calculation carried out before the experiment, it was considered that 11 min is appropriate. The MVD value in Appendix O is the MVD when cumulative mass of droplet reaches 0.5, shown in Figures 6 and 7.

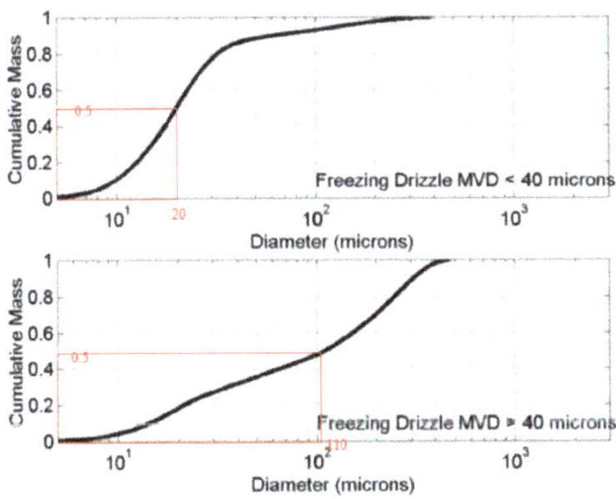

Figure 6. Appendix O—freezing drizzle, drop diameter distribution.

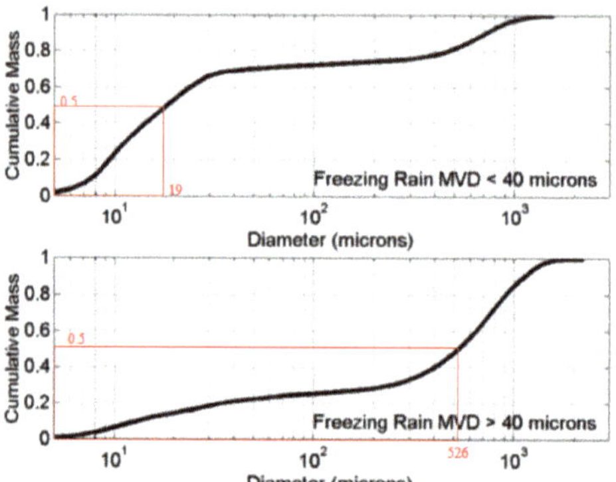

Figure 7. Appendix O—freezing rain, drop diameter distribution.

5. Test Results

The pressure distribution of the leading region in both flight states was matched in the wind tunnel prior to the icing test. The results of the pressure distribution matching results for the two states, AOA = 0° and AOA = 4.9°, are shown in Figures 8 and 9. In Figures 8 and 9, the abscissa x/c (dimensionless) represents relative position in the chordal direction (the c is the length of the airfoil chord). The ordinates Cp represent the dimensionless pressure coefficient.

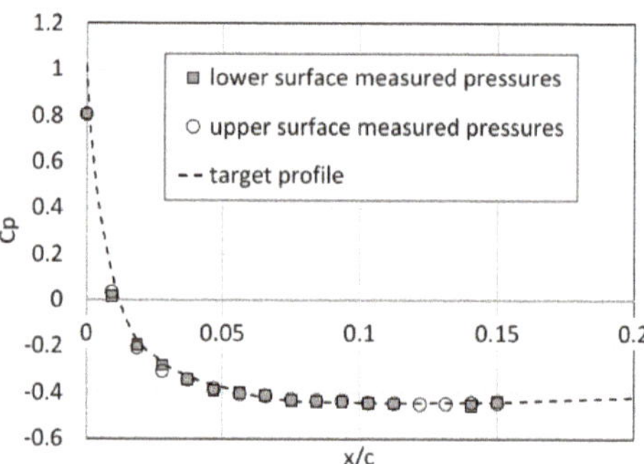

Figure 8. Pressure distribution of the leading−edge matching at AOA = 0°.

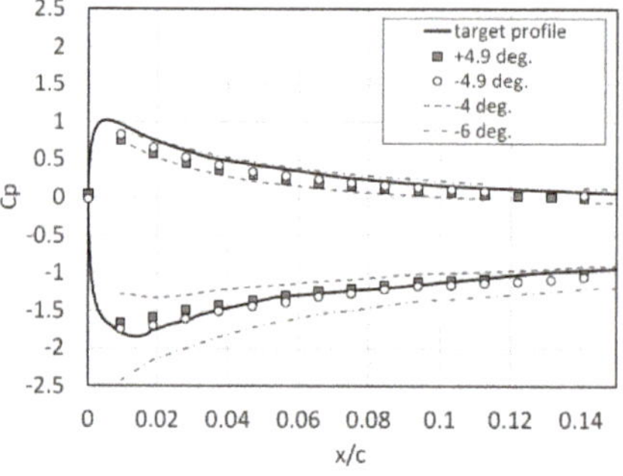

Figure 9. Pressure distribution of the leading−edge matching at AOA = 4.9°.

After pressure distribution matching, icing runs were conducted: first, the wind speed was adjusted to the target wind speed. Second, the air was cooled down to the target temperature. Third, the nozzle was turned on to spray supercooled droplets and this was timed. Fourth, when the icing time was up, the spray ended. Fifth, the temperature was reduced (to prevent ice from falling off). Sixth, after cooling, the wind was decreased to make the wind speed zero. Seventh, the hatch door was opened, and the "hot knife" (square copper plate, with which the shape of the leading edge of the airfoil was removed) was heated to melt the ice shape, forming a groove (as shown in Figure 10), so that the hot knife could fit with the airfoil. Eighth, the coordinate paper was stuck on the hot knife, and then the ice shape was drawn on the coordinate paper along the circumference of the ice shape. Ninth, the ice was heated to melt it quickly. Tenth, the airfoil surface was cleaned and the next test run was conducted.

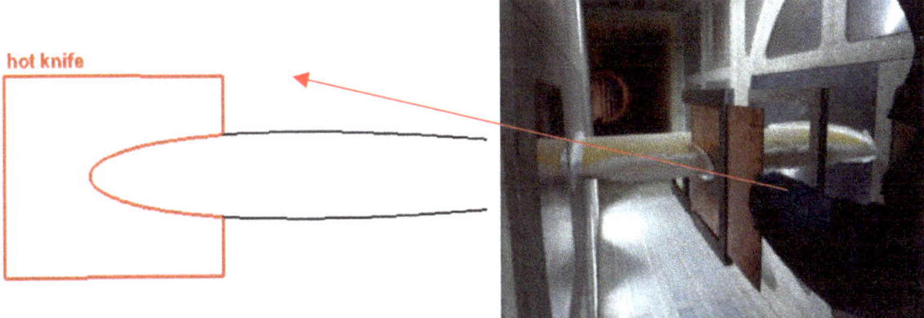

Figure 10. Hot knife used in many icing wind tunnels.

After ice accumulation, ice shapes were taken at three different span locations (center line and 100 mm on the left and right sides of the center of the model as shown in Figure 11 in red curves). The ice shapes at the three different sections coincide in position and height of the ice horn well. The ice shapes presented in this paper were all obtained at the centerline position of the model.

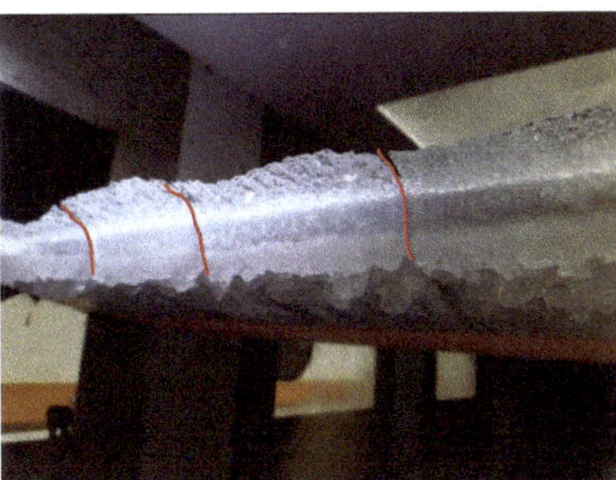

Figure 11. Ice shape at three different span locations.

In Figure 12, ice shapes with a single ice horn in different states are shown in different colors. The main ice horn of Appendix C is the highest, followed by FZDZ/L, FZDZ/G and FZRA/L and FZRA/G. Ice shape of Appendix C has the smallest range of the rough element, and the height of the rough element decreases sharply far away from the main ice horn; the range of the SLD ice shapes after the main ice horn are very large, and the rough element is almost the same in different states; the height of the rough element does not decrease significantly with the distance away from the main ice horn.

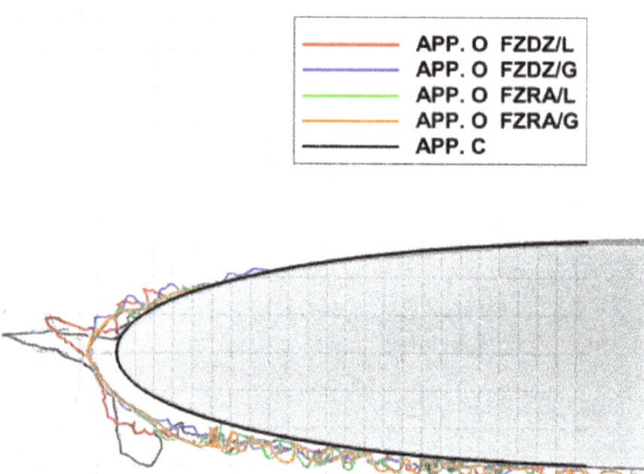

Figure 12. Ice shapes with single horn at AOA = 4.9°.

In Figure 13, ice shapes with double ice horns in different states are shown in different colors. The height of the main ice horns decreases gradually from Appendix C to FZDZ/L, FZDZ/G; while the main ice horn of FZRA/L and FZRA/G is not very obvious. Appendix C has minimized the range of the rough element, and the rough element height decreases sharply far away from the main ice horn; the range of the SLD ice shape after the main ice horn is very large, and the rough element is almost the same in different states; the height of the rough element does not decrease significantly with the distance away from the main ice horn.

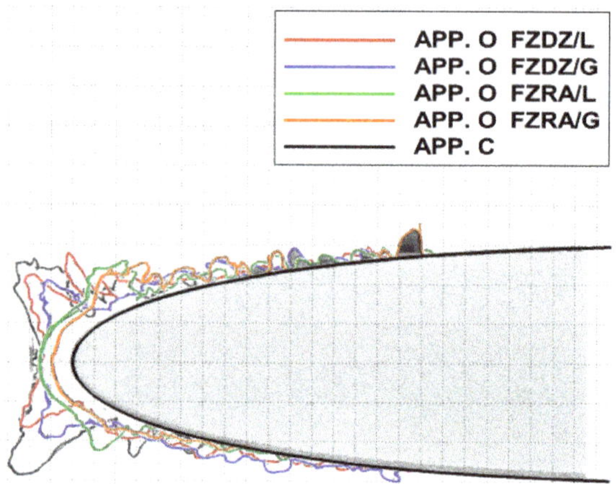

Figure 13. Ice shapes with double horns at AOA = 0°.

Through the comparative analysis of the above test results, it can be seen that a significant result of the SLD icing was ice accretions that formed downstream of the ice-protected surfaces [27]. The shape characteristics of ice in Appendix O are consistent with those in other references [13,19,28,29]. After observing the shape of the roughness element behind the main ice horn shown in Figure 14, the roughness cannot be formed by runback water, it can only be formed by reattachment of water droplet splashing or bouncing [30–33] as shown in Figure 15 because the roughness element particles are discontinuous, as shown in Figure 13.

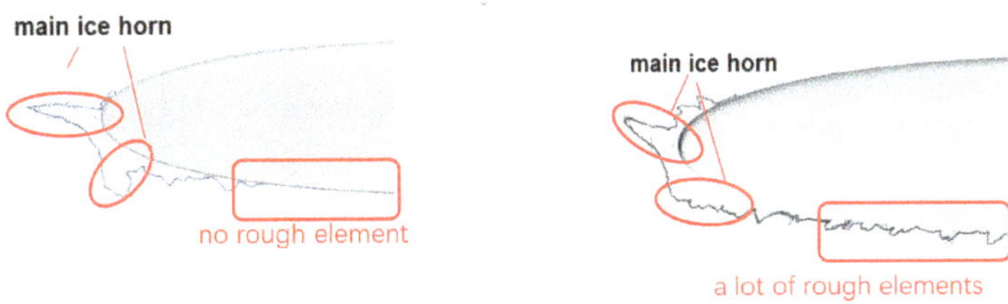

Figure 14. Ice shape of normal drop (**left**) and SLD (**right**).

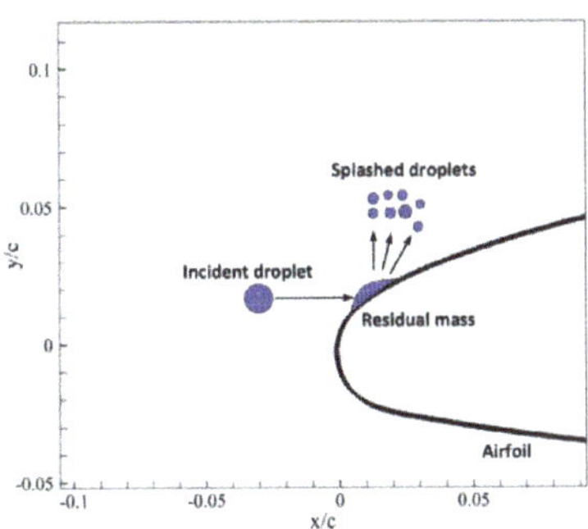

Figure 15. Simplification of droplet splashing for 2D simulation.

The shape characteristics of ice in Appendix O are consistent with those in other references. For example, the test results in NASA IRT and Chinese FL-61 icing wind tunnel. It can be found in the above icing wind tunnel from References [16,19,29,34] that after the main ice horn, the range of roughness element is wide, and there is little ice at the gap between the roughness elements, indicating that the formation of the roughness element here is not due to the runback water, but SLD splashing and bouncing.

6. Conclusions

In NRC's AIWT, the icing wind tunnel test of normal droplets and SLD was carried out. The differences from the comparison of ice shapes were consistent with the splashing and bouncing [30,31] theory of SLD:

(1) As the LWC of SLD is smaller than the LWC of normal droplets, the ice horn height of SLD is smaller than that of the normal droplets. At the same time, due to the splashing and rebound phenomenon of large droplets, the main ice horn of SLD ice will be further reduced, and the position of the main ice horn will move to the trailing edge, especially in the case of freezing rain.

(2) After splashing and rebounding, droplets will continue to fly to the trailing edge with the airflow. Due to the influence of gravity and airflow, some droplets will hit the airfoil twice, thus forming rough elements in a far range after the main ice horn of the leading edge. This rough element is mainly caused by the splashing and rebound of droplets, rather than the runback water, so the height of the rough element does not decrease significantly with the distance away from the main ice angle, and obvious discontinuities between rough elements will occur.

Author Contributions: Conceptualization and methodology, Z.H.; methodology, J.S.; resources and data curation, D.W.; writing—original draft preparation, Z.H.; writing—review and editing, J.S.; supervision, D.W.; project administration, Z.H.; funding acquisition, J.S. All authors have read and agreed to the published version of the manuscript.

Funding: This research was funded by the National Key R&D Program of China (Project No. 2020YFA0712000).

Conflicts of Interest: The authors declare that they have no known competing financial interest or personal relationships that could have appeared to influence the work reported in this paper.

References

1. Appiah-Kubbi, P.; Martos, B.; Atuahene, I.; William, S. U. S. inflight icing accidents and incidents, 2006 to 2010. In Proceedings of the Industrial and Systems Engineering Research Conference, San Juan, Puerto Rico, USA, 18–22 May 2013.
2. Johnson, C.L. Wing loading, icing and associated aspects of modern transport design. *J. Aeronaut. Sci.* **1940**, *8*, 43–54. [CrossRef]
3. Gulick, B.G. *Effects of Simulated Ice Formation on the Aerodynamic Characteristics of an Airfoil*; Technique Report L292; NACA: Washington, DC, USA, 1938.
4. Ashenden, R.; Lindberg, W.; Marwitz, J.D.; Hoxie, B. Airfoil performance degradation by supercooled cloud drizzle and rain drop icing. *AIAA J. Aircr.* **1996**, *33*, 1040–1046. [CrossRef]
5. Lee, S.; Bragg, M.B. Experimental investigation of simulated large-droplet ice shapes on airfoil aerodynamics. *AIAA J. Aircr.* **1999**, *36*, 844–850. [CrossRef]
6. Bragg, M.B.; Broeren, L.A.; Blumenthal, A.P. Iced-airfoil aerodynamics. *Prog. Aerosp. Sci.* **2005**, *41*, 323–362. [CrossRef]
7. Federal Aviation Administration. *Part 25: Airworthiness Standard: Transport Category Airplanes, Admt 25-140. Title 14: Aeronautics and Space. U.S.*; Federal Aviation Administration: Washington, DC, USA, 2014.
8. European Union Aviation Safety Agency. *CS-25: Certification Specifications and Acceptable Means of Compliance for Large Aeroplanes, Admt 27*; European Union Aviation Safety Agency: Cologne, Germany, 2021.
9. Bragg, M.B.; Perkins, W.; Sarter, N.; Basar, T.; Voulgaris, P.; Gurbacki, H.; Melody, J.; McCray, S. An interdisciplinary approach to inflight aircraft icing safety. In Proceedings of the 36th AIAA Aerospace Sciences Meeting and Exhibit, Reno, NV, USA, 12–15 January 1998.
10. Fu, C.; Song, W.; Peng, Q.; Liao, D.; Wang, C. An overview of supercooled large droplets icing condition simulation capacity in icing wind tunnels. *J. Exp. Fluid Mech.* **2017**, *31*, 1–7.
11. Judith, F.; Van, Z.; IDE, R.F. NASA Glenn Icing Research Tunnel: 2012 cloud calibration procedure and results. In Proceedings of the 4th AIAA Atmospheric and Space Environments Conference, New Orleans, LA, USA, 25–28 June 2012.
12. Steen, L.E.; IDE, R.F. NASA Glenn Icing Research Tunnel: 2014 and 2015 Cloud Calibration Procedures and Results. NASA/TM—2015-218758. 2015. Available online: https://ntrs.nasa.gov/citations/20150009300 (accessed on 24 November 2022).
13. Potapczuk, M.G.; Tsao, J.C.; King-Steen, L.E. Bimodal SLD ice accretion on a NACA 0012 airfoil model. In Proceedings of the 9th AIAA Atmospheric and Space Environments Conference, Denver, CO, USA, 5–9 June 2017.
14. Potapczuk, M.G.; Tsao, J. Bimodal SLD ice accretion on swept NACA0012 airfoil models. In Proceedings of the AIAA AVIATION 2020 FORUM, Virtual Event, 15–19 June 2020.
15. Esposito, B.M.; Ragni, A.; Ferrigno, F.; Vecchione, L. Cloud Calibration Update of the CIRA Icing Wind Tunnel. SAE Technical Paper Series. In Proceedings of the FAA In-Flight Icing/Ground De-icing International Conference & Exhibition, Chicago, IL, USA, 16–20 June 2003. [CrossRef]
16. Shu, J.; Xu, D.; Han, Z.; Li, S.; Huang, X. Study on the hybrid wing design of the icing wind tunnel SLD icing test. *Acta Aeronaut. Astronaut. Sin.* **2023**, *44*, 627182. (In Chinese) [CrossRef]
17. Fu, C.; Peng, Q.; Zhang, H.; Wang, C.; Wu, S. Preliminary research on spray nozzle atomization characteristics in icing wind tunnel environment. *J. Exp. Fluid Mech.* **2015**, *29*, 30–34.
18. Potapczuk, M. SLD Research at NASA: Basic Research. NASA Rep. 2015-0007676. Presented to WEZARD SLD Workshop Brussels, Belgium, 10 June 2013. Available online: https://ntrs.nasa.gov/api/citations/20150007676/downloads/20150007676.pdf (accessed on 24 November 2022).
19. William, B.W.; Mark, G.P.; Laurie, H.L. Comparison of LEWICE and GlennICE in the SLD regime. In Proceedings of the 46th AIAA Aerospace Sciences Meeting and Exhibit, Reno, Nevada, 7–10 January 2008.
20. Bidwell, C. Icing Simulation. NASA/CP-2009-215797, NASA Glenn Research Center. 2009; pp. 71–86. Available online: https://ntrs.nasa.gov/api/citations/20090030605/downloads/20090030605.pdf (accessed on 24 November 2022).
21. Oleskiw, M.M.; Hyde, F.H.; Penna, P.J. In-flight icing simulation capabilities of NRC's altitude icing wind tunnel. In Proceedings of the 39th Aerospace Sciences Meeting and Exhibit, Reno, NV, USA, 8–11 January 2001.
22. Orchard, D.; Clark, C.; Oleskiw, M. Development of a supercooled large droplet environment within the NRC Altitude Icing Wind Tunnel. In Proceedings of the SAE 2015 International Conference on Icing of Aircraft, Engines, and Structures, Prague, Czech Republic, 22–25 June 2015; SAE Technical Paper: Warrendale, PA, USA, 2015. [CrossRef]
23. Orchard, D.; Clark, C.; Chevrett, G. Measurement of liquid water content for supercooled large drop conditions in the NRC's altitude icing wind tunnel. In Proceedings of the International Conference on Icing of Aircraft, Engines, and Structures, Minneapolis, MN, USA, 17–21 June 2019; SAE Technical Paper: Warrendale, PA, USA, 2019. [CrossRef]
24. Orchard, D.; Szlider, K.; Davison, C. Design of an icing wind tunnel contraction supercooled large drop conditions. In Proceedings of the 2018 Atmospheric and Space Environments Conference, Atlanta, Georgia, 25–29 June 2018.
25. Heinrich, A.; Ross, R.; Zumwalt, G.; Provorse, J.; Padmanabhan, V. *Aircraft Icing Handbook*; ADA238040, DOT/FAA/CT-88/8-2; Federal Aviation Administration: Washington, DC, USA, 1991.
26. Cornell, D.; Donahue, C.A.; Keith, C. A Comparison of Aircraft Icing Forecast Models. Technical Note, AFCCC/TN-95/004. 1995. Available online: https://apps.dtic.mil/sti/pdfs/ADA303307.pdf (accessed on 24 November 2022).

27. Miller, D.R.; Addy, H.E., Jr.; Ide, R.F. A study of large droplet ice accretions in the NASA-Lewis IRT at near-freezing conditions. In Proceedings of the 34th Aerospace Sciences Meeting and Exhibit, Reno, NV, USA, 15–18 January 1996.
28. Tsao, J.C. Additional results of glaze icing scaling in SLD conditions. In Proceedings of the 8th AIAA Atmospheric and Space Environments Conference, Washington, DC, USA, 13–17 June 2016.
29. Potapczuk, M.G.; Tsao, J.C. Further examinations of bimodal SLD ice accretion in the NASA icing research tunnel. In Proceedings of the 2018 Atmospheric and Space Environments Conference, Atlanta, Georgia, 25–29 June 2018.
30. Wright, W.B.; Potapczuk, M.G. Semi-empirical modeling of SLD physics. In Proceedings of the 42nd AIAA Aerospace Sciences Meeting and Exhibit, Reno, Nevada, 5–8 January 2004.
31. Tan, S.C. A tentative mass loss model for simulating water droplet splash. In Proceedings of the 42nd AIAA Aerospace Sciences Meeting and Exhibit, Reno, Nevada, 5–8 January 2004.
32. Trujillo, M.F.; Mathews, W.S.; Lee, C.F.; Peters, J.E. Modeling and experiment of impingement and atomization of a liquid spray on a wall. *Int. J. Engine Res.* **2000**, *1*, 87–104. [CrossRef]
33. Bilodeau, D.R.; Habashi, W.G.; Fossati, M.; Baruzzi, G.S. Eulerian modeling of supercooled large droplet splashing and bouncing. *J. Aircr.* **2015**, *52*, 1611–1624. [CrossRef]
34. Potapczuk, M.G.; Tsao, J.C. The Influence of SLD Drop Size Distributions on Ice Accretion in the NASA Icing Research Tunnel. In Proceedings of the International Conference on Icing of Aircraft, Engines, and Structures, Minneapolis, MN, USA, 17 June 2019.

Article

Numerical and Experimental Research on Flight Control of a V-Tail Configuration for the Wind Tunnel Model of Aircraft

Jun Liu [1], Wei Qian [1,2,3], Yuguang Bai [1,2,3,*] and Xiaole Xu [1]

[1] School of Aeronautics and Astronautics, Dalian University of Technology, Dalian 116023, China
[2] State Key Laboratory of Structural Analysis for Industrial Equipment, Dalian University of Technology, Dalian 116023, China
[3] Advanced Technology for Aerospace Vehicles of Liaoning Province, Dalian University of Technology, Dalian 116023, China
*   Correspondence: baiyg@dlut.edu.cn

Abstract: The V-tail configuration has excellent stealth performance and has been using widely in the aerodynamic shape design of advanced aircraft. Many recent studies have focused on numerical simulation about V-tail configuration flight performance. The relative wind tunnel tests still need to be developed. This challenge is a focused aspect in such research. In the present experimental study, the role of flight control law was investigated in order to keep the test model in the target attitude and height. An effective design method of a full model of the aircraft with twin V-tails is proposed based on CFD evaluation. This model was manufactured based on the design of a two degrees of freedom support system via a Chinese wind tunnel. A longitudinal flight control law was proposed and simulated. Wind tunnel tests were employed to find the effectiveness of the model design and the control law. It is seen from the results that the proposed experimental method via a full model of the aircraft with twin V-tails and a novel longitudinal flight control law is effective. These test results can provide appliable contributions on the development of the support system for wind tunnel experiments. The proposed model design and test methods can be useful for applications in the aeroelastic wind tunnel tests of the full model aircrafts.

Keywords: full model aircraft; V-tail; wind tunnel test; flight control; pitch and plunge freedom

Citation: Liu, J.; Qian, W.; Bai, Y.; Xu, X. Numerical and Experimental Research on Flight Control of a V-Tail Configuration for the Wind Tunnel Model of Aircraft. *Aerospace* **2022**, *9*, 792. https://doi.org/10.3390/aerospace9120792

Academic Editors: Dan Zhao, Chenzhen Ji and Hexia Huang

Received: 29 October 2022
Accepted: 29 November 2022
Published: 3 December 2022

Publisher's Note: MDPI stays neutral with regard to jurisdictional claims in published maps and institutional affiliations.

1. Introduction

The V-tail configuration has excellent stealth performance and has been using widely in the aerodynamic shape design of advanced aircraft [1–4]. The research on V-tail aircraft is of more significance. However, the literature and papers about V-tails are very limited [5]. Malcolm J. Abzug [6] studied the stall problem of V-tails. Qiao et al. [7] put forward an adaptive back-stepping neural control (ABNC) method for the coupled nonlinear model of a novel type of embedded surface morphing aircraft, based on a large number of aerodynamic data for different V-tail configurations. Wang et al. [8] studied the three-axis static and dynamic stability characteristics of an example Blended-Wing-Body (BWB) aircraft with V-tail configuration. Leshikar et al. [9] developed an approach for generating linear time invariant state-space models of a small Unmanned Air System and verified the approach by an inverted V-tail aircraft model. Jin et al. [10] summarized the design technique of rigid/flexible hybrid model and testing methods of all-moving V-tail buffet wind tunnel test, and established a set of systematical theoretical analysis techniques, design criteria, and test methods for aircraft V-tail structure buffet dynamic strength design and test of advanced fighters and unmanned aerial vehicles. The research about V-tail configuration mainly concentrates on numerical simulations whereas the related wind tunnel tests are rare.

Compared with numerical simulations and flight tests, wind tunnel tests are important means of aeroelastic performance evaluation and verification of aerospace vehicles due to their advantages of high reliability [11–13]. The focus of this paper is to study the flight performance of V-tail configuration aircraft by means of wind tunnel tests.

According to different test requirements, it is important to select an appropriate wind tunnel test support system. With the development of wind tunnel test technology, various support systems have been developed. Many kinds of support system are used for wind tunnel test [14], such as tail support system, external/balance support system, side wall support system and wing tip support system. As a special wind tunnel for aeroelastic test, NASA Langley center Transonic Dynamics Tunnel (TDT) has a variety of test model sup-port methods, such as sidewall support, cable mount, and forced/free oscillation crossbar [15]. Russia TsAGI has developed a Floating Suspension System in a T-128 wind tunnel [16]. Typical facilities allow one or two degrees of freedom and some sophisticated testing facilities such as those operated by NASA and German Dutch Wind Tunnels (DNW) allow up to six degrees of freedom [17]. Many scholars made corresponding experimental studies based on different support systems. Gebbink et al. [18] conducted a wind tunnel test in the High-Speed Tunnel (HST) of DNW about a full-span scaled model mounted to a dorsal sting. Tang and Dowell [19] studied the effects of a free-to-roll fuselage on wing flutter and verified the proposed theory by wind tunnel test with a tail support. Allen et al. [20,21] carried out a dynamically scaled aeroelastic wind tunnel test in NASA's TDT wind tunnel on a sidewall support half span Truss-Braced Wing model.

Recently, in order to simulate the free flight condition accurately, a wind tunnel model support system with release of pitch and plunge degrees of freedom has been developed in China, which can simulate rigid body motion mode of aircraft [22]. The research of this paper is carried out based on a developed support system. The structural design needs to match the support system. The flight control law was designed to keep the aircraft model in steady level flight. The purpose of the paper is to verify the flight control law under the support system with release of pitch and plunge degrees of freedom and investigate the longitudinal flight performance of the V-tail configuration aircraft. This experimental research method provides technical support for the future wind tunnel test of elastic aircraft model.

The paper is organized as follows. Section 2 describes the design and manufacture process of the wind tunnel test model. Section 3 presents details on the design of flight control law, which contains the establishment of dynamic equations, the measurement of parameters and the simulation of flight control law. Section 4 presents analysis of the experimental results of the wind tunnel test. Conclusions are drawn in Section 5.

2. Model Design and Manufacture

The design of a wind tunnel test model needs to comprehensively consider aerodynamic shape, stiffness, strength, and flight control requirements. It can make it more feasible to simulate the steady level flight of the aircraft in the wind tunnel. The design process of the aircraft model is shown in Figure 1.

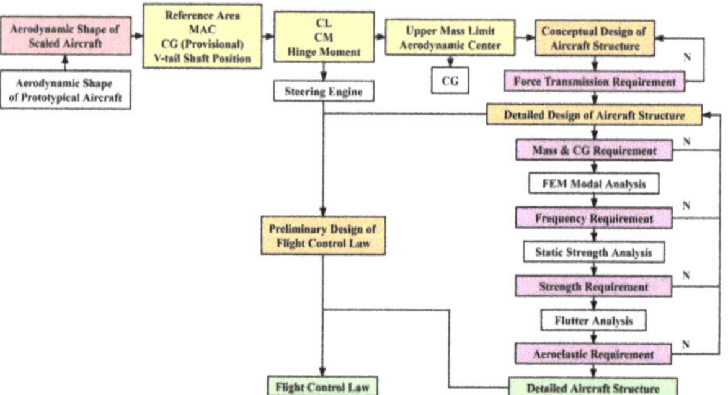

Figure 1. The design process of the aircraft model.

2.1. Aerodynamic Shape Design

A 3D full model aircraft is established as shown in Table 1. The wind tunnel test model is an 8%-length, full-span, rigidly scaled model of the original model. As shown in Figure 2, this aircraft was composed of a fuselage, wings, and V-tails. The aircraft is 1.82 m with 1.28 m wingspan. The mean dynamic chord (MAC) is 0.32 m and the dihedral angle of the V-tails is 40°.

Table 1. Model geometric parameters.

Parameters	Value
Scale ratio	8%
Length	1.82 m
Wingspan	1.28 m
MAC	0.32 m
Dihedral angle of V-tails	40°

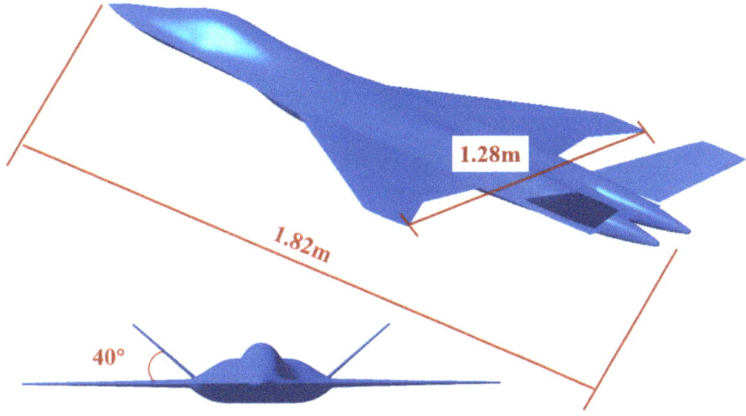

Figure 2. Aerodynamic shapes of the present aircraft.

2.2. Aerodynamic Parameters

Some aerodynamic parameters such as lift force, pitch moment, and the position of aerodynamic center (AC) should be considered before the structure design.

The aerodynamic parameters of the aircraft at 0.2 Mach can be obtained through high-precision Computational Fluid Dynamics (CFD) calculation [23]. The parameters included the lift coefficient, drag coefficient, pitch moment coefficient, and hinge moment under different angles of attack and V-tail deflection angles (see Figures 3 and 4).

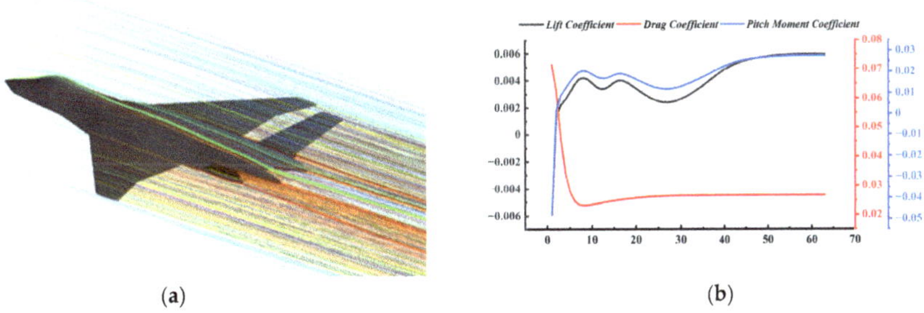

Figure 3. CFD calculation process: (**a**) the path lines and (**b**) the aerodynamic coefficient iteration process.

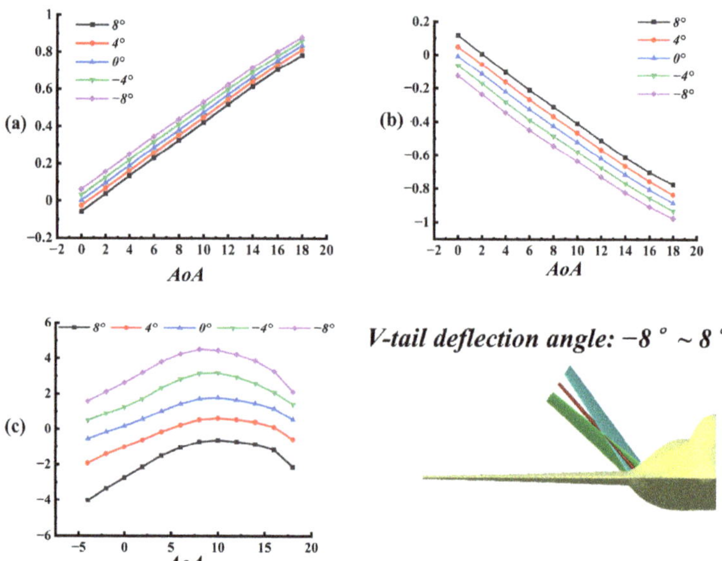

Figure 4. Aerodynamic coefficients calculated by CFD: (**a**) lift coefficient; (**b**) pitch moment coefficient; and (**c**) hinge moment.

2.3. Aircraft Trimming

As shown in Figure 4, the sign of the pitch moment coefficient changed from negative to positive with the increase in V-tail deflection angle. This indicated that the aerodynamic shape can be balanced. The pitch moment generated by V-tail deflection was enough to change the pitch angle of the aircraft.

2.4. Aerodynamic Center

The AC is the action point of aerodynamic increment of aircraft. The pitch moment at this position is always equal to the zero-lift moment. During the process of CFD calculation, the position of the reference center of gravity (CG) was given in advance. The moment balance equation is established according to the obtained aerodynamic parameters [24]. The position of CG is taken as the moment action point, see Figure 5.

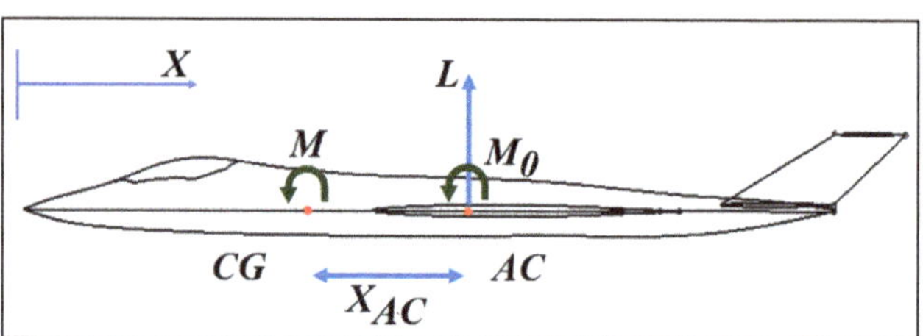

Figure 5. Force analysis of the aircraft.

$$M = -L \times X_{AC} + M_0 \tag{1}$$

$$M = \frac{1}{2}\rho V^2 S \bar{c} C_M \tag{2}$$

$$L = \frac{1}{2}\rho V^2 S C_L \tag{3}$$

$$M_0 = \frac{1}{2}\rho V^2 S\bar{c}C_{M_0} \tag{4}$$

According to the aerodynamic theory, the following equations can be obtained. Substitute Equations (2)–(4) into Equation (1), it is obtained that

$$C_{M_0} = \frac{X_{AC}}{\bar{c}}C_L + C_M \tag{5}$$

The lift coefficient curve and pitch moment coefficient curve (i.e., as shown in Figure 6) without V-tail deflection are linearly fitted, and the expressions are obtained.

$$C_L = A_1\alpha + B_1 \tag{6}$$

$$C_M = A_2\alpha + B_2 \tag{7}$$

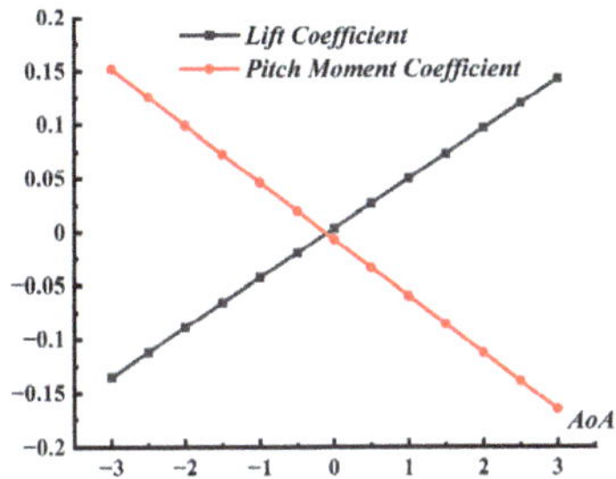

Figure 6. The lift coefficient and pitch moment coefficient curves.

Substitute Equations (6) and (7) into Equation (5), it is obtained that

$$C_{M_0} = \frac{(A_1\alpha + B_1)X_{AC}}{\bar{c}} + A_2\alpha + B_2 = \left(\frac{A_1 X_{AC}}{\bar{c}} + A_2\right)\alpha + \frac{B_1 X_{AC}}{\bar{c}} + B_2 \tag{8}$$

Since the zero-lift moment coefficient is independent of the angle of attack, so

$$\frac{A_1 X_{AC}}{\bar{c}} + A_2 = 0 \tag{9}$$

$$X_{AC} = -\frac{A_2 \bar{c}}{A_1} \tag{10}$$

The position of AC is thus obtained.

The stability of an aircraft is closely dependent to the position of its CG and AC [25]. The proposed CG is selected as 17.1% of MAC in front of AC by considering the model design and the assembly of wind tunnel test support system, as shown in Figure 7.

Figure 7. CG and AC positions of the aircraft.

The pitch moment coefficients based on the new CG with different V-tail deflection angles are shown in Figure 8, which is significant for the design of flight control law.

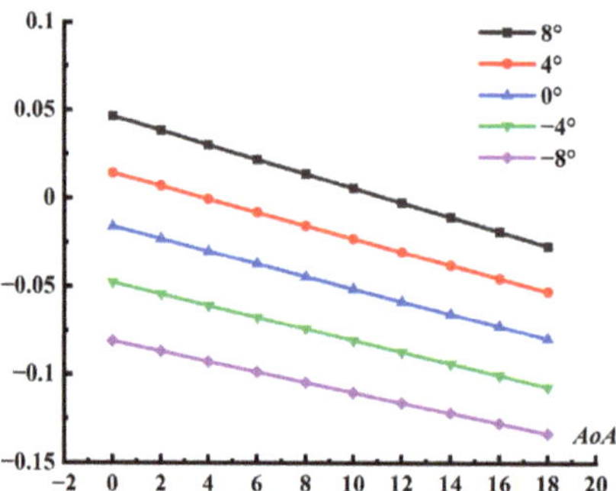

Figure 8. Pitch moment coefficient based on the new CG.

2.5. Detailed Structure Design of Full Model Aircraft

The concept of the wind tunnel support system is shown in Figure 9, in which a vertical beam and a carriage were included. It was installed inside the aircraft model and provides the pitch and plunge degrees of freedom to allow the model to "fly" in the wind tunnel test section.

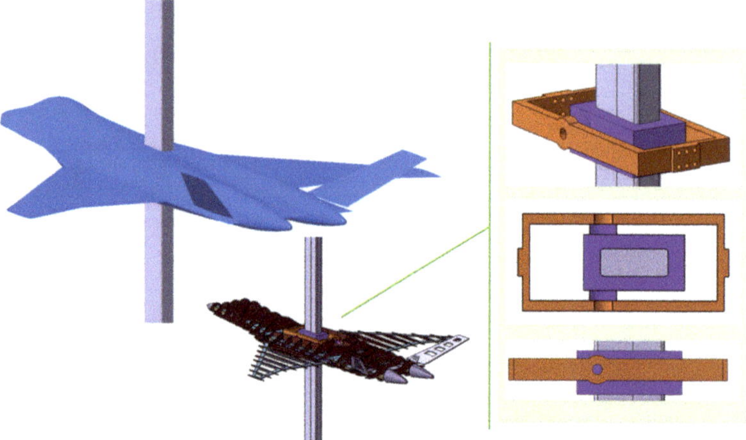

Figure 9. The aircraft model installed with the wind tunnel test support system.

The real aircraft structure is always composed of beams, shear webs, and skins. The aircraft in this paper has obvious thin fuselage. Considering assembly within the wind tunnel support system, it is a better choice to replace the beams with one core board. The shear webs were adhered to the core board vertically. In addition, the skins were stuck to the edge of shear webs.

During the process of structural design, it is necessary to ensure that CG coincides with the rotating axis of the support system to avoid additional pitch moment. This can be achieved by placing lead counterweight at the front of the model after the main structure of the model was assembled. An opening was set at CG to provide space for connection with support system. The front and rear metal joints were designed to ensure the rigid connection between the model and the support system. In addition, the metal joint at the back can slide back or forth to fit with support system, see Figure 10.

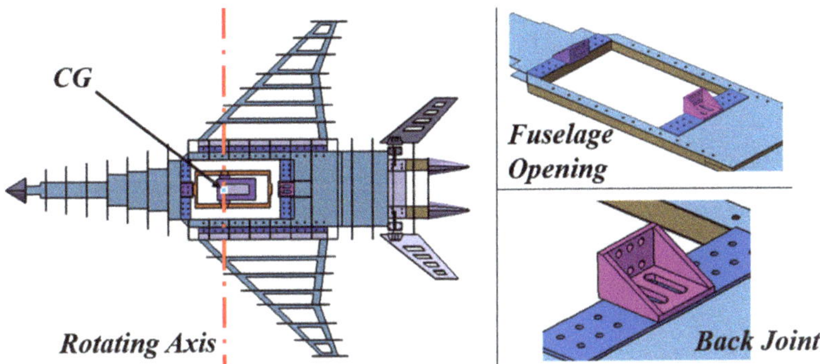

Figure 10. Structure design of the aircraft.

The fuselage included one core board, shear webs and skins. The core board has typical sandwich construction. The top and bottom layers are made of carbon fiber composites, and the core (i.e., the ② part shown in Figure 11) is balsa wood strip (i.e., it was wrapped with carbon fiber cloth cured with resin), which can satisfy the requirements of mass and stiffness at the same time. The shear webs were made of aviation laminate. In order to reduce the mass, the skins were made of balsa wood. Then the PVC heat shrinkable film was adhered to the surface of balsa to preserve smoothness.

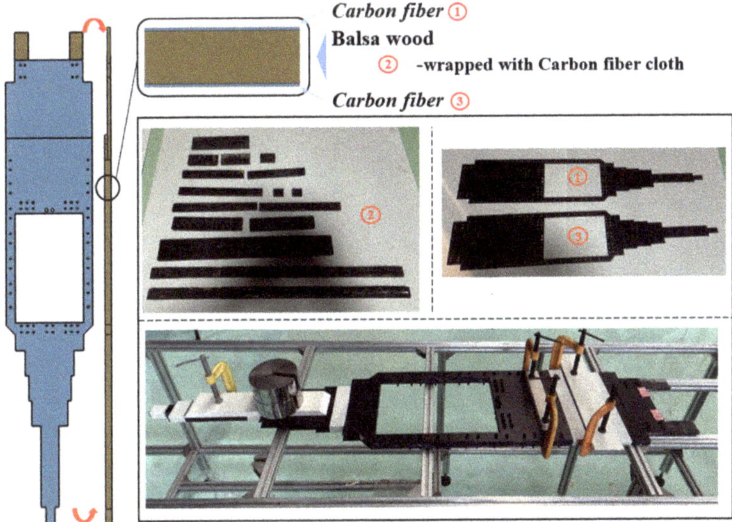

Figure 11. Structure design and assembly of the fuselage core board. The materials of the core board: ① carbon fiber; ② balsa wood; and ③ carbon fiber.

The structure of the present wing was similar with the fuselage, except that the core board was only made of Carbon fiber composites. The stringers made of aviation laminate were stuck to the edge of wing shear webs. The wing was connected to the fuselage by an aluminum alloy joint, as shown in Figure 12.

The shear webs were not included in the structure of V-tails due to narrow space. The leading and tailing edges were made of balsa wood and stuck to the edge of the core board, respectively. Aviation laminate was used to stick to the top and bottom surface of the core board. The aerodynamic shape was preserved by polishing. Similarly, the PVC heat shrinkable film was adhered to the surface. The structure and assembly of the V-tails are shown in Figure 13.

Figure 12. The wing structure and its joint with the fuselage.

Figure 13. Structure and assembly of the V-tails.

The full-span assembled aircraft model is shown in Figure 14.

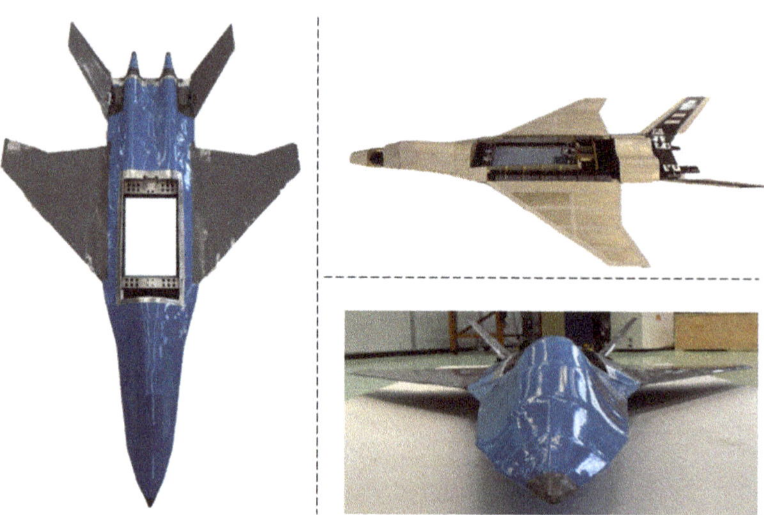

Figure 14. The final wind tunnel test aircraft model.

2.6. V-Tail Actuator Design

The key components in closed-loop active control are sensor, controller (optimization algorithm) and dynamic actuator [26]. The servo actuator was directly connected with the tail shaft to control the V-tail deflection. The aerodynamic loads of the V-tails are mainly transmitted to the fuselage in the form of bending and torque moment. The torque generated by the aerodynamic loads can be offset by the servo actuator. In order to ensure the normal operation of the servo actuator, it was not allowed to balance the bending moment by the servo actuator. Therefore, two bearings were added at the tail shaft to resist the bending moment generated by the aerodynamic force, see Figure 15.

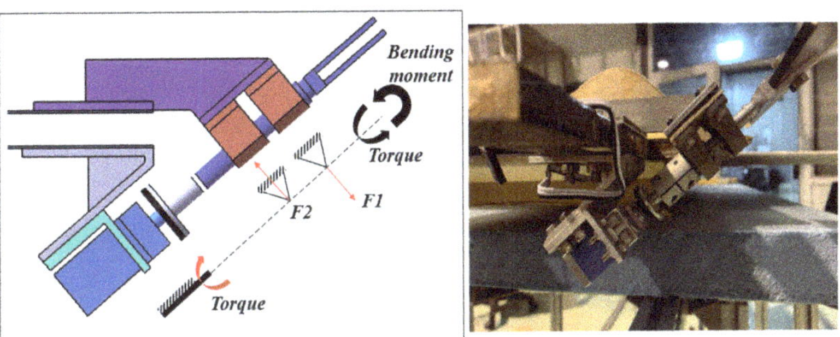

Figure 15. V-tail actuator and its force analysis process.

In order to guarantee that the rated torque of the servo actuator is big enough to resist the aerodynamic force, the chosen rated torque is 10 times that of the hinge moment calculated by CFD.

3. Flight Control Law

3.1. Dynamic Equation

Under the constraint of the wind tunnel support system, the force analysis of the wind tunnel model is shown in Figure 16. Based on the unsteady aircraft equations of motion for 2D flight [27], the longitudinal dynamic equations of the model are established while the longitudinal motion parameters are considered.

$$\frac{mdV}{dt} = Fcos(\gamma) - D - mgsin(\gamma) \tag{11}$$

$$\frac{mVd\gamma}{dt} = -Fsin(\gamma) + L - mgcos(\gamma) \tag{12}$$

$$\frac{I_y dq}{dt} = M_y \tag{13}$$

$$\frac{d\theta}{dt} = q \tag{14}$$

$$\frac{dh}{dt} = Vsin(\gamma) \tag{15}$$

where

$$\gamma = \theta - \alpha \tag{16}$$

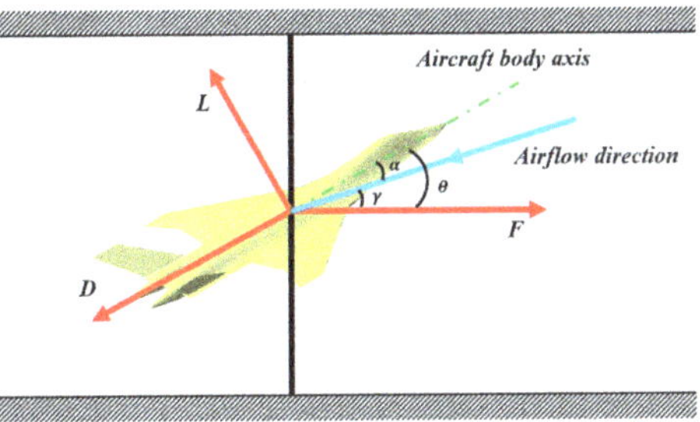

Figure 16. Force analysis of the aircraft installed in the present support system.

The variable F can be obtained with the analysis bellow.

The relative wind speed of the model is determined by the air velocity, which was steady during the wind tunnel test. In the ground coordinate system, the relative velocity of the model in the x direction is constant.

So

$$\frac{dV_x}{dt} = 0 \tag{17}$$

See Figure 16,

$$V_x = V cos(\gamma) \tag{18}$$

So

$$\frac{dV_x}{dt} = \frac{dV}{dt}cos(\gamma) - \frac{Vsin(\gamma)d\gamma}{dt} = 0 \tag{19}$$

Multiplying both sides of the equal sign of Equation (11) by $cos(\gamma)$ and Equation (12) by $sin(\gamma)$ at the same time and considering Equation (19), the expression of F can be obtained.

$$F = Dcos(\gamma) + Lsin(\gamma) \tag{20}$$

The dynamic equations of the aircraft model in the wind tunnel can be obtained after the variable F is derived.

3.2. Measurement of Moment of Inertia

Before simulation, the mass and moment of inertia must be measured. The compound pendulum method [28] is used in the measurement.

According to the compound pendulum model in Figure 17, it can be recognized as harmonic vibration when the swing angle β is very small. The moment around the rotating axis O is expressed as

$$M_O = -mglsin(\beta) \tag{21}$$

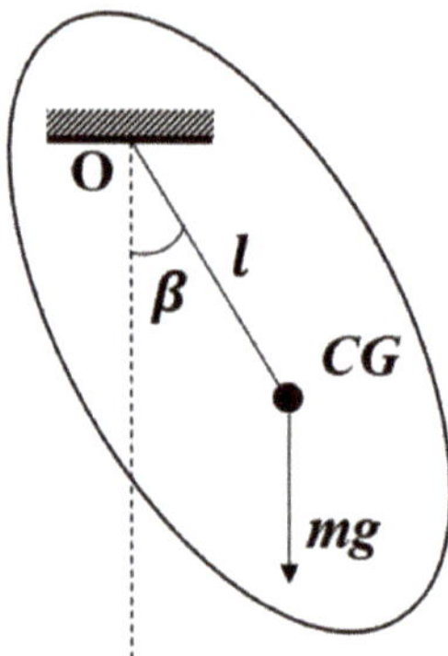

Figure 17. The compound pendulum.

When β is small, it can be approximately expressed as

$$M_O = -mgl\beta \tag{22}$$

According to Newton's second law of rotation

$$M_O = I_O\ddot{\beta} \tag{23}$$

Then

$$\ddot{\beta} = -\omega^2\beta = -\frac{mgl}{I_O}\beta \tag{24}$$

The vibration period of the compound pendulum

$$T = 2\pi\sqrt{\frac{I_O}{mgl}} \tag{25}$$

Based on parallel axis theorem [29]

$$I_O = I_{CG} + ml^2 \tag{26}$$

So

$$I_{CG} = \frac{mglT^2}{4\pi^2} - ml^2 \tag{27}$$

Once the vibration period of the aircraft model is measured, the moment of inertia about CG can be obtained with Equation (27).

The moment of inertia measurement test is shown in Figure 18.

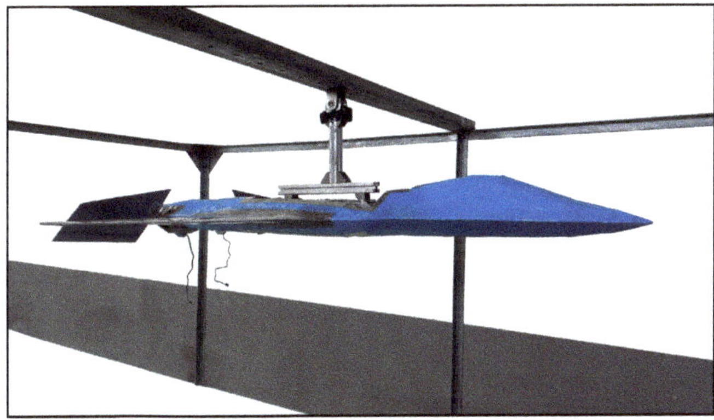

Figure 18. Moment of inertia measurement test using compound pendulum method.

3.3. Simulation of the Flight Control Law

Based on Equations (11)–(16), the dynamic model was established in Simulink. The flight control model is shown in Figure 19. According to the air velocity in the wind tunnel, the trim angle of attack and V-tail deflection angle are calculated and set as the initial state. The angular velocity, acceleration and altitude signals were collected, and the classical PID control method was adopted to keep the model in the target attitude and height by adjusting the deflection angle of the V-tails. The proportional gain, integral gain, and differential gain are tuned manually to ensure the response speed, transform time, and stability of the control system.

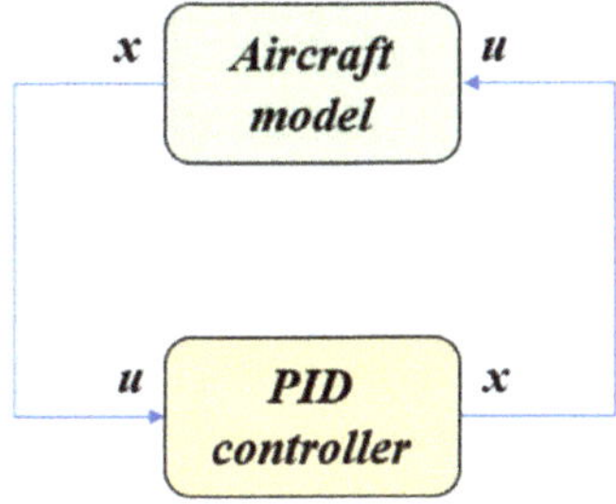

Figure 19. The flight control model.

The longitudinal PID controller is shown in Figure 20. The upper components form the pitch angle controller and the angular velocity signal input is used as the differential part. Similarly, the lower components form the height controller and the longitudinal velocity is taken as the differential part. The whole displayed controller in Figure 20 is used to realize the height control of the V-tail aircraft. The attitude control can be achieved by disconnecting the height control components and setting an attitude command to replace the theta control.

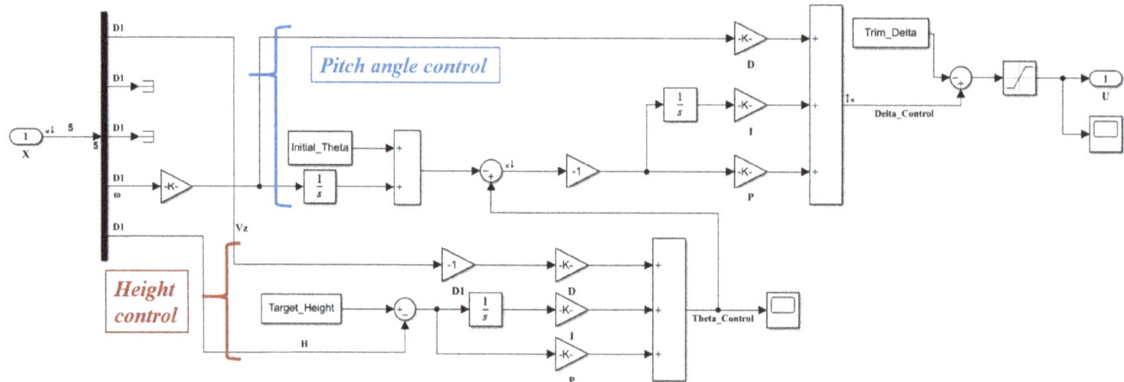

Figure 20. The longitudinal control law.

The simulation results are shown in Figure 21. Simulink results (a) and (c) are in good agreement with the semi-physical simulation results (b) and (d), respectively. Due to the delay of control desk measurement, the semi-physical simulation images lag behind the Simulink images by 1.8 s.

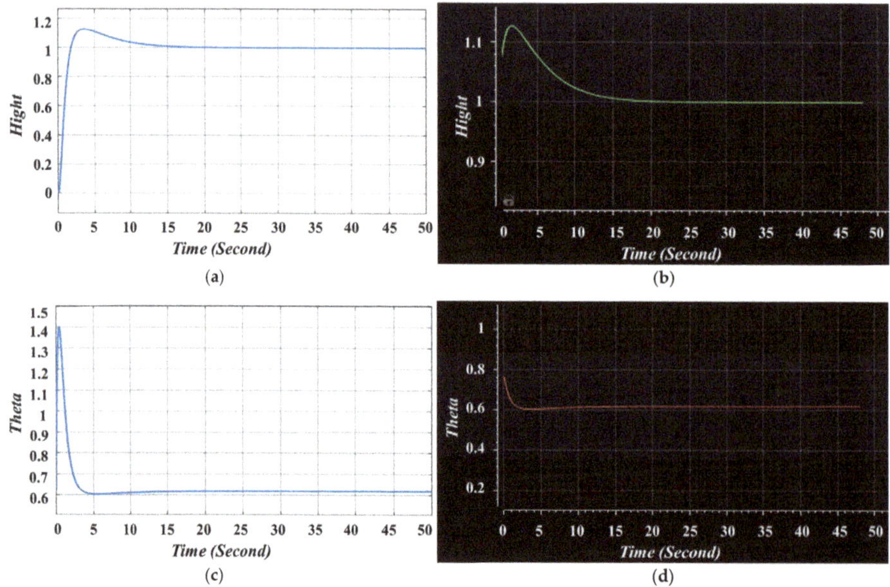

Figure 21. Simulation results of the flight control law: (**a**) the height result by Simulink; (**b**) the height result by semi-physical simulation; (**c**) the pitch angle result by Simulink; and (**d**) the pitch angle result by semi-physical simulation.

4. Wind Tunnel Test

A wind tunnel test is presented to verify the effectiveness of the flight control law. The full model aircraft assembled in the wind tunnel test section is shown in Figure 22. Three test conditions were arranged: (1) different wind speeds were set to verify the static stability of the full model aircraft when the control system is closed; (2) for attitude control, the tail deflection angle was controlled to change the pitch moment under the given wind speed, whereas the model can reach the target pitch angle and keep stable; and (3) for height control, the angle of attack was changed due to the change in pitch moment which occurs with variety of the tail deflection. Then the increment of lift force led to the ascending or descending of the model until it reached the target height.

Figure 22. The full model aircraft assembled in the wind tunnel test section.

In the present case of low wind speed and small angle of attack (i.e., the wind speed is no more than 30 m/s and the angle of attack is no more than 10°), the lift is difficult to be balanced with gravity for the present model. In order to ensure the safety and integrity of the test, a spring structure was installed in the wind tunnel to make up for the lack of lift at low wind speed and small angle of attack, see Figure 23. When the wind speed and the angle of attack reach a certain value, the lift is sufficient to support the model. At this time, it is the focus of the test to ensure that the model is stable at the target pitch angle and height under this flight state.

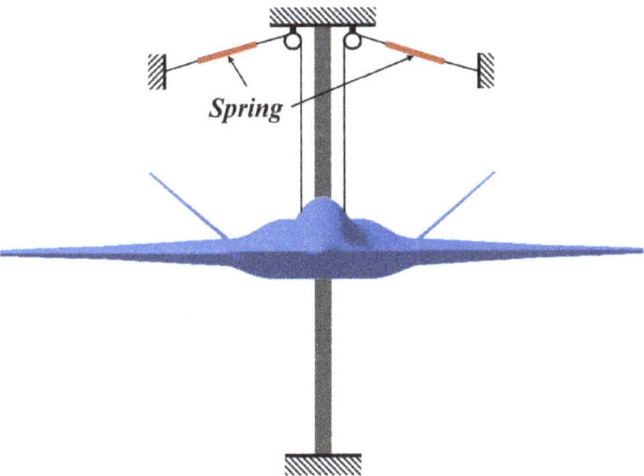

Figure 23. Auxiliary springs for suspending the present model.

4.1. Test Condition 1: Static Stability Verification

A brake cylinder was installed on the support system. At the beginning of the test, the brake system was turned on and the model was fixed at the middle height of the wind tunnel test section. The tail deflection angle was set to $0°$ and the flight control system was closed. During the test, the model kept stable all the time. In addition, it was able to restore to stable state in a few seconds after applying manual interference. It is seen from the test phenomenon that the model has good static stability. The aircraft model reached steady state at different wind speeds (16 m/s, 20 m/s, 24 m/s, 28 m/s, and 30 m/s). Trim angles of attack at different wind speeds can be seen in Figure 24.

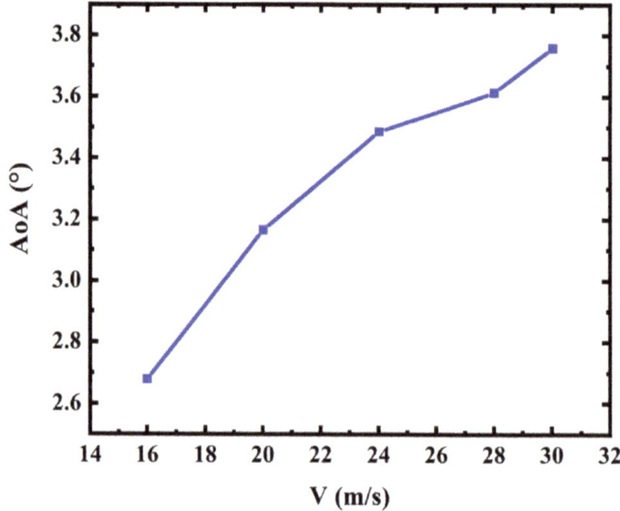

Figure 24. Trim angles of attack at different wind speeds.

4.2. Test Condition 2: Attitude Control

The wind speeds of the attitude control test are: 16 m/s, 20 m/s, and 24 m/s. Under each wind speed, the target pitch angles were set as $-5°$, $0°$, $5°$, and $10°$. The brake system was turned off and the model obtained pitch and plunge degrees of freedom at the same time. The pitch angle time history curves under different wind speeds are shown in Figure 25. Before opening the flight control system, the test model was stable at the static trim angle of attack, which was consistent with the results of test condition 1. After the target pitch angle was set and the flight control system was opened, the model reached and stabilized at the target state within 6 s. As shown in Figure 25c, the model has slight pitch oscillation after reaching the target attitude at 24 m/s and $10°$. It is speculated that there are two possible reasons: (1) the stability judgment error band of the control law is set too large; and (2) at large pitch angle and height, the flow field is unstable due to the increase in blockage percentage [30].

The states of the model in the wind tunnel are shown in Figure 26 after the model reached the target pitch angle.

The sign of tail deflection angle is determined by the pitch change in the aircraft driven by it. The head up is positive and the head down is negative. The change in pitch angle is in good agreement with the change in tail deflection angle when the flight control law was opened, e.g., the results shown in Figure 27. The aerodynamic pitch moment caused the aircraft to rise and the pitch angle increased when the tail deflection angle was downward (i.e., corresponding to the value increase in Figure 27a,b presents the similar regularity when the tail deflection angle was upward). When the pitch angle reached the target value and was stable, the tail deflection angle was also stable at a certain value.

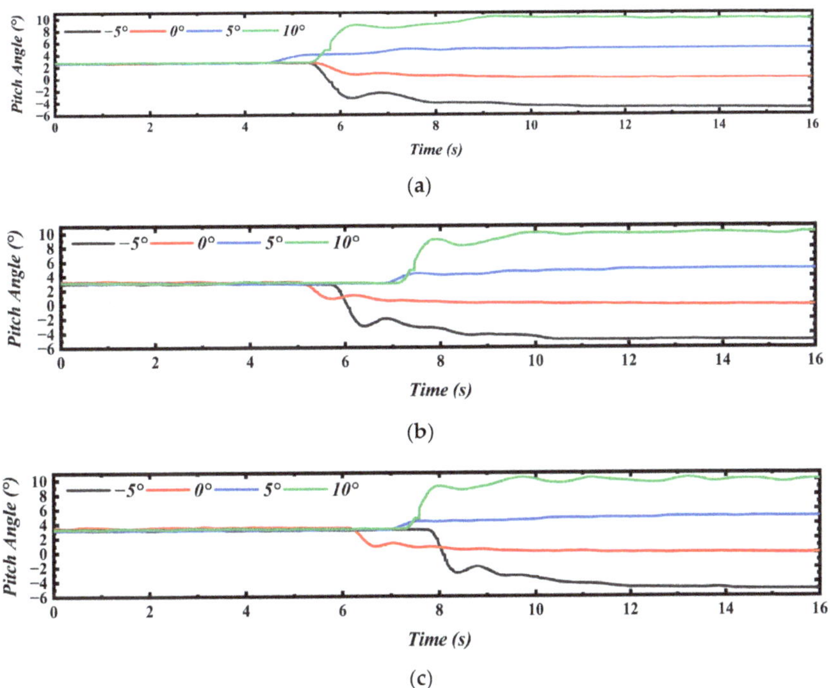

Figure 25. Pitch angle control time history curves at different wind speeds: (**a**) 16 m/s; (**b**) 20 m/s; and (**c**) 24 m/s.

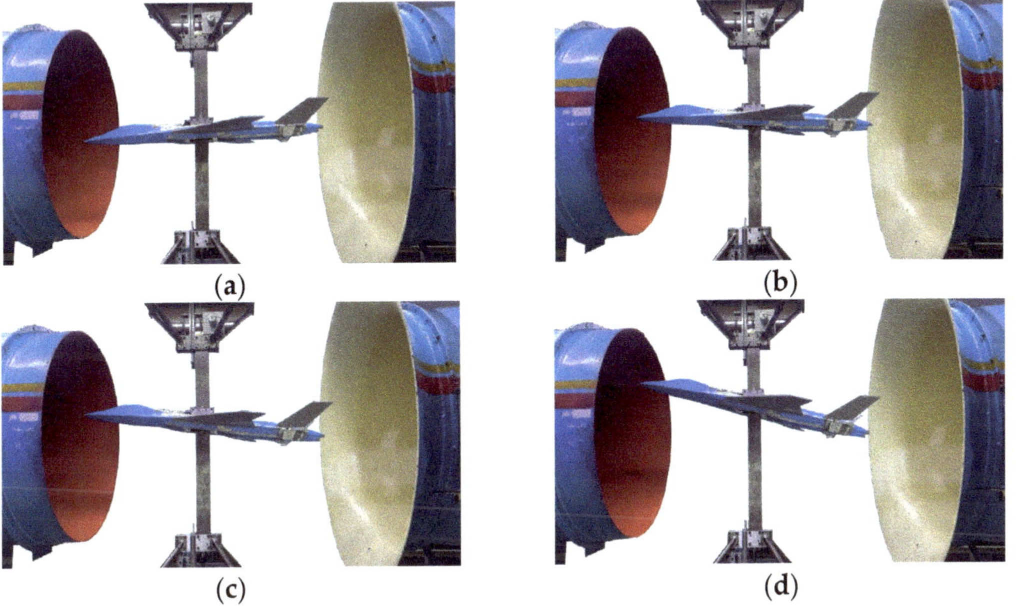

Figure 26. Captures when the model reached different target pitch angles: (**a**) −5°; (**b**) 0°; (**c**) 5°; and (**d**) 10°.

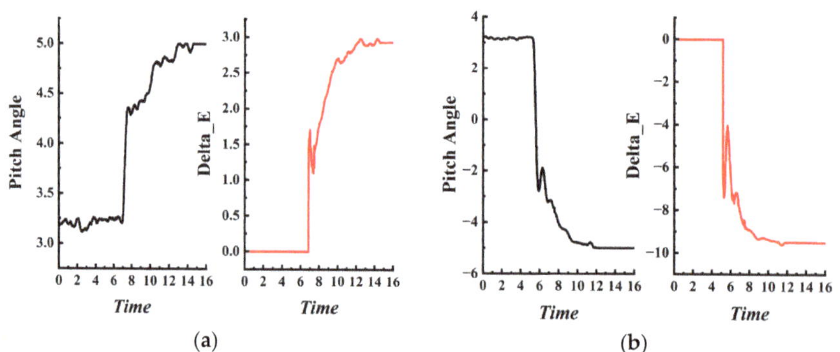

Figure 27. The pitch angle and tail deflection angle time history curves at wind speed 24 m/s: (**a**) 5° and (**b**) −5°.

4.3. Test Condition 3: Height Control

The data curves of the height control test are shown in Figure 28. The ascending and descending tests were carried out at the wind speeds of 24 m/s and 30 m/s. The time history curves of height, pitch angle, and tail deflection angle are shown in Figure 29. The target height was set to 110 mm. The balance position under the wind speed of 24 m/s was taken as the initial position of height control. After the wind speed was stable and the model was in static equilibrium state, the target height was set to 110 mm and the height control system was opened. The pitch angle increased due to the downward deflection of the V-tails. Then the aircraft model climbed to the target height and preserved stable. The captures of the aircraft model ascending process are shown in Figure 30. According to the test results, the test auxiliary springs were completely relaxed when the wind speed was 30 m/s and the target height was 125 mm. It means that the lifting force is exactly equal to gravity at this moment. This test state truly simulated the pitch and plunge free state of the aircraft in the wind tunnel.

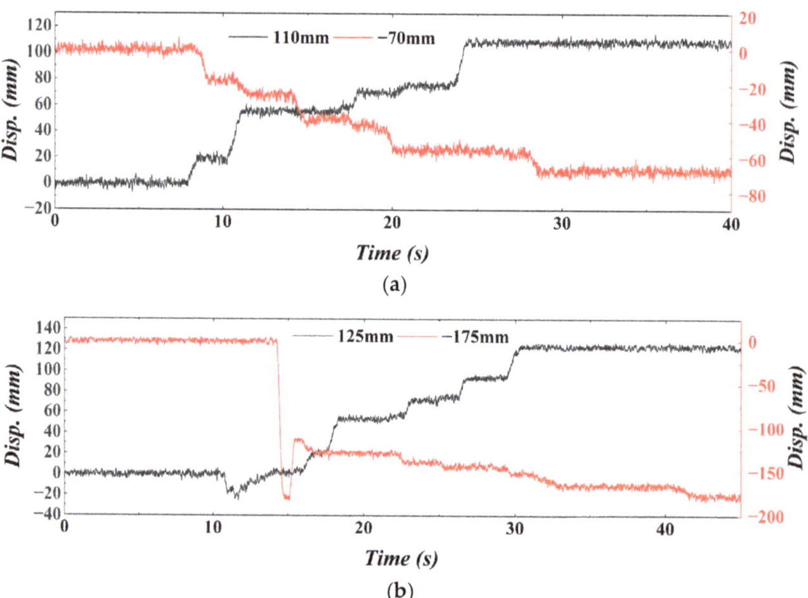

Figure 28. Height time history curves at different wind speeds: (**a**) 24 m/s and (**b**) 30 m/s.

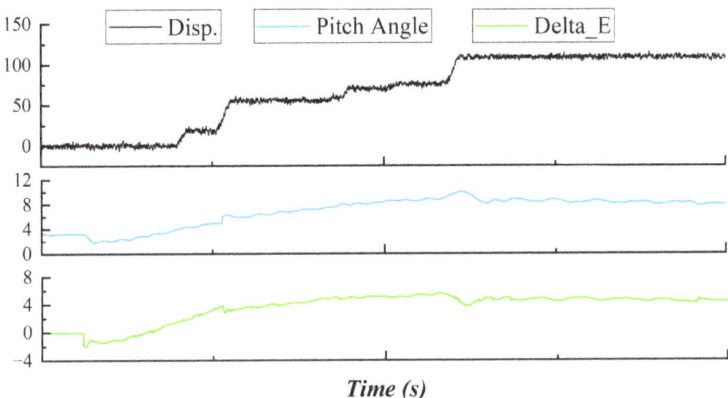

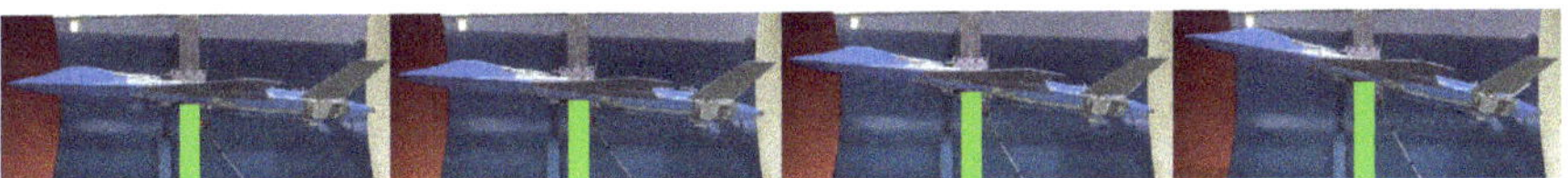

Figure 29. Height, pitch angle and tail deflection angle time history curves at wind speed 24 m/s and the target height was 110 mm.

Figure 30. Captures of the test when the model arose at wind speed 24 m/s to the height 110 mm.

It is seen from Figure 28 that the height time history curves show obvious step features, which are caused by the friction between the aircraft model and the model support system. Under the given wind speed, the mutual squeezing force between the model and the support system structure increased as the pitch angle increased. The lifting force has to overcome both gravity and nonlinear friction, which lead to the step feature curves of the height. From the results, it is summarized that the support system still needs to be developed to reduce the impact of friction on the aircraft motion.

The success of height control test shows great applicable potential of the flight control law in full model wind tunnel test. The height can be controlled to keep the model in a better position to avoid the negative influence due to the increase in the blockage percentage.

5. Conclusions

In this paper, a low-speed full model aircraft was designed and manufactured based on a wind tunnel test support system with release of pitch and plunge degrees of freedom. The flight control law was designed according to the model parameters and the support system features. Three test conditions were arranged. The static stability of the aircraft model was verified with test one. The attitude control of the model was realized with test two. The pitch angle control of $-5°, 0°, 5°$, and $10°$ was investigated under the wind speed of 16 m/s, 20 m/s, and 24 m/s. The model height control was studied through test three. The model achieved an ascent of 110 mm and a descent of 70 mm at a wind speed of 24 m/s. Meanwhile, the model achieved an ascent of 125 mm and a descent of 175 mm under the wind speed of 30 m/s. In addition, the height control of the model without springs was successfully achieved.

In this research, the effectiveness of the proposed flight control law is proved by wind tunnel test. The V-tail configuration aircraft model flies in the wind tunnel with pitch and plunge degrees of freedom successfully. The target pitch angle and height can be realized efficiently.

Wind tunnel tests with release of pitch and plunge freedom are usually conducted in the research of gust load alleviation and body freedom flutter of flying wing aircrafts [31–37]. Half model wind tunnel tests are the mainstream due to the difficulty of full model tests. Based on the present work, the wind tunnel test fully shows the great potential of the support system in the low-speed wind tunnel test of the full model aircraft. The proposed test method can make contributions to the full model tests of the research on gust load alleviation and body freedom flutter in the future.

Author Contributions: Conceptualization, W.Q. and J.L.; Methodology, J.L., W.Q. and Y.B.; Software, J.L.; Validation, J.L. and X.X.; Investigation, J.L.; Data curation, J.L.; Writing—original draft preparation, J.L. and Y.B.; Visualization, J.L.; Supervision, Y.B. All authors have read and agreed to the published version of the manuscript.

Funding: This research received no external funding.

Data Availability Statement: Not applicable.

Conflicts of Interest: The authors declare no conflict of interest.

Nomenclature

M	Pitch moment relative to center of gravity
L	Lift force
AC	Aerodynamic center
X_{AC}	Distance between center of gravity and aerodynamic center
M_0	Zero lift pitch moment (relative to aerodynamic center)
ρ	Air density
V	Air velocity
S	Reference area
$\bar{c}$	Mean aerodynamic chord
C_M	Pitch moment coefficient relative to center of gravity
C_L	Lift force coefficient
C_{M0}	Zero lift pitch moment coefficient (relative to aerodynamic center)
$A_1\ A_2\ B_1\ B_2$	Slope and intercept of fitting curve
α	Angle of attack
m	Mass
F	Horizontal reaction force
γ	Climb angle
D	Aerodynamic drag
I_y	Pitch moment of inertia
q	Pitch angular velocity
M_y	Moment about y-axis (pitch moment)
θ	Pitch angle
h	Height
V_x	Component of flight relative velocity in x direction
M_O	Moment relative to fix point O
CG	Center of gravity
l	Distance between CG and fix point O
β	Swing angle of the compound pendulum
I_O	Moment of inertia relative to point O
$\ddot{\beta}$	Swing angular acceleration
ω	Angular frequency
I_{CG}	Moment of inertia relative to CG
T	Swing period

References

1. Gong, W.; Xia, M.; Yue, L.; Zheng, J.; Zheng, S. Numerical Investigation on Flow Features and Static Stability Characteristics of the V-Tail Aircraft. *Flight Dyn.* **2021**, *39*, 19–24. [CrossRef]
2. Sanchez-Carmona, A.; Cuerno-Rejado, C. Vee-Tail Conceptual Design Criteria for Commercial Transport Aeroplanes. *Chin. J. Aeronaut.* **2019**, *32*, 595–610. [CrossRef]
3. Phillips, W.F.; Hansen, A.B.; Nelson, W.M. Effects of Tail Dihedral on Static Stability. *J. Aircr.* **2006**, *43*, 1829–1837. [CrossRef]
4. Metz, P. Flight Test of the YF-23A Advanced Tactical Fighter. In Proceedings of the Aerospace Design Conference, Irvine, CA, USA, 3–6 February 1992; American Institute of Aeronautics and Astronautics: Irvine, CA, USA, 1992.

5.	García-Hernández, L.; Cuerno-Rejado, C.; Pérez-Cortés, M. Dynamics and Failure Models for a V-Tail Remotely Piloted Aircraft System. *J. Guid. Control Dyn.* **2018**, *41*, 505–513. [CrossRef]

6.	Abzug, M.J. V-Tail Stalling at Combined Angles of Attack and Sideslip. *J. Aircr.* **1999**, *36*, 729–731. [CrossRef]

7.	Qiao, F.; Shi, J.; Qu, X.; Lyu, Y. Adaptive Back-Stepping Neural Control for an Embedded and Tiltable V-Tail Morphing Aircraft. *Int. J. Control Autom. Syst.* **2022**, *20*, 678–690. [CrossRef]

8.	Wang, L.; Zhang, N.; Liu, H.; Yue, T. Stability Characteristics and Airworthiness Requirements of Blended Wing Body Aircraft with Podded Engines. *Chin. J. Aeronaut.* **2022**, *35*, 77–86. [CrossRef]

9.	Leshikar, C.; Gosnell, S.; Gomez, E.; Moy, L.; Valasek, J. System Identification Flight Testing of Inverted V-Tail Small Unmanned Air System. In Proceedings of the AIAA SCITECH 2022 Forum, San Diego, CA, USA, 3–7 January 2022; p. 15.

10.	Jin, W.; Yang, Z.; Meng, D.; Chen, Y.; Huang, H.; Wang, Y.; He, S.; Chen, Y. Strength Desigh and Test of Advanced Fighter All-Moving Twin V-Tail Buffet. *Acta Aeronaut. Astronaut. Sin.* **2020**, *41*, 15. [CrossRef]

11.	Zhu, W. Models for Wind Tunnel Tests Based on Additive Manufacturing Technology. *Prog. Aerosp. Sci.* **2019**, *110*, 100541. [CrossRef]

12.	Fujii, K. Progress and Future Prospects of CFD in Aerospace—Wind Tunnel and Beyond. *Prog. Aerosp. Sci.* **2005**, *41*, 455–470. [CrossRef]

13.	Damljanović, D.; Vuković, Đ.; Ocokoljić, G.; Ilić, B.; Rašuo, B. Wind Tunnel Testing of ONERA-M, AGARD-B and HB-2 Standard Models at Off-Design Conditions. *Aerospace* **2021**, *8*, 275. [CrossRef]

14.	Ocokoljić, G.; Rašuo, B.; Kozić, M. Supporting System Interference on Aerodynamic Characteristics of an Aircraft Model in a Low-Speed Wind Tunnel. *Aerosp. Sci. Technol.* **2017**, *64*, 133–146. [CrossRef]

15.	Ivanco, T.G. Unique Testing Capabilities of the NASA Langley Transonic Dynamics Tunnel, an Exercise in Aeroelastic Scaling. In Proceedings of the AIAA Ground Testing Conference, San Diego, CA, USA, 24–27 June 2013; p. 23.

16.	Yang, X.; Liu, N.; Guo, C.; Zhang, Y.; Sun, J.; Zhang, G.; Yu, X.; Yu, J.; Hou, L. A Survey of Aeroelastic Wind Tunnel Test Techonlogy of Flight Vehicles. *Acta Aerodyn. Sin.* **2018**, *36*, 995–1008. [CrossRef]

17.	Raju Kulkarni, A.; La Rocca, G.; Veldhuis, L.L.M.; Eitelberg, G. Sub-Scale Flight Test Model Design: Developments, Challenges and Opportunities. *Prog. Aerosp. Sci.* **2022**, *130*, 36. [CrossRef]

18.	Gebbink, R.; Wang, G.; Zhong, M. High-Speed Wind Tunnel Test of the CAE Aerodynamic Validation Model. *Chin. J. Aeronaut.* **2018**, *31*, 439–447. [CrossRef]

19.	Tang, D.; Dowell, E.H. Effects of a Free-to-Roll Fuselage on Wing Flutter: Theory and Experiment. *AIAA J.* **2014**, *52*, 2625–2632. [CrossRef]

20.	Scott, R.C.; Allen, T.; Castelluccio, M.; Sexton, B.; Claggett, S.; Dykman, J.R.; Funk, C.; Coulson, D.; Bartels, R.E. Aeroservoelastic Wind-Tunnel Test of the SUGAR Truss Braced Wing Wind-Tunnel Model. In Proceedings of the 56th AIAA/ASCE/AHS/ASC Structures, Structural Dynamics, and Materials Conference, Kissimmee, FL, USA, 5–9 January 2015; American Institute of Aeronautics and Astronautics: Kissimmee, FL, USA, 2015.

21.	Allen, T.; Sexton, B.; Scott, M.J. SUGAR Truss Braced Wing Full Scale Aeroelastic Analysis and Dynamically Scaled Wind Tunnel Model Development. In Proceedings of the 56th AIAA/ASCE/AHS/ASC Structures, Structural Dynamics, and Materials Conference, Kissimmee, FL, USA, 5–9 January 2015; American Institute of Aeronautics and Astronautics: Kissimmee, FL, USA, 2015.

22.	Tang, J.; Wu, F.; Pu, L.; Zeng, X.; Zhang, H.; Zhang, L. Development of a Two Degrees of Freedom Support System for Full Model Gust Tests. *J. Exp. Fluid Mech.* **2021**, *35*, 94–99. [CrossRef]

23.	Gu, X.; Duc Vo, H.; Mureithi, N.W.; Laurendeau, E. Plasma Gurney Flap Flight Control at Low Angle of Attack. *J. Aircr.* **2022**, 1–18. [CrossRef]

24.	Courtland, W.J.D.; Perkins, D.; Hage, R.E. Airplane Performance, Stability and Control. *J. R. Aeronaut. Soc.* **1950**, *54*, 607–608.

25.	Cusati, V.; Corcione, S.; Ciliberti, D.; Nicolosi, F. Design Evolution and Wind Tunnel Tests of a Three-Lifting Surface Regional Transport Aircraft. *Aerospace* **2022**, *9*, 133. [CrossRef]

26.	Zhao, D.; Lu, Z.; Zhao, H.; Li, X.Y.; Wang, B.; Liu, P. A Review of Active Control Approaches in Stabilizing Combustion Systems in Aerospace Industry. *Prog. Aerosp. Sci.* **2018**, 35–60. [CrossRef]

27.	Jasa, J.P.; Brelje, B.J.; Gray, J.S.; Mader, C.A.; Martins, J.R.R.A. Large-Scale Path-Dependent Optimization of Supersonic Aircraft. *Aerospace* **2020**, *7*, 152. [CrossRef]

28.	Junos, M.H.; Mohd Suhadis, N.; Zihad, M.M. Experimental Determination of the Moment of Inertias of USM E-UAV. *AMM* **2013**, *465–466*, 368–372. [CrossRef]

29.	Hong, S.-I.; Hong, S.-C. Moments of Inertia of Spheres without Integration in Arbitrary Dimensions. *Eur. J. Phys.* **2014**, *35*, 025003. [CrossRef]

30.	Katz, J.; Walters, R. Effects of Large Blockage in Wind-Tunnel Testing. *J. Aircr.* **1995**, *32*, 1149–1152. [CrossRef]

31.	Lei, P.; Yu, L.; Chen, D.; Lyu, B. Influence of Flight Control Law on Body Freedom Flutter Characteristics: Experimental Study. *Acta Aeronaut. Astronaut. Sin.* **2021**, *42*, 124378. [CrossRef]

32.	Lei, P.; Lyu, B.; Yu, L.; Chen, D. Influence of Inertial Parameters on Body Freedom Flutter of Flying Wings. *Acta Aerodyn. Sin.* **2021**, *39*, 18–24.

33.	He, S.; Guo, S.; Liu, Y.; Luo, W. Passive Gust Alleviation of a Flying-Wing Aircraft by Analysis and Wind-Tunnel Test of a Scaled Model in Dynamic Similarity. *Aerosp. Sci. Technol.* **2021**, *113*, 106689. [CrossRef]

34. Shi, P.; Liu, J.; Gu, Y.; Yang, Z.; Marzocca, P. Full-Span Flying Wing Wind Tunnel Test: A Body Freedom Flutter Study. *Fluids* **2020**, *5*, 34. [CrossRef]
35. Yang, J.; Wu, Z.; Dai, Y.; Ma, C.; Yang, C. Wind Tunnel Test of Gust Alleviation Active Control for Flying Wing Configuration Aircraft. *J. Beijing Univ. Aeronaut. Astronaut.* **2017**, *43*, 184–192.
36. Ricci, S.; De Gaspari, A.; Riccobene, L.; Fonte, F. Design and Wind Tunnel Test Validation of Gust Load Alleviation Systems. In Proceedings of the 58th AIAA/ASCE/AHS/ASC Structures, Structural Dynamics, and Materials Conference, Grapevine, TX, USA, 9–13 January 2017; American Institute of Aeronautics and Astronautics: Grapevine, TX, USA, 2017.
37. Scott, R.; Coulson, D.; Castelluccio, M.; Heeg, J. Aeroservoelastic Wind-Tunnel Tests of a Free-Flying, Joined-Wing SensorCraft Model for Gust Load Alleviation. In Proceedings of the 52nd AIAA/ASME/ASCE/AHS/ASC Structures, Structural Dynamics and Materials Conference, Denver, CO, USA, 4–7 April 2011; American Institute of Aeronautics and Astronautics: Denver, CO, USA, 2011.

MDPI
St. Alban-Anlage 66
4052 Basel
Switzerland
www.mdpi.com

Aerospace Editorial Office
E-mail: aerospace@mdpi.com
www.mdpi.com/journal/aerospace